COLLEGE ALGEBRA

Fotios C. Paliogiannis
St. Francis College

Fotios C. Paliogiannis
Mathematics Department
St. Francis College
180 Remsen Street
Brooklyn Heights, NY 11201
USA
fpaliogiannis@sfc.edu

Library of Congress Control Number - LCCN: 2014911558

Paliogiannis, Fotios C.
COLLEGE ALGEBRA / Fotios C. Paliogiannis
Includes index
ISBN-13: 978-1-63048-043-1
ISBN-10: 1630480436

Printed in the United States of America

Preface

This book contains the fundamentals of Algebra. The prerequisite for reading this book is only a knowledge of basic Arithmetic. The primary goal of the book is to *teach Algebra in a thorough, clear and systematic way.* Since good algebra skills are important for a successful study of Pre-calculus and Calculus, the book provides students with a comprehensive and solid background in Algebra. The plethora of illustrative standard and non-standard examples, as well as, the numerous exercises make the book suitable for a course textbook and for self-study. Answers to the exercises can be found at the end of the book.

The book begins in Chapter 1, with a review of numbers and their operations. In Chapter 2, we study first-degree equations, ratios, percentages, proportions, direct and inverse variations, and several word problems. Chapter 3, deals with powers, roots and their properties. In Chapters 4 through 8, we give an extensive and detailed study of the main topics of Algebra, such as, polynomials and factorization, rational and radical expressions, algebraic equations, inequalities and systems of equations. Finally, in Chapter 9, we study exponentials and logarithms, their properties and their applications. We also study exponential and logarithmic equations, inequalities and systems. A star (*) indicates a more advanced topic which can be ommited at a first reading of the book. However, the more ambitious student desiring further undersdanding on the subject will wish to study this material as well.

I wish to thank Dr. Martin Moskowitz for valuable discussions and his encouragment during the process of the preparation of this book. I also thank many of my St. Francis College students for their comments and suggestions. In particular, I thank Le Xu for her help in catching errors and misprints. Of course, any errors and misprints are the author's responsibility. As a final note, the author will be very grateful to anybody who wants to inform him about errors or just misprints, or wants to express criticism or other comments. The e-mail address is: *fpaliogiannis@sfc.edu*

Fotios C. Paliogiannis, Ph.D.   New York, 2014.

Contents

Chapter 1

Numbers

In this chapter, we review the basic operations and properties of numbers.

1.1   Natural numbers

Our knowledge of numbers derives from our perception of the physical world. Counting is the first mathematical activity that we learn. The numbers obtained as a result of counting are called *natural numbers*. The natural numbers arranged in increasing order form the sequence of natural numbers:

$$1 < 2 < 3 < 4 < 5 < ...,$$

(the three dots mean *"and so on"*). The set of **natural numbers** is denoted by the symbol $\mathbb{N}$. That is,

$$\mathbb{N} = \{1, 2, 3, 4, 5, ...\}.$$

The set of natural numbers is infinite. It contains the least number 1, but has no greatest number. The numbers

$$0, 1, 2, 3, 4, 5, ...$$

are known as **whole number**.

The whole numbers are symbolized by the following ten digits: $0, 1, 2, 3, 4, 5, 6, 7, 8, 9$, for example, $10, 17, 56, 94, 129, 752, 5,842$, etc.

On the whole numbers we perform the four arithmetic operations of **addition**, **subtraction**, **multiplication** and **division**. We assume that the reader is familiar, from elementary school, with the four arithmetic operations of whole numbers. For example,

$$5 + 3 = 8, \quad 125 + 357 = 482, \quad 9 - 4 = 5, \quad 400 - 264 = 136,$$

$$5 \cdot 3 = 15, \quad 20 \cdot 17 = 340, \quad 36 \div 4 = 9, \quad 867 \div 17 = 51.$$

Note that
$$5 \cdot 3 = 3 + 3 + 3 + 3 + 3 = 15,$$

that is, *multiplication is repeated addition.*

We recall that subtraction and division are defined as the inverse operations to addition and multiplication, respectively. To **subtract** a number b from a number a means to find a number c, which when added to the number b, will give the number a, that is,

$$a - b = c \quad \text{if} \quad b + c = a.$$

The number c is called the *difference* of the numbers a and b.

For instance, $10 - 6 = 4$ because $6 + 4 = 10$.

To **divide** a number a by a number b means to find a number q such that the product $b \cdot q$ will yield the number a, that is,

$$a \div b = q \quad \text{if} \quad b \cdot q = a.$$

For instance, $18 \div 6 = 3$ because $6 \cdot 3 = 18$.
Remark. *Division by zero is not permitted.*

If $a \div b = q$, then we say the number a is *divisible by* b, the number a is called the *divident*, the number b is called the *divisor* and the number

q is called the *quotient* of the division of a by b.

A number is said to be **even** if it is divisible by 2, it is said to be **odd** if it is not divisible by 2. Thus, the even whole numbers are

$$0, 2, 4, 6, 8, 10, 12, \ldots$$

and the odd numbers are

$$1, 3, 5, 7, 9, 11, 13, \ldots$$

An even number n is of the general form is $n = 2k$, while an odd number m is of the form $m = 2k + 1$, where $k = 0, 1, 2, 3, \ldots$.

If $a = bq$, then we also say a is a **multiple** of b and b is a **factor** of the number a. For instance:

The multiples of 5 are: 5, 10, 15, 20, 25, 30, 35, 40, 45, 50, 55, ...

The factors of 24 are: 1, 2, 3, 4, 6, 8, 12 and 24.

Definition 1.1.1. A natural number p, with $p \neq 1$, is called **prime** if it has only two factors, 1 and itself.

Here are the first twenty prime numbers:

2, 3, 5, 7, 11, 13, 17, 19, 23, 29, 31, 37, 41, 43, 47, 53, 59, 61, 67, 71.

A natural number which is not prime is called a **composite number**. We state without proof the following theorem:

Theorem 1.1.2. *(**Fundamental Theorem of Arithmetic**). Every natural number, not equal to 1, can be expressed as a product of primes. Except for the order of the factors, this expression is unique.*

For example, the *prime factorization* of 60 is:

$$60 = 2 \cdot 2 \cdot 3 \cdot 5 = 2^2 \cdot 3 \cdot 5.$$

Common factors and multiples

A *common factor* of two or more natural numbers is a number which divides each of them. The largest of all common factors is called the **greatest common factor**, abbreviated as GCF.

If the GCF of the numbers a, b is equal to 1, then the numbers a and b are said to be *relatively prime*. For instance, it is easy to check that 8 and 15 are relatively prime. We wrire GCM$(8, 15) = 1$.

*To find the **GCF** of two or more numbers, we write the prime factorization of each number, and we form the product of the common factors taken each with the smallest exponent.*

Example 1.1.3. Find the GCF of 252 and 1080.

Solution. Writing the prime factorizations we get:

$$252 = 2^2 \cdot 3^2 \cdot 7 \quad \text{and} \quad 1080 = 2^3 \cdot 3^3 \cdot 5.$$

We have only two common factors, the prime factors 2 and 3. The smallest exponent of these factors is 2. Hence the GCF is $2^2 \cdot 3^2 = 36$. We write GCF$(252, 1080) = 36$.

A *common multiple* of two or more natural numbers is a number which is divisible by each of them. The smallest of the common multiples is called the **least common multiple**, abbreviated as LCM.

*To find the **LCM** of two or more natural numbers, we write the prime factorization of each number, and we form the product of the common and not common factors taken each with the greatest exponent.*

Example 1.1.4. Find the LCM of 252 and 441.

Solution. Factor into prime factors.

$$252 = 2^2 \cdot 3^2 \cdot 7 \quad \text{and} \quad 441 = 3^2 \cdot 7^2.$$

The common and not common prime factors are 2, 3 and 7. The greatest exponent is 2. Hence the LCM is $2^2 \cdot 3^2 \cdot 7^2 = 1,764$. We write LCM$(252, 441) = 1,764$.

1.2 Integer Numbers

The operations of addition and multiplication of whole numbers are always possible on the set of whole numbers (that is, the result of both of these operations is a whole number). However, subtraction is not always possible on the set of whole numbers. For instance, $8 - 3 = 5$, but $3 - 8$ is not possible. For subtraction to be always possible we must count backwards beyond the number zero. Being able to count backward as freely as forward, we consider the new numbers

$$-1, -2, -3, -4, -5, \ldots$$

To distinguish these new numbers as a class from the natural numbers, we call them **negative**, and the natural numbers **positive**. The numbers of both kinds and 0 are called **integer numbers**. The set of integers is denoted by the symbol $\mathbb{Z}$. That is,

$$\mathbb{Z} = \{\ldots, -4, -3, -2, -1, 0, 1, 2, 3, 4, \ldots\}.$$

Observe the *symmetry* of this ordered sequence of integers with respect to the number 0. As 3 is three units *after* 0, so -3 is three units *before* 0; and so on in general.

The number 0 is neither positive nor negative. For every nonzero number a the number $-a$ is called the **opposite** of a. Thus -5 is the opposite of 5, while 3 is the opposite of -3.

Equality and Inequality

To say that a and b are the *same* number, we use the notation $a = b$ and call it the **equality** of integers.

Equality has the following properties:

1. $a = a$ (**reflexive property**)

2. $a = b \Leftrightarrow b = a$ (**symmetric property**)

3. $a = b$ and $b = c$ implies $a = c$ (**transitive property**)

We say that an integer a is *smaller* than an integer b, and we write $a < b$, if b "precedes" a in the sequence of integer numbers. Similarly, we say that a is *greater* than b, and we write $a > b$, if a precedes b. These notations are the **inequalities** of integers. Thus, for example,

$$4 < 7, \quad 9 > 5, \quad 1 > 0, \quad -1 < 0, \quad -2 < -1, \quad -8 < -3.$$

Note that the natural numbers are the positive integers, that is, $n > 0$ for all $n \in \mathbb{N}$. Note also that $a > b$ is equivalent to $b < a$. We write

$$a > b \iff b < a.$$

Observe that changing the signs of any two numbers a and b *reverses the order* in which they occur in the numbers sequence. Thus, we have

$$-3 < -2, \quad \text{but} \quad 3 > 2; \quad -4 < 1, \quad \text{but} \quad 4 > -1.$$

The **absolute value** of a number a is denoted by $|a|$. The absolute value of a positive number, or 0, is the number itself. The absolute value of a negative number is its opposite. Thus,

$$|5| = 5, \quad |0| = 0, \quad \text{and} \quad |-5| = 5.$$

The absolute value is an important notion in mathematics. In Section 1.5.2, we shall study the absolute value of a number in more detail and we shall derive its basic properties.

1.2.1 Operations with Integer Numbers

1. Addition. (a) *To **add** two integers with the same sign, add the absolute values of the numbers and give the answer the common sign.*

Example 1.2.1. $7 + 5 = 12, \quad (-3) + (-5) = -8.$

(b) *To **add** two integers with the different signs, find the difference between their absolute values of the numbers and give the answer the sign of the number whose absolute value is greater.*

Example 1.2.2. $(-2) + 5 = 3, \quad (-3) + 1 = -2, \quad 4 + (-9) = -5.$

2. Subtraction. Let a, b be intergers. *To **subtract** b from a, we add the opposite of b to a.* That is,

$$a - b = a + (-b).$$

Example 1.2.3.

$$7 - 5 = 7 + (-5) = 2, \quad , \; 1 - 9 = 1 + (-9) = -8,$$

$$(-7) - (-3) = -7 + 3 = -4, \quad (-1) - (-1) = -1 + 1 = 0.$$

3. Multiplication. (a) *The product of two integers with the same sign is positive.*

Example 1.2.4. $4 \cdot 5 = 20, \quad (-4) \cdot (-5) = 20.$

(b) *The product of two integers with the opposite signs is negative.*

Example 1.2.5. $2 \cdot (-3) = -6, \quad (-5) \cdot 8 = -40.$

Thus, (**Rules of signs in multiplication**):

$$(+) \cdot (+) = + \quad (-) \cdot (-) = +$$

$$(-) \cdot (+) = - \quad (+) \cdot (-) = -$$

Properties of Operations

The operation of addition of integers satisfies a number of properties such as, for example:

$$1+2 = 2+1 = 3, \ (-8)+2 = 2+(-8) = -6, \ (-2)+(-7) = (-7)+(-2) = -9,$$

$$(-5+3) + 6 = (-5) + (3+6) = 4, \ 0+3 = 3, \ 2+(-2) = 0.$$

Multiplication of integers satisfies similar properties, for example:

$$2 \cdot 3 = 3 \cdot 2 = 6, \ (2 \cdot 5) \cdot 3 = 2 \cdot (5 \cdot 3) = 30, \ 1 \cdot 8 = 8, \ 0 \cdot (-5) = 0.$$

Moreover, multiplication is distributive with respect to addition, for example

$$3 \cdot (4 + 6) = 3 \cdot 4 + 3 \cdot 6 = 30.$$

Thus, in general, addition and multiplication of integer numbers a, b, c possess the following basic properties:

1. $a + b = b + a$ (**commutativity of addition**);

2. $(a + b) + c = a + (b + c)$ (**associativity of addition**);

3. $a + 0 = a$ (**identity property of zero in addition**)

4. $a + (-a) = 0$ (**additive inverse property**)

5. $ab = ba$ (**commutativity of multiplication**);

6. $(ab)c = a(bc)$ (**associativity of multiplication**);

7. $1a = a$ (**identity property of unity in multiplication**);

8. $0a = 0$ (**property of zero in multiplication**);

9. $a(b + c) = ab + ac$ (**distributive property**).

Zero products.

$$\text{If } a \cdot b = 0, \text{ then either } a = 0 \ \text{ or } \ b = 0.$$

4. Division. The rules of signs in division are the *same* as the rules of sign in multiplication. That is, **(a)** *The quotient of two integers with the same sign is positive.*

Example 1.2.6.

$$15 \div 3 = 5, \quad (-15) \div (-3) = 5.$$

(b) *The quotient of two integers with the opposite signs is negative.*

Example 1.2.7.

$$(-20) \div 5 = -4, \quad (12) \div (-4) = -3.$$

Thus, (**Rules of signs in division**):

$$(+) \div (+) = + \qquad (-) \div (-) = +$$

$$(-) \div (+) = - \qquad (+) \div (-) = -$$

We state without proof the division algorithm for integers.

Theorem 1.2.8. *(**Division Algorithm.**) For any two integers a and b, with $b > 0$, there are unique integers q and r such that*

$$a = bq + r, \quad with \ \ 0 \leq r < b.$$

The number q is called the quotient of the division $a \div b$, and the number r is called the remainder.

Example 1.2.9.

$$-30 = 6 \cdot (-5). \text{ Here, } q = -5 \text{ and } r = 0.$$

$$43 = 7 \cdot 6 + 1. \text{ Here, } q = 6 \text{ and } r = 1.$$

$$-25 = 4 \cdot (-7) + 3. \text{ Here, } q = -7 \text{ and } r = 3.$$

$$-18 = 4 \cdot (-5) + 2. \text{ Here, } q = -5 \text{ and } r = 2.$$

1.2.2   Exercises

1. Find the prime factorization of the numbers 96 and 2520.

2. Find the greatest common factor of the numbers 54 and 72.

3. Find the least common multiple of the numbers 45 and 108.

4. Perform the operations

   (a) $(-3) + (-2) + 4$

   (b) $(-3) - (-2) + 1$

   (c) $2 - 6 - 4 + 3$

   (d) $-19 + 8 - 3 + 15$

5. Do the operations

   (a) $(-8)(-3)(2)(-5)$

   (b) $(-5 + 2)(6 - 2) + (-7)(-2)$

6. Do the operations

   (a) $(3 - 9)(1 - 2) - (-7 + 5)(-1 - 2)$

   (b) $[(-3)(-5)(4)] \div [(-2)(-3)]$

1.3   Rational Numbers. Fractions

As we have seen there are four standard arithmetic operations: addition, subtraction, muliplication and division. In $\mathbb{N}$ only addition, and multiplication are possible in general, since it need not be the case, for natural numbers a and b, that $a - b$ or $a \div b$ are natural numbers. The set of integers is such that subtraction is possible, but it is still the case, that division may not work in $\mathbb{Z}$, in other words, for integer numbers a and b, the division $a \div b$ may not be an integer number. For division to be always possible we introduce the new numbers (**fractions**) of the form $\frac{a}{b}$ representing quotients of integers. That is,

$$\frac{a}{b} = a \div b.$$

The number a is called the **numerator** and the number b the **denominator** of the fraction. These numbers are called **rational numbers**. The set of rational numbers is denoted by the symbol $\mathbb{Q}$. That is,

$$\mathbb{Q} = \left\{ \frac{a}{b} : a, b \in \mathbb{Z}, \ b \neq 0 \right\}.$$

When the numerator of a fraction has smaller absolute value than the denominator, we call the fraction a **proper fraction**. For example, $\frac{1}{2}$, $\frac{3}{4}$, and $-\frac{6}{7}$ are all proper fractions. When the absolute value of the numerator is greater than or equal to that of the denominator, we call the fraction an **improper fraction**. For example, $\frac{3}{2}$, $\frac{12}{7}$, and $-\frac{9}{5}$ are all improper fractions. A **mixed number** consists of an integer number and a fraction. For example, $4\frac{2}{3}$ and $-3\frac{4}{5}$ are mixed numbers. It is important to understand that a mixed number is really an addition without the $+$ sign: $5\frac{1}{2} = 5 + \frac{1}{2}$. A mixed number can be written as an improper fraction and vice versa. For instance,

$$1\tfrac{1}{2} = \tfrac{3}{2} \quad \text{and} \quad 5\tfrac{2}{3} = \tfrac{17}{3}.$$

Definition 1.3.1. We say that two fractions $\frac{a}{b}$ and $\frac{c}{d}$ are **equal** if and only if $ad = cb$. That is,

$$\frac{a}{b} = \frac{c}{d} \quad \Leftrightarrow \quad ad = cb.$$

It is important to note that, since $ab = cd \Leftrightarrow abk = cdk$, for $k \neq 0$, the definition of equality of fractions implies that

$$\frac{a}{b} = \frac{ak}{bk}.$$

Hence there follows the fundamental property of fractions: *if the numerator and denominator of a fraction are multiplied (or divided) by one and the same number not equal to zero, then we obtain a fraction equal to the given fraction*, that is,

$$\frac{ak}{bk} = \frac{a}{b}.$$

This fundamental property allows us to reduce (change) a fraction to lower or higher terms.

Reducing Fractions

To reduce a fraction to *lowest terms* we factor both the numerator and the denominator and we cancel the common factors. For example

$$\frac{3}{6} = \frac{1 \cdot 3}{2 \cdot 3} = \frac{1}{2}, \qquad \frac{18}{30} = \frac{3 \cdot 6}{5 \cdot 6} = \frac{3}{5}.$$

The reduction to lowest terms may be accomplished gradually or at once using the GCF (as we did above). If the numerator and denominator of a fraction are relatively prime, then the fraction is automatically in its lowest terms.

To reduce a fraction to *higher terms*, or change a fraction to another denominator, we multiply both the numerator and denominator by the same number. For example

$$\frac{1}{4} = \frac{1 \cdot 2}{4 \cdot 2} = \frac{2}{8}, \qquad \frac{5}{7} = \frac{5 \cdot 3}{7 \cdot 3} = \frac{15}{21}.$$

Frequently, one has to change two or several fractions to a *common denominator*. This is done in the following way: we find the least common multiple of the denominators, and then each fraction is reduced to this denominator. This common denominator is called the **least common denominator** or **LCD**.

Example 1.3.2. Reduce the fractions $\frac{5}{6}$ and $\frac{7}{9}$ to the least common denominator.

Solution. The LCD of 6 and 9 is 18. Now we change the fractions so that they both have denominator 18. We have

$$\frac{5}{6} = \frac{5 \cdot 3}{6 \cdot 3} = \frac{15}{18}$$

and

$$\frac{7}{9} = \frac{7 \cdot 2}{9 \cdot 2} = \frac{14}{18}.$$

1.3.1 Operations with Rational Numbers

The *addition* or *subtraction* of fractions are defined by the following rules:

1. *If the denominators of the fractions are the same, we add or subtract the numerators, and keep the denominator the same.*

Example 1.3.3.

$$\frac{5}{8} + \frac{7}{8} = \frac{5 + 7}{8} = \frac{12}{8} = \frac{3}{2} = 1\frac{1}{2}.$$

Example 1.3.4.

$$\frac{1}{6} - \frac{5}{6} = \frac{1 - 5}{6} = \frac{-4}{6} = -\frac{4}{6} = -\frac{2}{3}.$$

2. *If the denominators differ, first reduce the fractions to the least common denominator, and then we add or subtract their numerators.*

Example 1.3.5.

$$\frac{2}{3} + \frac{1}{5} = \frac{2 \cdot 5}{3 \cdot 5} + \frac{1 \cdot 3}{5 \cdot 3} = \frac{10}{15} + \frac{3}{15} = \frac{13}{15}.$$

Example 1.3.6.

$$\frac{1}{4} - \frac{5}{6} = \frac{1 \cdot 3}{4 \cdot 3} - \frac{5 \cdot 2}{6 \cdot 2} = \frac{3}{12} - \frac{10}{12} = -\frac{7}{12}.$$

Example 1.3.7.

$$\frac{1}{3} - 1 = \frac{1}{3} - \frac{3}{3} = \frac{1 - 3}{3} = \frac{-2}{3} = -\frac{2}{3}.$$

*To **multiply** fractions, multiply the numerators and the denominators separately.* That is,

$$\frac{a}{b} \cdot \frac{c}{d} = \frac{a \cdot c}{b \cdot d}.$$

If there are mixed numbers, convert them to improper fractions before multiplying. Also, before multiplying cancel any common factors in the numerator and the denominator.

Example 1.3.8. Find each product, and write the answer in lowest terms.

$$(a) \quad \frac{3}{5} \cdot \left(-\frac{7}{8}\right), \quad (b) \quad \frac{4}{9} \cdot \frac{21}{30} \quad (c) \quad 2\frac{2}{5} \cdot 3\frac{3}{4}$$

Solution.

$$(a) \quad \frac{3}{5} \cdot \left(-\frac{7}{8}\right) = -\frac{3 \cdot 7}{5 \cdot 8} = -\frac{21}{40}.$$

$$(b) \quad \frac{4}{9} \cdot \frac{21}{30} = \frac{4 \cdot 21}{9 \cdot 30} = \frac{2 \cdot 2 \cdot 3 \cdot 7}{3 \cdot 3 \cdot 2 \cdot 15} = \frac{2 \cdot 7}{3 \cdot 15} = \frac{14}{45}.$$

$$(c) \quad 2\frac{2}{5} \cdot 3\frac{3}{4} = \frac{12}{5} \cdot \frac{15}{4} = \frac{12 \cdot 15}{5 \cdot 4} = \frac{3 \cdot 4 \cdot 3 \cdot 5}{4 \cdot 5} = \frac{3 \cdot 3}{1 \cdot 1} = \frac{9}{1} = 9.$$

Definition 1.3.9. For every nonzero fraction $\frac{a}{b}$, with $a \neq 0$ and $b \neq 0$, the fraction $\frac{b}{a}$ is called the **reciprocal** of $\frac{a}{b}$. In particular, the reciprocal of any nonzero integer $a = \frac{a}{1}$ is the number $\frac{1}{a}$.

The reciprocal $\frac{b}{a}$ is also called the *multiplicative inverse* of $\frac{a}{b}$, because

$$\frac{a}{b} \cdot \frac{b}{a} = 1.$$

*To **divide** two fractions, multiply the first by the reciprocal of the second.* That is,

$$\frac{a}{b} \div \frac{c}{d} = \frac{a}{b} \cdot \frac{d}{c} = \frac{ad}{bc}.$$

Example 1.3.10.

$$\frac{2}{3} \div \frac{4}{15} = \frac{2}{3} \cdot \frac{15}{4} = \frac{2 \cdot 15}{3 \cdot 4} = \frac{2 \cdot 3 \cdot 5}{3 \cdot 2 \cdot 2} = \frac{5}{2} = 2\frac{1}{2}.$$

Properties of Operations

Note that we have extended the numbers from the natural numbers to integers, and from the integers to rational numbers. The natural numbers are the positive integers. Of course, every integer a can be written trivially as a rational number of the form $\frac{a}{1}$. For instance, $-3 = \frac{-3}{1}$. Therefore, the set of natural numbers is contained in the set of integers, and the set of integers is contained in the set of rational numbers. In set notation this means

$$\mathbb{N} \subset \mathbb{Z} \subset \mathbb{Q}.$$

The basic properties of addition and multiplication of integers are easily extended to rational numbers. Thus, addition is *commutative, associative*, $0 = \frac{0}{1}$ is the *identity of addition*, and every rational $\frac{a}{b}$ has an *opposite* $-\frac{a}{b}$. Multiplication is *commutative, associative*, $1 = \frac{1}{1}$ is the *identity of multipication*, and the *distributive property* hold. An additional property that the rational numbers satisfy is that any nonzero rational $\frac{a}{b}$ has a *reciprocal* $\frac{b}{a}$.

Rational numbers as Decimals

Rational numbers can be written as decimals by dividing the numerator by the denominator. Some examples of rational numbers are:

$$\frac{1}{2} = 0.5, \quad \frac{5}{8} = 0.625, \quad \frac{1}{3} = 0.333... = 0.\overline{3}, \quad \frac{7}{11} = 0636363... = 0.\overline{63}.$$

As we see from these examples, the decimal forms of rational numbers are either *terminating decimals* or *repeating decimals*.

1.3.2 Exercises

1. Perform the operations

   (a) $-2 + \frac{1}{2}$

   (b) $-\frac{1}{3} + 1$

2. Perform the operations

   (a) $-\frac{2}{5} - \left(-\frac{3}{1}\right)$

   (b) $\frac{3}{8} - \frac{4}{3} + 1$

3. Perform the operations

   (a) $-\frac{4}{5} + \frac{3}{10} - 3\frac{1}{2} + 1$

   (b) $3\frac{3}{8} - 5\frac{1}{6} + 3\frac{1}{3} - 1$

4. Perform the operations

   (a) $-15 + 15.5 - \frac{1}{2} + 2.3 - 0.6$

   (b) $-\frac{5}{8} + \left(-\frac{7}{4}\right) + \frac{11}{2} + \frac{12}{16} - 0.25$

5. Perform the operations

   (a) $2\left(-\frac{3}{5}\right)\left(-\frac{1}{2}\right)\left(\frac{3}{4}\right)$

   (b) $\left[\left(-\frac{2}{7}\right)\left(-\frac{3}{8}\right)(-5)\right] \cdot \frac{56}{6}$

6. Perform the operations

   (a) $\left[\left(-\frac{3}{4}\right) + \left(-\frac{2}{3}\right)\right] \cdot \left[\left(-\frac{4}{3}\right) - \left(-\frac{3}{2}\right)\right]$

   (b) $\left(-1 + \frac{3}{2} - \frac{5}{3}\right) \cdot \left(-2 + \frac{1}{2}\right)$

7. Perform the operations

   (a) $\left(-\frac{4}{5}\right) \div \left(-\frac{8}{10}\right)$

   (b) $(-6) \div \frac{2}{3}$

8. Do the operations

   (a) $\left[\left(-\frac{1}{2}\right)\left(-\frac{3}{5}\right)\left(-\frac{2}{7}\right] \div \left(-\frac{3}{5}\right)\right.$

   (b) $\left(-\frac{3}{4} - \frac{6}{2} + 1\right) \div \left(-\frac{1}{2}\right)$

1.3.3 Powers and Roots of Rational Numbers

In addition to the fundamental arithmetic operations $+, -, \cdot, \div$ with rational numbers, we introduce the operations of *raising to a power* and *taking the root* for rational numbers.

Powers with Natural Exponent

Here we study powers of rational numbers with a *natural exponent*. Powers with integer or rational exponents will be studied in Chapter 3.

Let a be a rational number. The product $a \cdot a$ is represented by a^2, read *"a square"* ; the product $a \cdot a \cdot a$ by a^3, read *"a cube"* or *"the 3rd power of a"*. In general, we have the following definition.

Definition 1.3.11. Let a be a rational number and let n be a natural number. The product of n factors each of which is equal to a is denoted by a^n and is called the **power** of a with a natural **exponent** n. That is,

$$a^n = \underbrace{aa \cdots a}_{n \ times}.$$

The number a repeated as a factor is called the **base** of the power.

Of course, $a^1 = a$. We also define $a^0 = 1$, for $a \neq 0$. The expression 0^0 is considered to be *meaningless*.

It follows from the above definition that:

1. $0^n = 0$ and $1^n = 1$.

2. $a^n > 0$ if $a > 0$, ie, any power of a positive number is a positive number.

3. any even power of a negative number is a positive number; for instance, $(-2)^2 = (-2)(-2) = 4 > 0$.

4. any odd power of a negative number is a negative number; for instance, $(-2)^3 = (-2)(-2)(-2) = -8 < 0$.

In particular, for $k = 0, 1, 2, 3, \dots$ we have

$$(-1)^{2k} = 1, \quad (-1)^{2k+1} = -1.$$

Here are some examples of powers:

Example 1.3.12.

$$3^2 = 3 \cdot 3 = 9, \quad \left(-\frac{1}{5}\right)^2 = \left(-\frac{1}{5}\right) \cdot \left(-\frac{1}{5}\right) = \frac{1}{25}, \quad \left(\frac{2}{3}\right)^3 = \frac{2}{3} \cdot \frac{2}{3} \cdot \frac{2}{3} = \frac{8}{27}.$$

Rules of Exponents

Theorem 1.3.13. *Let a, b be rational numbers and m, n natural numbers. Then*

1. $a^n \cdot a^m = a^{n+m}$.

2. $(a^m)^n = a^{mn}$.

3. $(a \cdot b)^n = a^n \cdot b^n$.

4. $\frac{a^m}{a^n} = a^{m-n}$, *where $a \neq 0$, and $m > n$.*

5. $\left(\frac{a}{b}\right)^n = \frac{a^n}{b^n}$, *where $b \neq 0$.*

Proof.   1. $a^n \cdot a^m = \underbrace{(aa \cdots a)}_{n \text{ times}} \underbrace{(aa \cdots a)}_{m \text{ times}} = \underbrace{aa \cdots a}_{(n+m) \text{ times}} = a^{n+m}$.

Thus,
$$a^3 \cdot a^2 = (aaa)(aa) = aaaaa = a^5 = a^{3+2}.$$

2. $(a^m)^n = \underbrace{a^m a^m \cdots a^m}_{n \text{ times}} = a^{\overbrace{m + m + \dots + m}^{n \text{ times}}} = a^{mn}$.

Thus,
$$(a^3)^4 = a^3 \cdot a^3 \cdot a^3 \cdot a^3 = a^{3+3+3+3} = a^{4 \cdot 3} = a^{12}.$$

3. $(a \cdot b)^n = \underbrace{(ab)(ab)\cdots(ab)}_{n\,times} = (\underbrace{aa\cdots a}_{n\,times})(\underbrace{bb\cdots b}_{n\,times}) = a^n \cdot b^n.$

Thus,

$$(ab)^3 = (ab)(ab)(ab) = aaa \cdot bbb = a^3 \cdot b^3.$$

4. $\dfrac{a^m}{a^n} = \dfrac{\overbrace{aa\cdots a}^{m\,times}}{\underbrace{aa\cdots a}_{n\,times}}.$ Reducing the fraction we obtain

$$\frac{a^n}{a^m} = \frac{\overbrace{aa\cdots a}^{(m-n)\,times}}{1} = \frac{a^{m-n}}{1} = a^{m-n}.$$

Thus,

$$\frac{a^5}{a^3} = \frac{aaaaa}{aaa} = \frac{aa}{1} = a^2 = a^{5-3}.$$

5. $\left(\dfrac{a}{b}\right)^n = \underbrace{\dfrac{a}{b} \cdot \dfrac{a}{b} \cdots \dfrac{a}{b}}_{n\,times} = \dfrac{\overbrace{aa\cdots a}^{n\,times}}{\underbrace{bb\cdots b}_{n\,times}} = \dfrac{a^n}{b^n}.$

Thus,

$$\left(\frac{a}{b}\right)^4 = \frac{a}{b} \cdot \frac{a}{b} \cdot \frac{a}{b} \cdot \frac{a}{b} = \frac{aaaa}{bbbb} = \frac{a^4}{b^4}.$$

$\square$

Example 1.3.14. Compute

$$\frac{15^3 \cdot 21^2}{35^2 \cdot 3^4}$$

Solution.

$$\frac{15^3 \cdot 21^2}{35^2 \cdot 3^4} = \frac{(3 \cdot 5)^3 \cdot (3 \cdot 7)^2}{(5 \cdot 7)^2 \cdot 3^4} = \frac{3^3 \cdot 5^3 \cdot 3^2 \cdot 7^2}{5^2 \cdot 7^2 \cdot 3^4} = \frac{3^5 \cdot 5^3 \cdot 7^2}{3^4 \cdot 5^2 \cdot 7^2} = 3 \cdot 5 = 15.$$

Example 1.3.15. Compute

$$\frac{(-2) \cdot (-3)^{17} - (-3)^{16}}{9^7 \cdot 15}.$$

Solution.

$$\frac{(-2)\cdot(-3)^{17}-(-3)^{16}}{9^7\cdot 15}=\frac{(-2)\cdot(-3^{17})-3^{16}}{9^7\cdot 15}=\frac{2\cdot 3^{17}-3^{16}}{9^7\cdot 15}=$$

$$\frac{3^{16}(2\cdot 3-1)}{(3^2)^7\cdot 3\cdot 5}=\frac{3^{16}(6-1)}{3^{14}\cdot 3\cdot 5}=\frac{3^{16}\cdot 5}{3^{15}\cdot 5}=\frac{3^{16}}{3^{15}}=3.$$

Roots

For a natural number $n\geq 2$, we shall define the notion of the nth-root of a rational number. We shall see that *taking the root of a number is an operation inverse to raising to a power.*

Definition 1.3.16. An nth-**root** of a number a is a number b, which when it is raised to the nth power, produces the number a. The nth-root of a is symbolized as $\sqrt[n]{a}$. Thus,

$$\sqrt[n]{a}=b\ \Leftrightarrow\ b^n=a.$$

The symbol $\sqrt{\ }$ is called the **radical sign**, the natural number n is called the *index* of the root, and the numbers a is said the **radicand**.

Clearly, $\sqrt[n]{0}=0$ and $\sqrt[n]{1}=1$.

Even Roots

If $n=2$, the root is called the **square root**; in this case the index 2 is usually ommited. For instance, instead of $\sqrt[2]{5}$ we write $\sqrt{5}$. Thus,

$$\sqrt{a}=b\ \Leftrightarrow\ b^2=a.$$

Note that, since $b^2\geq 0$, the expression $\sqrt{a}$ has sense only for $a\geq 0$. Note also that since $(-b)^2=b^2=a$, the square root of $a>0$ will have two values $\pm b$. We agree to consider only the positive value of $\sqrt{a}$. For instance, regardless that $(-5)^2=5^2=25$, we consider only $\sqrt{25}=5$. In other words, by the square root of a number $a>0$, we mean the *positive square root* of a. Thus, $\sqrt{a}\geq 0$.

The same holds for all even roots, $\sqrt[4]{a}$, $\sqrt[6]{a}$, $\sqrt[8]{a}$ and so on.

Thus, even roots $\sqrt[2k]{a}\geq 0$, and make sense only for nonegative numbers $a\geq 0$.

Odd Roots

When $n = 3$, the third root is called the **cube root**. Here the definition tells us

$$\sqrt[3]{a} = b \iff b^3 = a.$$

Since b^3 can be either positive or negative, the numbers a and $\sqrt[3]{a}$ can also be either positive or negative. For instance,

$$\sqrt[3]{8} = 2 \iff 2^3 = 8,$$

and

$$\sqrt[3]{-8} = -2 \iff (-2)^3 = -8.$$

The same holds for all odd roots, $\sqrt[5]{a}$, $\sqrt[7]{a}$, $\sqrt[9]{a}$, and so on.

Thus, odd roots $\sqrt[2k+1]{a}$ can be either positive or negative depending whether a is positive or negative number.

Here are some examples of roots.

Example 1.3.17.

$$\sqrt{9} = 3 \iff 3^2 = 9, \quad \sqrt{100} = 10 \iff 10^2 = 100.$$

$$\sqrt{\frac{1}{36}} = \frac{1}{6} \iff \left(\frac{1}{6}\right)^2 = \frac{1}{36}, \quad \sqrt{\frac{4}{9}} = \frac{2}{3} \iff \left(\frac{2}{3}\right)^2 = \frac{4}{9}.$$

$$\sqrt[3]{-27} = -3 \iff (-3)^3 = -27, \quad \sqrt[3]{64} = 4 \iff 4^3 = 64.$$

$$\sqrt[3]{-\frac{64}{125}} = -\frac{4}{5} \iff \left(-\frac{4}{5}\right)^3 = -\frac{64}{125}.$$

$$\sqrt[4]{81} = 3 \iff 3^4 = 81, \quad \sqrt[5]{-32} = -2 \iff (-2)^5 = -32.$$

1.3.4   Exercises

1. Simplify

   (a) $(-2)^3(-2)^2$

   (b) $\left(-\frac{1}{2}\right)^4\left(-\frac{2}{3}\right)^3$

   (c) $\left(-\frac{3}{5}\right)^5 \div \left(-\frac{3}{5}\right)^3$

   (d) $\left[\left(-\frac{1}{2}\right)^2\right]^3$.

2. Simplify

   (a) $(0.03)^3$

   (b) $(-1.2)^2$

   (c) $(-1.4)^3 \cdot \left(3\frac{4}{7}\right)^3$

   (d) $\sqrt[3]{\frac{1}{27}} - \frac{5}{6}\sqrt[3]{27}$.

3. Perform the operations

   (a) $\sqrt{64} - \sqrt[4]{81} - \sqrt[5]{-32}$

   (b) $\frac{2}{3}\sqrt{81} - \frac{7}{3}\sqrt{\frac{36}{49}}$.

1.4   Irrational numbers

The set of rational numbers has some mathematical limitations. For example, *there is no rational number whose square is* 2. Let us prove this.

Proposition 1.4.1. $\sqrt{2}$ *is not a rational number.*

Proof. Suppose the contrary that $\sqrt{2}$ is a rational number. That is, suppose $\sqrt{2} = \frac{a}{b}$, where $a, b \in \mathbb{N}$ and the fraction $\frac{a}{b}$ is in its *lowest terms*.

According to the definition of the square root, we have

$$\sqrt{2} = \frac{a}{b} \Leftrightarrow \left(\frac{a}{b}\right)^2 = 2 \Leftrightarrow a^2 = 2b^2.$$

This means that a^2 is an even number. But a^2 is even only if a is even (if a is not dividible by 2, then a^2 is not divisible by 2 either). Since a is even, there is an integer k such that $a = 2k$. Then $(2k)^2 = 2b^2$ or $b^2 = 2k^2$. That is, b is also an even number, say $b = 2m$. Consequently, the fraction

$$\frac{a}{b} = \frac{2k}{2m} = \frac{k}{m}$$

is reducible, which contradicts the hypothesis. From the obtained contradiction it follows that $\sqrt{2}$ is not a rational number. $\qquad\square$

A need therefore arises of introducing new numbers, distinct from rational numbers, such for instance, the number $\sqrt{2}$. There are many other numbers which are not rational. It can be proved that the square root of any non-perfect square number is not rational. For example, the numbers

$$\sqrt{3}, \quad \sqrt{5}, \quad \sqrt{6}, \quad \sqrt{7}, \quad \sqrt{8}, \sqrt{10} \quad \text{are not rational}.$$

Also the cube root of a any non-perfect cube number is not rational, and so on. Thus, for example

$$\sqrt[3]{2}, \quad \sqrt[3]{4}, \quad \sqrt[3]{7}, \quad \sqrt[3]{9}, \quad \sqrt[4]{2}, \quad \sqrt[4]{4}, \quad \sqrt[5]{25} \quad \text{are not rational}.$$

These numbers are called ***irrational numbers***.[1]

In contrast to rational numbers, which when are written as decimals either terminate, or if not, they repeat a decimal digit or a block of digits, the *irrational numbers written as decimals contain an infinite number of decimal digits and do not repeat a digit or blocks of digits.*

---

[1]Irrational numbers do not arise only by taking roots. Most values of logarithms of positive numbers are irrational, and most values of trigonometric functions are also irrational.

For example, $\sqrt{2} \approx 1.411421356237...$ and $\sqrt{5} \approx 2.236067978...$ The number $\pi \approx 3.141592654....$ used to denote the ratio of the circumferance of a circle to its diameter is also irrational.

The four arithmetic operations $+, -, \cdot, \div$ can be performed with irrational numbers, and the basic properties of arithmetic operations with rational numbers hold for the irrational numbers as well. A rigorous proof of these operations and their properties is given in higher mathematics.

1.5   Real numbers

The set of rational numbers together with set of irrationals form the set of **real numbers**. Thus, a real number means either rational or irrational number. The set of real numbers is denoted by $\mathbb{R}$.

Every real number can be approximately replaced by a terminating decimal.

Example 1.5.1. 1. The terminating decimals -0.8, -0.87 are approximate values for the number $-\frac{7}{8} = -0.875$.

2. The terminating decimals 0.1, 0.16, 0.166, 0.1666 are approximate values for the number $\frac{1}{6} = 0.1\overline{6}$.

3. The terminating decimals 1.4, 1.41, 1.411, 1.4114, 1.41142 are approximate values for the number $\sqrt{2}$.

4. The terminating decimals 3.1, 3.14, 3.141, 3.1415, 3.14159 are approximate values for the number π.

The terminating decimals we use to replace an irrational number a are termed as *rational approximating representantives* of a. The more decimal digits a rational representative q of a contains, the *better the rational q approximates a*. We write $a \approx q$

Real numbers are subject to the four arithmetic operations, with the aid of their approximating representatives (a rigorous development of this subject is beyond the level of the present book, and practically for us here does not have any purposefulness). To illustrate the method, we consider an example: Let $a = \sqrt{3}$ and $b = \sqrt{5}$. Even though in mathematics there is no need to convert $\sqrt{3}$, $\sqrt{5}$ and $\sqrt{3} + \sqrt{5}$ into decimal form, in order to find approximately the sum $\sqrt{3} + \sqrt{5}$, we write $\sqrt{3}$ and $\sqrt{5}$ using rational approximating representatives with the *same* number of decimal digits. For instance, $\sqrt{3} = 1.73$ and $\sqrt{5} = 2.23$. Then we find the sum $1.73 + 2.23 = 3.96$ and we say that *the number 3.96 is an approximating representative of the sum $a + b$.* In practice, we say that the sum $\sqrt{3} + \sqrt{5}$ is *about* 3.96, and we write

$$\sqrt{3} + \sqrt{5} \approx 3.96.$$

Properties of Real Numbers

The basic properties of addition and multiplication hold for real numbers. Thus, if a, b and c are real numbers, then

1. $a + b = b + a \qquad\qquad ab = ba$
   (*commutativity of addition and multiplication*);

2. $(a + b) + c = a + (b + c) \qquad\quad (ab)c = a(bc)$
   (*associativity of addition and multiplication*);

3. $0 + a = 0 \qquad\qquad 1 \cdot a = a$
   (*identity of addition and identity of multiplication*);

4. $0 \cdot a = a$ (*property of zero in multiplication*);

5. $a + (-a) = 0, \qquad\qquad a \cdot \frac{1}{a} = 1$ for $a \neq 0$
   (*opposite and reciprocal of a nonzero number*);

6. $a(b + c) = ab + ac$ (*distributive property*);

The following generalization of the distributive property also holds:

$$a(b_1 + b_2 + b_3 + ... + b_n) = ab_1 + ab_2 + ab_3 + ... + ab_n.$$

1.5.1   The Real Number Line

We shall now give a *geometric interpretation* to real numbers. Assume
we are given a horizontal straight line. It has two opposite directions,
one positive and the other negative. For definiteness we take the direction to the right (as we look at the picture) as the positive direction.
We fix a point O on that line and we call it the *origin*. The point O
divides the line into two parts called *rays*. The ray directed to the right
is a *positive* ray and that directed to the left a *negative* ray. Assume we
are given a line segment accepted as a *unit* of length. In that case we
say that a *scale* is introduced on the directed line. Such a straight line
is known as a **number line** or **number axis**.

Any real number can be represented by a point on the number line.
Indeed, *there is a correspondence between every real number and a point
of the number line* which obeys the following rule: The number zero
corresponds to the origin. Every positive number a is associated with a
point **A** on the positive ray, where the length of the segment OA is a.
Every negative number b is associated with a point **B** on the negative
ray, where the length of the segment OB is $|b|$.

$$
\begin{array}{ccccccccc}
 & \mathbf{B} & & & \mathbf{O} & & & \mathbf{A} & \\
 & b & & -2 \; -1 & 0 & 1 \; 2 & & a &
\end{array}
$$

We have thus put *every real number into a correspondance with a unique
point of the number line (to the chosen scale), and vice versa.*

It is important to note that real numbers are often identified with
points on the number line associated with them. The number 0 is identified with the point **O**, the number a with the point **A**, and the number
b with point **B** on the number line. The number a is called the *coordinate* of the point **A**. The coordinate of the point **B** is the number b.
The coordinate of the origin **O**, is of course, the number 0.

Using this fact, we can easily say which of two real numbers is
greater: *the number which is to the right of the other number on the
number line is greater.*

This definition can be written in other form using comparison of real
numbers with the number zero, namely, the number a is *greater* than

the number b if and only if the difference $a - b$ is positive, ie,

$$a > b \iff a - b > 0.$$

To say that a is *greater or equal* to b we write $a \geq b \iff a - b \geq 0$. Similarly, we say a is *smaller* that b if and only if $a - b$ is negative, ie,

$$a < b \iff a - b < 0.$$

and to say a is *less or equal* to b we write $a \leq b \iff a - b \leq 0$.

1.5.2   The Absolute Value

Definition 1.5.2. If a is a real number, the **absolute value** of a is defined by

$$|a| = \begin{cases} a & \text{if } a \geq 0, \\ -a & \text{if } a < 0. \end{cases}$$

Example 1.5.3.

$$|-4| = 4, \quad \left|-\frac{1}{2}\right| = \frac{1}{2}, \quad -\left|-\frac{3}{5}\right| = -\frac{3}{5}, \quad |-\sqrt{3}| = \sqrt{3}, \quad |-\pi| = \pi.$$

Let us state some immediate consequences of the definition: First, note that since $(-a)^2 = (-a)(-a) = a^2$, we have

$$|a|^2 = a^2.$$

Note also that

$$-|a| \leq a \leq |a|.$$

By virtue of the the definition of the square root, we also have

$$\sqrt{a^2} = |a|.$$

Moreover, for $k > 0$ and any $x \in \mathbb{R}$, we have (see, Section 7.2.3)

$$|x| < k \iff -k < x < k \tag{1.1}$$

Theorem 1.5.4. *Let a and b be real numbers. Then*

1. $|a| \geq 0$ *and* $|a| = 0 \iff a = 0$;

2. $|ab| = |a| \cdot |b|$ *and* $\left|\frac{a}{b}\right| = \frac{|a|}{|b|}$ *when* $b \neq 0$;

3. $|a + b| \leq |a| + |b|$ *(triangle inequality)*.

Proof.   1. This is obvious from the definition.

2. Since for any $a \in \mathbb{R}$, we have $\sqrt{a^2} = |a|$, it follows

$$|ab| = \sqrt{(ab)^2} = \sqrt{a^2 b^2} = \sqrt{a^2} \cdot \sqrt{b^2} = |a| \cdot |b|.$$

Moreover, for $b \neq 0$, we have

$$\left|\frac{a}{b}\right| = |a \cdot \frac{1}{b}| = |a| \cdot \frac{1}{|b|} = \frac{|a|}{|b|}.$$

3. We have $-|a| \leq a \leq |a|$ and $-|b| \leq b \leq |b|$. Adding these inequalities side by side, we get

$$-(|a| + |b|) \leq a + b \leq |a| + |b|.$$

By (1.1), this implies $|a + b| \leq |a| + |b|$.

$\square$

By property (2) we also have $|-a| = |(-1)a| = |-1| \cdot |a| = 1 \cdot |a| = |a|$. In particular, $|b - a| = |-a + b| = |-(a - b)| = |a - b|$. Thus,

$$|b - a| = |a - b|.$$

Exercise 1.5.5. Let a and b be real numbers. Show that

$$||a| - |b|| \leq |a - b|.$$

We now give the following definition, which tells us the geometrical meaning of the absoloute value.

Definition 1.5.6. Let a and b be real numbers represented on the number line by the points **A** and **B**, respectively. The ***distance*** d beteween the point **A** with coordinate a and the point **B** with coordinate b is given by

$$d = |b - a|.$$

Since $|a| = |a - 0|$, we see that the absolute value a is its *distance from the origin*. That is, the absolute value of a number is a nonnegative number which tells us how many units away from 0 the given number lies regardless of direction (below or above zero).

Example 1.5.7. Find the distance on the number line between the numbers: **(1)** 5 and 8, **(2)**. 3 and -4, **(3)**. -1 and -6 **(4)** $\frac{1}{2}$ and $\frac{1}{3}$ **(5)**. π and 2.

Solution. (1). $d = |8 - 5| = |3| = 3.$

(2). $d = |-4 - 3| = |-4 + (-3)| = |-7| = 7.$

(3). $d = |-6 - (-1)| = |-6 + 1| = |-5| = 5.$

(4). $d = \left|\frac{1}{3} - \frac{1}{2}\right| = \left|\frac{2}{6} - \frac{3}{6}\right| = \left|-\frac{1}{6}\right| = \frac{1}{6}.$

(5). $d = |2 - \pi| = -(2 - \pi) = \pi - 2.$

Intervals

The set of real numbers x for each of which the two-sided inequality $-2 < x < 3$ holds true is usually designated as $(-2, 3)$ and is called the *open interval* $(-2, 3)$. The set of real numbers satisfying the two-sided inequality $-2 \leq x \leq 3$ is designated as $[-2, 3]$ and is called the *closed interval* $[-2, 3]$. The interval $[-2, 3)$ of all real numbers satisfying $-2 \leq x < 3$ is usually called the *half-open interval* $[-2, 3)$, while the interval $(-2, 3]$ of all real numbers satisfying $-2 < x \leq 3$ is usually called the *half-closed interval* $(-2, 3]$.

In the general case, when a, b are two real numbers with $a < b$. We have the following types of **intervals**:

1. $(a, b) = \{x : a < x < b\}$ *open interval*;

2. $[a, b] = \{x : a \leq x \leq b\}$ *closed interval*;

3. $[a, b) = \{x : a \leq x < b\}$ *half-open interval*;

4. $(a, b] = \{x : a < x \leq b\}$ *half-closed interval*.

On the number line, we also have the following types of *unbounded intervals*:

1. $(a, \infty) = \{x : a < x\}$

2. $[a, \infty) = \{x : a \leq x\}$

3. $(-\infty, b) = \{x : x < b\}$

4. $(-\infty, b] = \{x : x \leq b\}$

5. $\mathbb{R} = (-\infty, \infty) = \{x : -\infty < x < \infty\}$.

The symbol ∞ or $+\infty$ is called **infinity** and denotes an *arbitrarily large* quantity (it is not a number, but a notion). The symbol $-\infty$ is read as negative infinity.

1.6 Numerical Expressions

A *numerical expression* is an expression composed of numbers and signs indicating the operations[2] to be performed with them.

For example,

$$30 \div 6 + 4, \quad 2 \cdot (3^2 - 7), \quad 1 - \sqrt{5^2 - 4^2}, \quad \tfrac{3}{5} + \tfrac{1}{4} - 1,$$

are numerical expressions. If all the operations indicated in a numerical expression can be performed the number obtained as a final result is called the *numerical value* of the given numerical expression. Let us find the values of the above numerical expression:

$$30 \div 6 + 4 = 5 + 4 = 9, \quad 2 \cdot (3^2 - 7) = 2 \cdot (9 - 7) = 2 \cdot 2 = 4.$$

$$1 - \sqrt{5^2 - 4^2} = 1 - \sqrt{25 - 16} = 1 - \sqrt{9} = 1 - 3 = -2.$$

$$\frac{3}{5} + \frac{1}{4} - 1 = \frac{12}{20} + \frac{5}{20} - \frac{20}{20} = -\frac{3}{20}.$$

Frequently, we use grouping symbols to separate operations. The most common such symbol is parenthesis (...). In more involved expressions we also use brackets [...] and braces {...}.

Order of Operations

When several operations occur in an expression, they should be done in the following order to get the correct result:

1. Perform the operations inside the grouping symbols, braces, brackets, parentheses, according to the steps 2, 3 and 4 below.

2. Find the value of any power.

3. Perform all multiplications and/or divisions, working from left to right.

---

[2]Throughout the book, it is assumed that the indicated operations in all expressions are applied a finite number of times.

4. Perform all additions and /or subtractions, working from left to right.

Example 1.6.1. Evaluate the numerical expression.

$$\frac{3\left[5 - 2(4-7)^2 + (-2)^3 - 3\right]}{6 + 10(-8+5) \div 2}.$$

Solution. We operate using the order of operations, and we have

$$\frac{3\left[5 - 2(4-7)^2 + (-2)^3 - 3\right]}{6 + 10(-8+5) \div 2} = \frac{3\left[5 - 2(-3)^2 + (-8) - 3\right]}{6 + 10(-3) \div 2}$$

$$= \frac{3\left[5 - 2(9) + (-8) - 3\right]}{6 + 10(-3) \div 2} = \frac{3\left[5 - 18 + (-8) - 3\right]}{6 + (-30) \div 2}$$

$$= \frac{3\left(-13 - 8 - 3\right)}{6 + (-15)} = \frac{3\left(-21 - 3\right)}{-9} = \frac{3\left(-24\right)}{-9} = \frac{-24}{-3} = 8.$$

Example 1.6.2. Compute.

$$\frac{\left(2\frac{1}{3} - 1\frac{2}{9}\right) \div 4 - 1\frac{2}{3}}{3\frac{2}{7} - \frac{3}{14} \div \frac{1}{6} - 1\frac{1}{2}}.$$

Solution. We have

$$\frac{\left(2\frac{1}{3} - 1\frac{2}{9}\right) \div 4 - 1\frac{2}{3}}{3\frac{2}{7} - \frac{3}{14} \div \frac{1}{6} - 1\frac{1}{2}} = \frac{\left(\frac{7}{3} - \frac{11}{9}\right) \div 4 - \frac{5}{3}}{\frac{23}{7} - \frac{3}{14} \cdot \frac{6}{1} - \frac{3}{2}} = \frac{\left(\frac{21}{9} - \frac{11}{9}\right) \div 4 - \frac{5}{3}}{\frac{23}{7} - \frac{9}{7} - \frac{3}{2}}$$

$$= \frac{\left(\frac{10}{9}\right) \cdot \frac{1}{4} - \frac{5}{3}}{\frac{14}{7} - \frac{3}{2}} = \frac{\frac{5}{18} - \frac{5}{3}}{\frac{2}{1} - \frac{3}{2}} = \frac{\frac{5}{18} - \frac{30}{18}}{\frac{4}{2} - \frac{3}{2}} = \frac{-\frac{25}{18}}{\frac{1}{2}} = -\frac{25}{18} \div \frac{1}{2} = -\frac{25}{18} \cdot \frac{2}{1} = -\frac{25}{9}.$$

1.7 Review Exercises

1. Simplify
$$(12 - 15) \div (-3) + (23 - 3) \div (-4)$$

2. Simplify
$$6 - (-5) \cdot (-2) + (-14) \div (-7) + 7$$

3. Simplify
$$12 - 6 \cdot (-3) + 7 - 15 \div (-3) + 18 - 16 \div (-4) + 1$$

4. Simplify
$$(-3 + \frac{3}{5}) \cdot (-\frac{5}{2}) + (2 - \frac{2}{5}) \div (-\frac{4}{5})$$

5. Do the operations
$$\left(\frac{1}{6} - \frac{1}{3} \right) \div \left(\frac{1}{3} - \frac{1}{6} \right) - \left(\frac{3}{2} - \frac{5}{4} \right) \div \left(-\frac{5}{2} + \frac{1}{4} \right)$$

6. Do the operations
$$\left(2 - \frac{1}{2} + \frac{3}{5} \right) \cdot (5) - \left(\frac{1}{3} + 4 - \frac{5}{6} \right) \div (-3)$$

7. Simplify
$$\frac{(3 \cdot 4)^3}{(2 \cdot 3 \cdot 4)^2}$$

8. Simplify
$$\frac{63^4 \cdot 35^3}{5^5 \cdot 21^7}$$

9. Simplify
$$(3 \cdot 2^{20} + 7 \cdot 2^{19}) \cdot 52 \div (13 \cdot 8^4)^2$$

10. Simplify
$$24 \cdot \left(-\frac{1}{3} \right)^3 \cdot \left(4\frac{1}{2} \right)^2 .$$

Chapter 2

Introduction to Algebra

In this chapter we introduce variables, algebraic expressions, first degree equations, rates, ratios, percentages, proportions and variations.

2.1 Algebraic Expressions

The subject of algebra involves the study of equations and a number of other issues that developed from the theory of equations. The roots of algebra lie in antiquity. In Babylonia, Greece, China, India, the Middle East, and eventually in Italy and Medieval Europe, algebra reaches the form that we use today.

Constants and Variables

When studying various phenomena of nature and in our everyday practical activity, we come across various quantities, such as length, area, volume, mass, temperature, time and so on. Depending on concrete conditions, some quantities have constant numerical values, while others vary. These quantities are called **constant** and **variable**, respectively.

In algebraic discussions we find it convenient to use letters from the Latin alphabet such as $a, b, c, d, e, ..., x, y, z, w$, or from the Greek alphabet such as $\alpha, \beta, \gamma, \delta, \epsilon, ..., \varphi, \chi, \psi, \omega$. A letter may be used to denote a quantity which *varries* or is *unknown*, that is, to represent a *variable*. A letter can also be used to denote any given number whatsoever having a particular value, which it is to retain throughout the discussion, that is, to represent a *constant*.

It customary to represent variables or unknowns by the later letters x, y, z, and constants by the early letters of the alphabet a, b, c. Besides simple letters we sometimes use letters with subscripts, for instance, a_0, $a_1, a_2, ..., a_n$ read "a sub-null", "a sub-one", "a sub-two", ..., "a sub-n".

Algebraic Expression

An **algebraic expression** is an expression involving numbers (or constants), variables, operation signs $+, \ -, \ \cdot, \ \div,$ *powers*, and/or *roots*. For example,

$$2x + 5, \quad a^2 - b^2, \quad x^5y^4 - 8x^2y^3, \quad \frac{x^2-1}{5x+3}, \quad \frac{x^2-y^2+z^2}{xyz}, \quad \sqrt{x^2+1}, \quad \frac{xy}{\sqrt{x^2+y^2}},$$

are algebraic expressions.

The algebraic expressions[1] are classified according to their complexity as: *monomials*, *polynomials*, *rational expressions* and *radical expressions*. In Chapter 4, we shall study these expressions in detail, and we will learn how to simplify and perform operations with them.

We shall use the notation such as $A = A(x)$ to denote an algebraic expression in a single variable x. An expression may contain more than one variable, say x, y, z. The numerical values of the variables x, y, z for which all operations indicated in the expression make sense are called *permissible values*. These values form the *domain of definition* of the expression.

---

[1]An algebraic expression is a particular case of a mathematical expression. Other mathematical expressions which we study here will involve logarithms and exponentials. Trigonometric expressions are part of Trigonometry.

For instance, the value $x = 0$ is *not* a permissible value for the expression $\frac{1}{x}$ since division by zero is not permitted. Thus, $\frac{1}{x}$ is defined for all $x \neq 0$. Similarly, the expression $\frac{x}{x-3}$ is defined for $x \neq 3$.

2.1.1 Evaluating Algebraic Expressions

To **evaluate an algebraic expression** for given values of its variables, means to substitute these values in the expression and find the numerical value of the resulting numerical expression.

Example 2.1.1. Evaluate

$$5x - 3y$$

when $x = -2$ and $y = -6$.

Solution. Setting $x = -2$ and $y = -6$ in the expression $5x - 3y$ we get

$$5(-2) - 3(-6) = -10 + 18 = 8.$$

Example 2.1.2. Evaluate

$$x^3 + x^2 - x - 1$$

when $x = -\frac{1}{2}$.

Solution. Substituting $x = -\frac{1}{2}$ we get

$$\left(-\frac{1}{2}\right)^3 + \left(-\frac{1}{2}\right)^2 - \left(-\frac{1}{2}\right) - 1 = -\frac{1}{8} + \frac{1}{4} + \frac{1}{2} - 1 = -\frac{3}{8}.$$

Example 2.1.3. Evaluate

$$\frac{x^3 - xy^4}{x - 3y}$$

when $x = -1$ and $y = -2$.

Solution. Substituting $x = -1$ and $y = -2$ we get

$$\frac{(-1)^3 - (-1)(-2)^4}{-1 - 3(-2)} = \frac{-1 + 16}{-1 + 6} = \frac{15}{5} = 3.$$

2.1.2 Exercises

1. Evaluate
$$x^2 - 5x + 6$$
when $x = 3$.

2. Evaluate
$$3x^2 - 13x - 10$$
when $x = -\frac{2}{3}$.

3. Evaluate
$$C = \frac{5}{9}(F - 32)$$
when $F = -4$.

4. Evaluate
$$4x^2 - 3xy + 4y$$
when $x = -3$ and $y = -2$.

5. Evaluate
$$\frac{x^2 - 6x - 7}{x + 3}$$
   (a) when $x = 0$
   (b) when $x = -1$.

6. Evaluate
$$\frac{x - y}{x + y}$$
when $x = \frac{1}{3}$ and $y = \frac{1}{2}$.

7. Evaluate
$$\frac{y - x^2 z^3}{2z^2 - 3y^2}$$
when $x = -2$, $y = 0$ and $z = 3$.

2.2 First Degree Equations

In general, an **equation** is an equality $A(x) = B(x)$ of two variable expressions A and B. A **solution** or **root** of an equation is a value of the unknown x which when substituted into the equation results into a true equality. In other words, a solution is a value of x which satisfies the equation. To **solve** an equation means to find all its solutions.

Two equations are called **equivalent** if they have the same solutions. We will use the symbol $\Leftrightarrow$ to indicate equivalent equations.

Here are some basic statements which make equivalent transition possible, and are frequently used in the process of solving equations:

1. $A = B \Leftrightarrow A - B = 0$.

2. $A = B \Leftrightarrow A + k = B + k$, for any real number k.

3. $A = B \Leftrightarrow kA = kB$, for any real number $k \neq 0$.

The simplest equations are the first degree equations. A **first degree equation** in one unknown x is an equation of the form

$$ax + b = 0,$$

where a, b are given real numbers, with $a \neq 0$. The numbers a, b are called the *coefficients* of the equation[2].

Theorem 2.2.1. *The first degree equation*

$$ax + b = 0 \quad (a \neq 0)$$

has a unique solution given by

$$x = -\frac{b}{a}.$$

---

[2]Note that if $a = 0$, then the equation is $0x + b = 0$ or $b = 0$. which when $b \neq 0$ is impossible (and we say, the equation has no solution).
If $a = b = 0$, then the equation becomes $0x + 0 = 0$, that is, $0 = 0$, which is true, and there is nothing to solve.

Proof. We first separate knowns from unknowns, by adding the opposite $-b$ of b to both sides of the equation to obtain equivalently

$$ax + b = 0 \quad \Leftrightarrow \quad ax + b - b = 0 - b \quad \Leftrightarrow \quad ax = -b.$$

Now, since $a \neq 0$, we divide both sides of $ax = -b$ by a (or equivalently we multiply both sides by $\frac{1}{a}$) and we find the unique solution

$$x = -\frac{b}{a}.$$

Conversely, substituting $x = -\frac{b}{a}$ into the equation $ax + b = 0$, we obtain

$$a\left(-\frac{b}{a}\right) + b = -b + b = 0.$$

$\square$

Example 2.2.2. Solve the equation

$$2x + 6 = 0.$$

Solution. We separate knowns from unknowns

$$2x + 6 = 0 \quad \Leftrightarrow \quad 2x = -6 \quad \Leftrightarrow \quad x = \tfrac{-6}{2} \quad \Leftrightarrow \quad x = -3.$$

Example 2.2.3. Solve the equation

$$3x - 7 = x + 5.$$

Solution. We separate knowns from unknowns

$$3x - 7 = x + 5 \quad \Leftrightarrow \quad 3x = x + 5 + 7 \quad \Leftrightarrow \quad 3x - x = 12 \quad \Leftrightarrow \quad 2x = 12.$$

Now we divide both sides by 2 to get $x = \frac{12}{2}$ or $x = 6$. To check the solution $x = 6$, means to substitute 6 for x into the original equation to see whether we get a true statement. Indeed, $3(6) - 7 = 6 + 5 \Rightarrow 11 = 11$.

Example 2.2.4. Solve the equation

$$2(x - 1) + 1 = 3 - (1 - 2x).$$

Solution. Here we must simplify first and then solve the equation. Removing the parentheses, we have

$$2x-2+1 = 3-1+2x \iff 2x-1 = 2+2x \iff 2x-2x = 2+1 \iff 0x = 3$$

That is, $0 = 3$, which impossible, and the equation has no solution.

Example 2.2.5. Solve the equation

$$\frac{x}{5} - \frac{3}{5} = \frac{x}{2} + \frac{1}{2}.$$

Solution. Here the coefficients are fractions. In such a case, it is preferable to multiply both sides of the equation by the least common denominator, so as to clear the equation of fractions. The LCD of the denominators is 10. Multiplying both sides by 10, we have

$$10\left(\frac{x}{5} - \frac{3}{5}\right) = 10\left(\frac{x}{2} + \frac{1}{2}\right) \iff 2x - 6 = 5x + 5$$

$$\iff 2x - 5x = 5 + 6 \iff -3x = 11 \iff x = -\frac{11}{3}.$$

Example 2.2.6. Solve the equation $ax = x + 1$ for x.

Solution. We have

$$ax = x + 1 \iff ax - x = 1 \iff (a-1)x = 1$$

If $a - 1 \neq 0$, ie, $a \neq 1$, then we divide both sides by $a - 1$ to find the solution $x = \frac{1}{a-1}$. If $a - 1 = 0$, ie, $a = 1$, then the equation is $0x = 1$ or $0 = 1$ which is impossible and the equation has no solution.

Frequently, we are interested in solving an equation for one variable in terms of others. Here is an example:

Example 2.2.7. (*Converting to Fahrenheit from Celsius*). Solve for F:

$$C = \frac{5}{9}(F - 32).$$

Solution. Multiplying by 9, we have

$$C = \tfrac{5}{9}(F - 32) \iff 9C = 5(F - 32) \iff 9C = 5F - 160 \iff$$

$$9C + 160 = 5F \iff 5F = 9C + 160 \iff F = \tfrac{9}{5}C + 32$$

2.2.1   Exercises

Solve the equations.

1. $7x - 4 = -2x + 5$

2. $3x + 5 = 3 - x$

3. $2(3 - 2x) = 3(x - 5)$

4. $8x - (2x + 1) = 3x - 10$

5. $x - \frac{2x-1}{3} = \frac{3(x+1)}{4}$

6. $\frac{1}{2}x - 2 = \frac{3}{4}x$

7. $2.5x = 1 + 1.25x$

8. $5(x - 2) - 2(3 - x) = 3x - 4$

9. $\frac{2x+1}{3} + \frac{1}{2} = 3x$

10. $\frac{x+1}{3} + \frac{x-2}{4} = 1$

11. $\frac{2x-1}{7} + \frac{x}{3} = x - 7$

12. $x + \frac{2x-7}{3} = 1 - \frac{x-5}{2}$

2.3   Ratios and Rates

When we compare two numbers, sometimes it is necessary to find out how many times one number is greater (or smaller) than the other or to express one number as a fraction of the other. For instance, 5 is $\frac{1}{6}$ of 30 or 30 is six times greater than 5. In this case we often say: "the ratio of 5 to 30 is equal to $\frac{1}{6}$ " or "the ratio of 30 to 5 is equal to 6 ".

In general, the ***ratio of a number*** a ***to a number*** b is defined as the quotient $\frac{a}{b}$. The ratio of a to b is denoted by $a : b$ or $\frac{a}{b}$.

For example,

1). The ratio of 12 to 10 is equal to 1.2, since $\frac{12}{10} = 1.2$.

2). The ratio 40 to 60 is $40 : 60 = \frac{40}{60} = \frac{2}{3}$.

By a ratio we do not mean only the result of division of one number by the other, but also the expression itself. The numbers a and b forming a ratio $a : b$ are called the *terms* of the ratio. The number a is called the *antecedent* and the number b the *consequent*. Note that, since ratios such as $a : b$ are fractions, their properties are the properties of fractions.

Example 2.3.1. If $\frac{a}{b} = 2$, find the ratio

$$\frac{a + b}{2a - b}.$$

Solution. We divide the terms of the ratio $\frac{a+b}{2a+b}$ by b, and we have

$$\frac{a + b}{2a - b} = \frac{\frac{a}{b} + \frac{b}{b}}{\frac{2a}{b} - \frac{b}{b}} = \frac{\frac{a}{b} + 1}{2\frac{a}{b} - 1} = \frac{2 + 1}{2 \cdot 2 - 1} = \frac{3}{4 - 1} = \frac{3}{3} = 1.$$

Rates

In mathematics, physics, and other sciences we frequently use ratios of homogeneous quantities. Sometimes it is possible to consider ratios of heterogeneous quantities, say, for example, the ratio $s : t$ of the distance travelled s to time t. Such ratios are knwon as **rates** and represent a new quantity. The rate $\frac{s}{t}$ represents the **velocity** (or *speed*) denoted by v. Thus

$$v = \frac{s}{t} \quad \Leftrightarrow \quad s = v \cdot t.$$

That is,

$$distance = velocity \cdot time.$$

Objects that are moving in accordance to the above formula are said to be in **uniform motion**.

Rates depend on the choice of the unit of measurement. For instance, let us find the travelling speed of a train if the latter covered 350 km during 5 hours. We have

$$\frac{350}{5} = \frac{70}{1} = 70\,(km/h).$$

If the distance travelled and time are given in other units, say meters and seconds, then 350 km = 350,000 meters and 5 h = 18,000 seconds. Consequently, we obtain a different rate

$$\frac{350,000}{18,000} = \frac{19\frac{4}{9}}{1} = 19\frac{4}{9}\,(m/sec).$$

The rates $\frac{70\,km}{1\,h}$ and $\frac{19\frac{4}{9}\,m}{1\,sec}$ are called **unit rates**. Unit rates are used to compare rates expressed in the same units. We illustrate this with an example:

Example 2.3.2. Which is the best buy when a 12 ounce package of cream cheese costs \$4.29 and an 8 ounce package of cream cheese costs \$3.09?

Solution. We must express the given rates $\frac{\$4.29}{12\,oz}$ and $\frac{\$3.09}{8\,oz}$ into unit rates. We have

$$\frac{4.29}{12} = \frac{4.29 \div 12}{12 \div 12} = \frac{0.3575}{1} = 0.3575 = 36\ cents\ per\ oz.$$

$$\frac{3.09}{8} = \frac{3.09 \div 8}{8 \div 8} = \frac{0.38625}{1} = 0.38625 = 39\ cents\ per\ oz.$$

Hence, the best buy is the 12 ounce package of cream cheese.

2.4   Percentages

A ratio of two numbers is often expressed in hundredths, ie, as a part of a hundred. A part of 100 is called a **percent** and is denoted by the symbol %. Thus, $1\% = \frac{1}{100} = 0.01$, $50\% = \frac{50}{100} = 0.5$, $100\% = \frac{100}{100} = 1$. In general,

$$r\% = \frac{r}{100}.$$

Example 2.4.1. The ratio $\frac{2}{5} = 0.4 = 0.40 = \frac{40}{100}$ or 40% . We say that *the number* 2 *is* 40% *of the number* 5.

The Basic Percent Equation

To express a ratio $\frac{a}{b}$ of two numbers a and b as a percent, we must multiply the value of this ratio by 100. Indeed, let the ratio $\frac{a}{b} = r\%$. This means $\frac{a}{b} = \frac{r}{100}$, and multiplying both sides by 100 we get $\frac{a}{b} \cdot 100 = r$. Note also that by multiplying both sides of the equality

$$\frac{r}{100} = \frac{a}{b}$$

by b we obtain equivalently

$$\frac{r}{100} \cdot b = a.$$

This is called the **basic percent equation**. It reads

$$r \ \textbf{\textit{percent of}} \ b \ \textbf{\textit{is}} \ a.$$

The number b is said the **base** *of the percent*.

According to which of the three numbers r, a, and b in the basic percent equation is unknown, we have the following three fundamental problems on percentages:

Problem 1. Finding the percent of a given number.

Example 2.4.2. Find 45% of 180.

Solution. Here $r = 45$, the base $b = 180$, and $a = x$ is unknown. Substituting in the percent equation $\frac{r}{100} \cdot b = a$, we have

$$x = \frac{45}{100} \cdot 180 = 0.45 \cdot 180 = 81.$$

Hence, 45% of 180 is 81. Note that all it takes here is to multiply the percent by the number.

Example 2.4.3. In a class of 25 students 84% were present at the lesson. How many students were present at the lesson?

Solution. Here the question is "what is 84% of 25?". Since $r = 84$, $b = 25$, and $a = x$, we obtain

$$x = \frac{84}{100} \cdot 25 = 0.84 \cdot 25 = 21.$$

Hence, 21 students were present in the lesson.

Problem 2. Finding the number by its percent.

Example 2.4.4. 15% of what number is 24.

Solution. Here $r = 15$, the base $b = x$ is unknown, and $a = 24$. Substituting in the percent equation $\frac{r}{100} \cdot b = a$, we have

$$\frac{15}{100} \cdot x = 24 \;\Rightarrow\; x = \frac{100}{15} \cdot 24 = \frac{2,400}{15} = 160.$$

Thus, 24 is 15% of 160 .

Example 2.4.5. During an hour a car travelled 54 miles which is 18% of the total distance. What is the entire distance?

Solution. Here the question is "18% of what number is 54?". Since $r = 18$, $b = x$, and $a = 54$, we obtain

$$\frac{18}{100} \cdot x = 54 \;\Rightarrow\; x = \frac{100}{18} \cdot 4 = \frac{5,400}{18} = 300.$$

Thus, the entire distance is 300 miles.

Problem 3. **Finding a percent from a portion**.

Example 2.4.6. 72 is what percent of 192?.

Solution. Here $r = x$ is unknown, the base $b = 192$, and $a = 72$. Substituting in the percent equation $\frac{r}{100} \cdot b = a$, we have

$$\frac{x}{100} \cdot 192 = 72 \;\Rightarrow\; 192x = 7,200 \;\Rightarrow\; x = \frac{7,200}{192} = 37.5.$$

Thus, 72 is 37.5% of 192 .

Example 2.4.7. The price of a book was reduced from \$175 to \$140. Find the percent reduction in price.

Solution. The total reduction in price is $175 - 140 = 35$ dollars. Now the question is "what percent of 175 is 35?". Since $b = 175$, $a = 35$, and $r = x$ is unknown, we have

$$\frac{x}{100} \cdot 175 = 35 \;\Rightarrow\; 175x = 3,500 \;\Rightarrow\; x = \frac{3,500}{175} = 20$$

Hence, the price was cut by 20%.

2.4.1   General Problems on Percentages

Problem 2.4.8. A student paid \$462.00 for a computer which was marked down by 23%. How much money did the student save?

Solution. To find the total discount, we need to find first the original price of the computer. Because of the 23% discount, the student paid $100\% - 23\% = 77\%$ of the original price. Now the question is "77% of what is 462?". The basic percent equation $\frac{r}{100} \cdot b = a$, with $r = 77$, $b = x$ and $a = 462$, gives

$$\frac{77}{100} \cdot x = 462 \;\Rightarrow\; 77xx = 46200 \;\Rightarrow\; x = \frac{46200}{77} = 600.$$

Hence, the original price was \$ 600. Since the student paid \$ 462, he/she saved the difference $600 - 462 = 138$ dollars.

Problem 2.4.9. The price of an item was reduced by 20%. By what percent must the new price be increased to obtain the original price of the item?

Solution. Let p be the original price of the item. The total reduction in price is $0.20p$, and the new price is $p - 0.20p = 0.80p$. Now the question is "what percent of $0.80p$ is $0.20p$?". Substituting in the percent equation $\frac{r}{100} \cdot b = a$, with $r = x$, $b = 0.80p$ and $a = 0.20p$, we have

$$\frac{x}{100} \cdot 0.80p = 0.20p \;\Rightarrow\; (0.80p)x = 20p \;\Rightarrow\; x = \frac{20p}{0.80p} = 25.$$

Hence, the new price must increase by 25%.

Problem 2.4.10. A merchant buys a merchandise for \$ 85.00. He has expences 12% (on the buying price) and he sells it with a profit 15% (on the cost price). What is the selling price? By what percentage was the buying price increased?

Solution. The total cost is 12% of 85 dollars. That is,

$$(.12)(85) = 10.2.$$

The cost price is then $85 + 10.2 = 95.2$. Now the selling price is 115% of the cost price. That is,

$$(1.15)(95.2) = 109.48.$$

To answer the second question of the problem, we must first find the total increase of the buying price. The total increase is

$$109.48 - 85 = 24.48.$$

Now the question is "what percent of 85 is 24.48? " The basic percent equation gives

$$\frac{x}{100} \cdot 85 = 24.48 \;\Rightarrow\; 85x = 2448, \;\Rightarrow\; x = \frac{2448}{85} = 28.88.$$

Hence, the buying price was increased by 28.88%.

2.4.2 Mixture Problems

Problem 2.4.11. 15 liters of water were added to 20 liters of solution containing 4% of salt. What is the concentration of salt in the new solution?

Solution. The total amount of salt before we add the 15 liters of water in the solution is 4% of 20, that is $(0.04)(20) = 0.8$. Let $x\%$ be the concentration of salt in the new 35 liters solution. Since the total amount of salt *remains the same* before and after adding 15 liters of water, we have

$$0.8 = \frac{x}{100} \cdot 35 \;\Rightarrow\; 80 = 35x \;\Rightarrow\; x = \frac{80}{35} = \frac{16}{7} = 2\frac{2}{7}.$$

Hence, the concentration of salt in the new solution is $2\frac{2}{7}\%$.

Problem 2.4.12. In a chemistry laboratory the concentration of one solution is 60% hydrochloric acid (HCI) and that of a second solution is 10% HCI. How many milliliters(mL) of each should be mixed to obtain 90 mL of a 40% HCI solution?

Solution. Let x be the number of milliliters of the 60% solution. Then $90 - x$ equals the number of milliliters of the 10% solution.

The resulting solution of 90 mL with concentration 40% HCI contains $(0.40)(90) = 36$ mL of pure HCI. Since the total amount of HCI *remains the same before and after the mixing*, we obtain the equation

$$0.60x + 0.10(90 - x) = 36.$$

Solving, we have

$$0.60x + 9 - 0.10x = 36 \;\Rightarrow\; 0.50x = 27 \;\Rightarrow\; x = 54.$$

Thus, 54 mL of the 60% acid solution, when mixed with 36 mL of the 10% acid solution, yields 90 mL of a 40% acid solution.

2.4.3   Simple Interest Problems

Interest is a charge for borrowing money or an income from investing money. The amount of money borrowed or invested is called the **principal**. An **interest rate** r is a percent which denotes the interest paid or received per 100 dollars. Interest rates affect our lives. They affect the national economy and they affect one's ability to borrow money for big purchases. For this reason students should be able to solve problems involving interest.

There are two basic types of interest: simple and compound[3]. The **simple interest** I is computed by multiplying the principal P times the rate of interest r times the period of time t over which the amount of money is borrowed or invested (usually measured in years unless otherwise stated). That is,

$$I = Prt.$$

Simple Interest Formula

If a principal P dollars is deposited in an account earning simple interest at an annual rate r, the amount A in the account after t years is given by

$$A = P + I \iff A = P + Prt \iff A = P(1 + rt).$$

Example 2.4.13. Find the interest on $\$3,000$ invested at a simple interest rate of 8% for one year.

Solution. Here the principal $P = 3,000$ dollars, the interest rate $r = 8\% = 0.08$ and the period of time $t = 1$ year. The simple interest formula gives

$$I = (3000)(0.08)(1) = 240.$$

Thus, the interest received for investing $\$3,000$ for one year at a simple interest rate of 8% is $\$240$.

---

[3]Most financial institutions use compounded interest in their transactions. Problems involving compound interset are solved by using exponential functions (See, Section 9.6).

Example 2.4.14. A company invested last year $ 600,000 in two different accounts. Part in an account earning 11% simple interest and the rest in an account earning 7% simple interest. At the end of the year, the company had earned $50,000 in interest. How much did it invest in each account?

Solution. Let x be the principal invested in the first account . Then $600,000 - x$ is the principal invested in second account. The first account earned interest $0.11x$, and the second account $.07(600,000 - x)$. Since the total interest earned from both accounts is $50,000$, we obtain the equation
$$0.11x + .07(600,000 - x) = 50,000.$$
Solving we have
$$0.11x + 42,000 - 0.07x = 50,000 \Rightarrow 0.04x = 8,000 \Rightarrow x = 200,000.$$
Thus, the company invested $200,000$ at 11% and $400,000$ at 7%.

2.4.4 Exercises

1. What is 15% of 40?

2. 63 is 35% of what number?

3. What percent of 75 is 12?

4. (a) Find 110% of 47 dollars and 20 cents.

   (b) Find 80% of 1 hour and 15 minutes.

5. An item at a discount of 12% was sold for $44. What was the original price of the item?

6. A television was sold with discount 30% for $455. What was the original price and was the total discount?

7. The price of Peter's new car is 6% more than the price of a similar model last year. He paid $15,900 for his car this year. What would a similar model have cost last year?

8. After two consecutive price reductions by one and the same percent the price of an item was reduced from $25 to $16. By how many percent was the price reduced each time?

9. The price of goods was first cut by 20%, then the new price was reduced by another 15%, and, finally, after recomputation, it was decreased once again-this time by 10%. By how many percent was the initial price cut in all?

10. Improved organization of labour raised the labour productivity by 10% and rationalization raised the labour productivity once again by 20%. By what percent does the labour productivity increase finally as compared with the original one?

11. A worker manufactured 480 parts and thus fulfilled his assignment by 120%. How many parts would be manufactured by the worker if he fulfilled his assignment by 110%?

12. How much water should be added into 20 ounces of a 15% solution of alcohol to dilute it to a 10% solution?

13. If a bottle holds 3 liters of milk containing $3\frac{1}{2}\%$ butterfat, how much skinned milk must be added to dilute the milk to 2% butterfat?

14. How many liters of water must evaporate to turn 12 liters of a 24% salt solution into a 36% solution?

15. $10,000 were invested partly at 9% and the rest at 14%. If the annual income from these investments is $1,275, how much was invested at each rate?

16. During the last two years, the price of an apartment was increased by 15% in the first year and consequently decreased by 20% the second year. The apartment was finally sold for $391,000. What was its original price two years ago?

2.5 Proportions

Let a, b, c, d be nonzero numbers. An equality of two ratios of the the form

$$\frac{a}{b} = \frac{c}{d}$$

is called a **proportion**. It can also be denoted as

$$a : b = c : d.$$

The proportion is read as follows: "a is to b as c is to d". The numbers a and d are called the **extremes** and the numbers b and c are called the **means**.

The **principal property of proportions** is the following: *In any proportion the product of the extremes is equal to the product of the means.* That is,

$$\frac{a}{b} = \frac{c}{d} \iff ad = bc.$$

To prove this property, we multiply both sides of the proportion by bd to obtain

$$(bd) \cdot \frac{a}{b} = (bd) \cdot \frac{c}{d} \implies \frac{bda}{b} = \frac{bdc}{d} \implies ad = bc.$$

Conversely, if $ad = bc$, then dividing both sides by bd, we get

$$\frac{ad}{bd} = \frac{cb}{bd} \implies \frac{a}{b} = \frac{c}{d}.$$

This property of proportions is known as "**cross multiplying**".

More Properties of Proportions

1. The following proportions are equivalent:

$$\frac{a}{b} = \frac{c}{d} \iff \frac{a}{c} = \frac{b}{d} \iff \frac{d}{b} = \frac{c}{a} \iff ad = bc.$$

Here is a more general property of proportions:

2. If $kn \neq ml$, Then

$$\frac{a}{b} = \frac{c}{d} \quad \Leftrightarrow \quad \frac{ka + lb}{ma + nb} = \frac{kc + ld}{mc + nd},$$

where $ka + lb \neq 0$, $ma + nb \neq 0$, $kc + ld \neq 0$, $mc + nd \neq 0$.

Indeed, using $ad = bc$, it is easy to check that

$$(ka + lb) \cdot (mc + nd) = (ma + nb) \cdot (kc + ld).$$

Assigning distinct values to the numbers k, l, m, n, we obtain particular cases of proportions. For example, for $k = l = 1$ and $m = 0$, $n = 1$, we get

$$\frac{a}{b} = \frac{c}{d} \quad \Leftrightarrow \quad \frac{a + b}{b} = \frac{c + d}{d},$$

for $k = 1\, l = -1$ and $m = 0$, $n = 1$, we get

$$\frac{a}{b} = \frac{c}{d} \quad \Leftrightarrow \quad \frac{a - b}{b} = \frac{c - d}{d},$$

for $k = l = 1$ and $m = 1$, $n = -1$, we get

$$\frac{a}{b} = \frac{c}{d} \quad \Leftrightarrow \quad \frac{a + b}{a - b} = \frac{c + d}{c - d}.$$

Another property of proportions is the following:

$$\frac{a}{b} = \frac{c}{d} = \frac{a + c}{b + d}.$$

Indeed, if $\frac{a}{b} = k$, then $\frac{c}{d} = k$ and so $a = bk$ and $c = dk$. Therefore

$$a + c = bk + dk \Rightarrow a + c = (b + d)k \Rightarrow \frac{a + c}{b + d} = k.$$

More generally, we have

$$\frac{a_1}{b_1} = \frac{a_2}{b_2} = \dots = \frac{a_n}{b_n} = \frac{a_1 + a_2 + \dots + a_n}{b_1 + b_2 + \dots + b_n}.$$

2.5.1 Solving Proportions

Example 2.5.1. Solve the proportion

$$\frac{9}{x} = \frac{3}{5}.$$

Solution. We cross multiply, and we have

$$\frac{9}{x} = \frac{3}{5} \ \Rightarrow \ 3x = 9 \cdot 5 \ \Rightarrow \ 3x = 45 \ \Rightarrow \ x = \frac{45}{3} \ \Rightarrow \ x = 15.$$

Example 2.5.2. Solve the proportion

$$\frac{4}{7} = \frac{x-5}{2x+3}.$$

Solution.

$$\frac{4}{7} = \frac{x-5}{2x+3} \ \Rightarrow \ 4(2x+3) = 7(x-5) \ \Rightarrow \ 8x+12 = 7x-35 \ \Rightarrow$$

$$8x - 7x = -35 - 12 \ \Rightarrow \ x = -47.$$

Example 2.5.3. Separate 253 into four proportional parts to 2,5,7, 9.

Solution. Let x, y, z and w be the required numbers.
Then $x + y + z + w = 253$. Furthermore, since they are proportional to 2,5,7 and 9, respectively, we have

$$\frac{x}{2} = \frac{y}{5} = \frac{z}{7} = \frac{w}{9}.$$

By the properties of proportions, we also have

$$\frac{x}{2} = \frac{y}{5} = \frac{z}{7} = \frac{w}{9} = \frac{x+y+z+w}{2+5+7+9} = \frac{253}{23} = 11.$$

Hence, $\frac{x}{2} = 11 \ \Rightarrow \ x = 22$, $\frac{y}{5} = 11 \ \Rightarrow \ y = 55$, $\frac{z}{7} = 11 \ \Rightarrow \ z = 77$, and $\frac{w}{9} = 11 \ \Rightarrow \ w = 99$. Thus, the four parts are 22, 55, 77, 99.

Various problems are solved with the aid of proportions.

Problem 2.5.4. A model of a building made to the scale 3 : 140 is 75 cm high. What will be the actual height of the building?

Solution. Let x cm be the actual height of the building. Since the scale shows the ratio of the height in the model to its true height, we obtain a proportion

$$\frac{3}{140} = \frac{75}{x}.$$

Solving the proportion

$$\frac{3}{140} = \frac{75}{x} \quad \Rightarrow \quad 3x = 140 \cdot 75 \quad \Rightarrow \quad 3x = 10,500 \quad \Rightarrow \quad x = 3,500.$$

Hence $x = 3,500 \, cm$, that is, $x = 35 \, m$.

Problem 2.5.5. A tank has a supply pipe A which will fill it in 3 hours, and a waste pipe B which will empty it in 3 hours and 40 minutes. If the tank be empty when both pipes are opened, how long will it take for the tank to be full?

Solution. Let t denote the number of hours required. Then $\frac{1}{t}$ is the part of the tank filled in one hour when both A and B are open. If were A alone open, then the part filled in one hour would be $\frac{1}{3}$. But were B alone open (and water in the tank) the part emptied in one hour would be $\frac{1}{3\frac{2}{3}} = \frac{3}{11}$. Hence

$$\frac{1}{t} = \frac{1}{3} - \frac{3}{11} \quad \Rightarrow \quad \frac{1}{t} = \frac{2}{33} \quad \Rightarrow \quad t = \frac{33}{2}.$$

Thus, $t = \frac{33}{2}$ hours, or 16 hours and 30 minutes.

Problem 2.5.6. The speed of a boat itself is 20 (mi/h) and the rate of the flow of the river is 4 (mi/h). Moving against the stream the boat went 80 miles. What distance will the boat cover during the same time going with the stream?

Solution. Let x miles be the distance covered by the boat going with the stream, ie, going down the river. In this case its speed is

$20 + 4 = 24$ (mi/h). The speed of the boat going against the stream, ie, up the river, is $20 - 4 = 16$ (mi/h). Since the time t is the same, we have $\frac{x}{24} = t = \frac{80}{16}$. Hence we obtain the proportion

$$\frac{x}{24} = \frac{80}{16} \;\Rightarrow\; 16x = 24 \cdot 80 \;\Rightarrow\; x = \frac{24 \cdot 80}{16} = 120\,(mi/h).$$

2.5.2 Exercises

1. If $\frac{a}{b} = -\frac{1}{2}$, find the ratios:

   $(a) \quad \dfrac{b}{a} \qquad (b) \quad \dfrac{b-a}{a+b} \qquad (c) \quad \dfrac{a+2b}{2a-b}.$

2. If $\frac{a}{b} = -2$, find the ratios:

   $(a) \quad \dfrac{2a+b}{a+3b} \qquad (b) \quad \dfrac{2ab-b^2}{a^2-b^2} \qquad (c) \quad \dfrac{a^2+b^2}{a^2-b^2}.$

3. If $(a+b):(a-b) = 5:2$, find $a:b$

4. Solve the proportions.

   $(a) \quad \dfrac{5}{4} = \dfrac{10}{x} \qquad (b) \quad \dfrac{6}{x} = \dfrac{3}{8}.$

5. Solve the proportions.

   $(a) \quad \dfrac{2x-7}{3} = \dfrac{x-5}{2} \qquad (b) \quad \dfrac{28-x}{x} = \dfrac{2}{5}.$

6. Two numbers have the ratio $3:4$. If 4 is added to each number the resulting ratio is $4:5$. Find the numbers.

7. Find two numbers whose sum is 560 and their ratio is $\frac{2}{5}$.

8. Find two numbers whose difference is 200 and their ratio is $\frac{7}{5}$.

9. The ratio of women to men in a mathematics class is $3:5$. How many men are in the class if there are 18 women?

10. The distance of two cities on a map is equal to 3cm . What is the true distance between the two cities if the map is made to the scale $1 : 2,000,000$?

11. A line segment of length 120 inches is divided into three parts whose lengths are propositional to 3, 4, 5. Find the legths of the parts.

12. Separate 81 into three propotional parts to 2, 3, 4.

13. A principal of $\$1,000$ was invested for 6 months, and another principal $\$900$ was invested for 10 months with the same interset rate. If both investments yielded total interset of $\$500$, how much interest corresponds to each principal? What was the interest rate?

14. Two barrels A and B are filled with two kinds of Sherry mixed in A in the ratio $3 : 5$, and in B in the ratio $3 : 7$. What amount must be taken from each barrel to form a mixture which shall consist of 6 gallons of one kind and 124 gallons of the other kind?

2.6   Direct and Inverse Variations

If two variables y and x are so related that however their values may change their ratio remains constant, then y is said to **vary directly** as x or y and x are said to **vary proportionally**. We write

$$\frac{y}{x} = k \quad \Leftrightarrow \quad y = kx,$$

where k is a constant.
The constant k is called the *constant of proportionality*.

Note that, if we are given a pair of corresponding values of x and y, we may find k. The equation $y = kx$ connecting y and x is then known, and from it we may compute the value of y which corresponds to any given value of x. Let us see some examples.

Example 2.6.1. If y varies directly with x, and $y = 12$ when $x = 3$, what is the value of y when $x = 9$?

Solution. We have $y = kx$, and by hypothesis, this equation is satisfied when $y = 12$, $x = 3$. Hence $12 = k \cdot 3$, that is $k = 4$.

Therefore $y = 4x$. Hence when $x = 9$ we have $y = 4 \cdot 9 = 36$.

Example 2.6.2. For a certain gas enclosed in a container of fixed volume, the pressure P (in Newtons per square meter) varies directly with temperature T (in Kelvins). If the pressure is found to be 100 Newtons per square meter at a temperature of 300 Kelvins, find a formula that relates pressure P to temperature T. What is the pressure P when $T = 360$ Kelvins?

Solution. Since P varies directly with T, there is some constant k such that
$$P = kT.$$
Because $P = 100$ when $T = 300$, we have
$$100 = k(300) \quad \Leftrightarrow \quad k = \frac{1}{3}.$$
Therefore,
$$P = \frac{1}{3}T.$$
In particular, when $T = 360$, we find
$$P = \tfrac{1}{3}(360) = 120 \text{ Newtons per square meter.}$$

Inverse Variation

If y varies as $\frac{1}{x}$, we say that y **varies inversely** as x. In this case their product is constant. We write
$$xy = k \quad \Leftrightarrow \quad y = \frac{k}{x},$$

where k is a constant.

Example 2.6.3. The intensity I of illumination from a light source varies inversely with the square of the distance d from the source. If the intensity of a light source is 125 Lumens at a distance of 8 feet, what is the intensity at 16 feet?

Solution. Since I varies inversely with the square d^2 of the distance, d, there is some constant k such that

$$I = \frac{k}{d^2}.$$

Because $I = 125$ when $d = 8$, we have

$$125 = \frac{k}{8^2} \quad \Leftrightarrow \quad k = 8,000.$$

Therefore,

$$I = \frac{8,000}{d^2}.$$

In particular, when $d = 16$, we find

$$I = \frac{8,000}{16^2} = \frac{8,000}{256} = 31.25 \text{ Lumens per square feet.}$$

2.6.1 Joint variation

Let x and y denote variables which are independent of one another. If a third variable z varies as the product xy, so that

$$z = kxy,$$

we say that z **varies jointly with x and** y. Moreover. if z varies as the quotient $\frac{x}{y}$, so that

$$z = k \cdot \frac{x}{y},$$

we say that z **varies directly as x and inversely as** y.

Example 2.6.4. In Physics the kinetic energy E of an object varies jointly with its mass m and the square of its velocity v. A 30 gram mass is moving at the rate of 40 centimeters per second has a kinetic energy of $12,000$ dyne-centimeters. Find the kinetic energy of a 5 gram mass that is moving with velocity 60 centimeters per second.

Solution. Since E varies jointly with mass m and the square v^2 of the velocity v, there is some constant k such that

$$E = kmv^2.$$

To find the constant k, we substitute the data $E = 12,000$, $m = 30$ and $v = 40$, and we get

$$12,000 = k(30)(40)^2 \iff 12,000 = 48,000k \implies k = \frac{1}{4}.$$

Therefore,

$$E = \frac{1}{4}mv^2.$$

In particular, when $m = 5$ and $v = 60$, we find

$$E = \tfrac{1}{4}(5)(60)^2 = 4,500 \text{ dyne-centimeters.}$$

.

Example 2.6.5. The wages of 3 men for 4 weeks being $\$\,6,720$, how many weeks will 5 men work for $\$\,8,400$?

Solution. The wages z vary directly with the number of men x and the number of weeks y, that is, z varies jointly with x and y. Therefore, there is a constant k such that

$$z = kxy.$$

To find k, we substitute $z = 6,720$, $x = 3$, $y = 4$, and we get

$$6,720 = k(3)(4) \quad \Leftrightarrow \quad k = \frac{6,720}{12} = 560.$$

Hence,

$$z = 560xy.$$

Now for $z = 8,400$, $x = 5$, we get

$$8,400 = 560(5)y \quad \Leftrightarrow \quad 8,400 = 2800y \quad \Leftrightarrow \quad y = \frac{8400}{2800} = 3.$$

Thus, 5 men for $ 8,400 will work 3 weeks.

.

Example 2.6.6. It takes 90 workers 5 weeks to construct 3 miles of road. How long it will take 50 workers to construct 10 miles of road?

Solution. The time it takes to construct a road varies directly with the length of the road but inversely with the number of workers. Let t denote the time in weeks, l denote the length in miles, and w denote the number of workers. The relationship of these three varibles is expressed by the equation

$$t = k \cdot \frac{l}{w}.$$

To find the constant k, we substitute $t = 5$, $l = 3$, $w = 90$, and we get

$$5 = k\frac{3}{90} \quad \Leftrightarrow \quad k = 150.$$

Therefore,

$$t = 150 \cdot \frac{l}{w}.$$

Now, when $l = 10$ and $w = 50$, we find

$$t = 150 \cdot \frac{10}{50} = 30.$$

Thus, it will take 30 weeks for 50 workers to construct 10 miles of the road.

.

2.6.2 Exercises

1. If y varies directly as x, and $y = -2$ when $x = 5$, what is the value of y when $x = 8$?

2. If y varies inversely as x^2, and $y = 1$ when $x = 2$, for what value of x will $y = 3$?

3. z varies jointly as x and y. If $z = 16$ when $x = 5$ and $y = -8$, find z when $x = 2$ and $y = -10$.

4. P varies jointly as r^2 and $\sqrt{t}$. If $P = 24$ when $r = 2$ and $t = 4$, find P when $r = 5$ and $t = 9$.

5. The cost of $\frac{3}{4}$ of a work is $900. What is the cost for $\frac{5}{6}$ of the same work?

6. For $\frac{8}{3}$ hours a car keeps the same speed of 67.5 miles per hour. How many miles will it travel with the same speed for $\frac{32}{9}$ hours?

7. A train traveling with a constant speed covers a distance of 182 miles in $3\frac{1}{4}$ hours. In how many hours will cover the same distance if it reduces its speed by $\frac{1}{4}$?

8. The distance that an object will fall in t seconds varies directly with the square of t. An object falls 16 feet in 1 second. How long it will take the object to fall 144 feet?

9. A work project was scheduled to be completed in 25 days. If 6 workers complete $\frac{1}{2}$ of the project in 10 days, how many workers must be used to finish the project according to scheduled time?

10. 8 workers working 7 hours per day complete a work in 12 days. How many days it will take 18 workers working 8 hours per day to complete the triple of the same work? (assume equal productivity per worker).

2.7   Equations. Word Problems

Algebra with the aid of equations provides a general method for solving a variety of application problems. In these problems it is required to find the value of certain *unknown* number from given relations, called *conditions of the problem*, connecting the given numbers with the unknown number.

Strategy for Solving Word Problems

For the given problem we denote the unknown number by a letter, as x. The given conditions of the problem then enable us to translate the word problem in terms of x and to form an **equation** connecting the expressions thus obtained. *This equation is the statement of the problem in algebraic symbols.* We solve it for x. Having solved the equation in x, we must check whether the result is a number of the *kind required* before we accept it as a solution of the problem. If it is not, then we say that the problem is an impossible one.

We illustrate with a number of problems.

Problem 2.7.1. The sum of three consecutive even numbers is 72. Find the numbers.

Solution. Any even number is of the form $2x$, where x is an integer. Since consecutive even numbers[4] differ by 2, let

$$2x, \ 2x + 2, \ 2x + 4$$

be the three consecutive even numbers. It is given that their sum is 72. Adding them up, we obtain the equation

$$2x + (2x + 2) + (2x + 4) = 72.$$

Solving this equation, we have

$$6x + 6 = 72 \ \Leftrightarrow \ 6x = 66 \ \Leftrightarrow \ x = 11.$$

Thus, the required even numbers are 22, 24, 26.

Problem 2.7.2. The sum of the digits of a certain number of two digits is 14. If the order of the digits is reversed, the number is increased by 18. What is the number?

Solution. Let $x =$ the tens digit. Then, $14 - x =$ the units digit. The value of the number is $10x + (14 - x)$. The value of the number obtained when we reverse the order of the digits is $10(14 - x) + x$. By the remaining condition of the problem, we have

$$10(14 - x) + x = 10x + (14 - x) + 18.$$

Solving this equation, we find $x = 6$, which being an integer less than 10, is an admissible solution to the problem. The same is true for $14 - 6 = 8$. Thus, the required number is 68.

---

[4]There are similar problems which involve consecutive integers, or concecutive odd integers. Three consecutive integers are represented by x, $x+1$, $x+2$. Three consecutive odd integers are represented by $2x + 1$, $2x + 3$, $2x + 5$.

Problem 2.7.3. A father is now 61 years old. His three children are 24, 21 and 18 years old, repsectively. When does the age of the father will be or was three times of the sum of ages of his children?

Solution. Suppose this will happen after x years from now. Then the ages of these 4 persons will be

$$61 + x, \ 24 + x, \ 21 + x, \ 18 + x.$$

The sum of the ages of the children is

$$(24 + x) + (21 + x) + (18 + x) = 63 + 3x.$$

Three times this, viz, $3(63 + 3x)$ must equal the age of the futher $61 + x$.

Hence, we obtain the equation

$$3(63 + 3x) = 61 + x \tag{2.1}$$

In (2.1) x must be a number within the logical limits of human life.

If $x > 0$, then the required event will happen in the future. If $x = 0$, then this happens now. If $x < 0$, then this happened in the past. In this case for the problem to make sense we must have $18 + x \geq 0$, for otherwise the younger child wouldn't be born.

Let us solve equation (2.1).

$$3(63+3x) = 61+x \ \Leftrightarrow \ 189+9x = 61+x \ \Leftrightarrow \ 8x = -128 \ \Leftrightarrow \ x = -16.$$

Thus, this happened 16 years ago when the ages were: father 45, children 8, 5, 2.

Problem 2.7.4. Two cars A and B are traveling along the same road in the same direction with velocities 45 miles per hour and 38 miles per hour, respectively. Car B is now 28 miles in advance of car A. Will they ever meet and if so, when?

Solution. Let x be the number of hours when the two cars will meet. Car A will then have traveled a distance of $45x$ miles, and car B a distance of $38x$ miles. Since car B is now 28 miles in advance of car A, we have

$$45x = 28 + 38x,$$

which is the equation of the problem.

Solving we find

$$45x - 38x = 28 \iff 7x = 28 \iff x = 4.$$

Thus, the two cars will meet after 4 hours.

Problem 2.7.5. A pipe fills a swimming pool in 4 hours. A second pipe fills the pool in 12 hours. In how many hours the two pipes will fill the pool if they operate simultaneously?

Solution. Suppose V is the volume of the swimming pool.

Let x be the number of hours that takes both pipes to fill the pool. Since the first pipe fills the pool in 4 hours, in 1 hour will fill $\frac{1}{4}V$, and in x hours $\frac{x}{4}V$. Similarly, the second pipe in x hours will fill $\frac{x}{12}V$. Now the equation of the problem is

$$\frac{x}{4}V + \frac{x}{12}V = V.$$

Cancelling V (ie, multiplying both sides of the equation by $\frac{1}{V}$), we get

$$\frac{x}{4} + \frac{x}{12} = 1.$$

Solving, we have

$$\frac{x}{4} + \frac{x}{12} = 1 \iff \frac{4x}{12} = 1 \iff 4x = 12 \iff x = 3.$$

Thus, it will take 3 hours for both pipes to fill the pool.

Problem 2.7.6. At what time between three and four o'clock do the hour and minute hands of the clock point

1. in the same direction?

2. in opposite directions?

Solution.

1. Let x be the number of minutes past three o'clock at which the hands of the clock point in the same direction.

   Since the minute hand starts at XII it will then have traversed x minute spaces. The hour hand starts at III or 15 minute spaces in advance of the minute hand, but it moves only $\frac{1}{12}$ as fast as the minute hand.

   Therefore when the minute hand is x minute spaces past XII, the hour hand is at $15 + \frac{x}{12}$ minute spaces past XII. Since the minute hand and the hour hand point in the same direction, we must have

   $$x = 15 + \frac{x}{12}.$$

   Solving

   $$x = 15 + \frac{x}{12} \iff 12x = 180 + x \iff 11x = 180 \iff x = \frac{180}{11} = 16\frac{4}{11}.$$

   Thus, the hands point in the same direction at $16\frac{4}{11}$ minutes after three o'clock.

2. When the minute hand and the hour hand point in opposite directions, the minute hand is 30 minute spaces in advance of the hour hand. Therefore the equation of the problem in this case is

   $$x = \left(15 + \frac{x}{12} \right) + 30.$$

Solving

$$x = 45 + \frac{x}{12} \iff 12x = 540 + x \iff 11x = 540 \iff x = \frac{540}{11} = 49\frac{1}{11}.$$

Thus, the hands point in opposite directions at $49\frac{1}{11}$ minutes after three o'clock.

Problem 2.7.7. It takes three days for a tourist traveling on a bicycle to cover the entire route. On the first day he traveled 30% of the entire path, on the second day 60% of the remaining part; after the first two days it remains for the tourist to travel by 1 mile less than the path he covered during the first day. What is the length of the entrire route?

Solution. Let x (miles) be the length of the entire route. Then during the first day the tourist covered 30% of x, which is $0.3x$ miles. After the first day the remainder of the path is $x - 0.3x = 0.7x$. During the second day the tourist traveled 60% of $0.7x$, which is $(0.6)(0.7x) = 0.42x$. During the third day, it remains for the tourist to travel $0.7x - 0.42x = 0.28x$ miles. Now by the hypothesis, during the third day the tourist covered distance by 1 mile less than that covered during the first day. Hence, we obtain the equation

$$0.28x = 0.3x - 1.$$

Multiplying both sides by 100 (to eliminate the decimals), we get

$$28x = 30x - 100 \iff 2x = 100 \iff x = 50.$$

Thus, the entire route is 50 miles.

2.7.1 Exercises

1. The sum of three consecutive odd positive integer numbers is 129. Find the numbers.

2. Three times a number is eight more than the number. What is the number?

3. If we add $\frac{1}{3}$ of a certain number to itself, we find the number 19 decreased by $\frac{1}{4}$ of the required number. What is the number?

4. The sum of the digits of a certain number of two digits is 10. If the order of the digits be reversed, the number is increased by 18. What is the number?

5. Find two positive integer numbers whose difference is 401, the quotient of the greater by the smaller is 6 and the remainder 6.

6. Mary returned from the grocery store with only $2.80. in change. She had two more dimes than nickels and twice as many quarters as nickels. How many nickels, dimes and quarters did she have?

7. A tank can be filled by one pipe in 3 hours, by a second in 6 hours, and emptied by a third in 4 hours. How long it will take to fill the tank if all three pipes are opened?

8. Four years ago father was six times older than his son; in sixteen years father will be twice as old as his son. How old is each of them?

9. At what time between eight and nine o'clock do the hands of a watch point in the same direction? in opposite direction?

10. In a fruit basket there are 24 pieces of fruit. There are twice as many apples as pears and four fewer peaches than pears. How many of each fruit are in the basket?

11. A baseball club bought 70 balls for $120. The balls used for games cost $2.25 and the balls used for practice cost $1.50. How many of each kind were bought?

12. Two cities A and B are 90 miles apart. A train leaves city A at 10:00 AM with a speed of 45 miles per hour. At 10:30 AM another train leaves from city B going to city A with a speed of 50 miles per hour. When will the two trains pass each other, and at what distance from city B?

2.8   Review Exercises

1. Evaluate the expressions $(a+b)^3$ and $a^3 + 3a^2b + 3ab^2 + b^3$ for $a = 2$ and $b = -4$. What do you observe?

2. Evaluate the expressions $x^2 + (a+b)x + ab$ and $(x+a)(x+b)$ for $x = -2$, $a = -3$ and $b = 4$. What do you observe?

3. Solve the equations.

   (a) $6 - [2x - (3x - 4) - 1] = 0$
   (b) $5 - 4(x - 3) = x - 2(x - 1)$

4. Solve the proportions.

   (a) $\frac{5x+1}{7} = \frac{2x-3}{3}$
   (b) $\frac{2x-1.5}{5} = \frac{0.8x-1}{2}$

5. Solve the equation.

$$\frac{3x-2}{8} - \frac{13x+3}{27} + 9 = \frac{5x-12}{18} - \frac{2-5x}{4}.$$

6. Solve the equation

$$\frac{3x}{4} - \frac{5}{17}(2x + 1) = x - 1 + \frac{7x - 5}{51} - \frac{2 - x}{2}.$$

7. What is a best buy 100 aspirin tablets for \$2.75 or 200 aspirin tablets for \$4.79?

8. The length of a rectangle is 7 meters longer than the width. The perimeter of the rectangle is 86 meters. Find the dimensions of the rectangle.

9. A tank can be filled by one pipe in 3 hours, and emptied by a second in 4 hours , and by a third in 2 hours. How long it will take to empty the tank if all three pipes operate simultaneously?

10. The digit of tens of a two digit number is four more than the digit of the ones. If we add $\frac{1}{5}$ of the number to itself we obtain 114. What is the number?

11. How soon after four o'clock are the hands of a watch at right angles?

12. Two workers A and B can do a piece of work in 10 days; but at the end of the seventh day A falls sick and B finishes the piece by working alone for 5 days. How long would it take each worker to do the entire piece, working alone?

13. A ship leaves from New York with speed 19.5 miles per hour. After 2 hours a second ship leaves in the same direction with speed 22.5 miles per hour. How long it will take the second ship to meet the first?

14. Thomas drove to Boston, Massachussets, in 5 hours. On the return trip, there was less traffic, and the trip took only 3

hours. If Thomas averaged 26 miles per hour faster on the return trip, how fast did he drive each way?

15. How much gallons of a 5% alcohol solution must be mixed with 90 gallons of 1% solution to obtain a 2% solution?

16. If a certain quantity of water be added to a gallon of a given liquid, it contains 30% of alcohol; if twice this qiuantity of water be added, it contains 20% of alcohol. How much water is added each time, and what percentage of alcohol did the original liquid contain?

17. A man invested $5000 partly at 4% and partly at 6%, so that he earned $275 total interest. What amount did he invest at each rate?

18. A traveler set out on a journey with a certain sum of money in his pocket and each day spent $\frac{1}{2}$ of what he began the day with and $2 besides. At the end of the third day his money was exhausted. How much had he at the outset?

19. An oil tank can be emptied by the main pump in 4 hours. An auxiliary pump can empty the tank in 9 hours. If the main pump started at 9:00 AM when should the auxiliary pump be started so that the tank is emptied by noon?

20. 20 workers working 8 hours per day complete the $\frac{2}{5}$ of a road in 14 days. How many hours per day must work 16 workers to complete the rest of the road in 30 days?

21. The volume of a gas in a ballon varies directly with the temperature and inversely with the pressure. When the temperature of a certain gas is 350°, the pressure is 50 pounds per square inch and the volume is 20 cubic inches. What is the volume when the temperature decreases to 330° and the pressure increases by 10 pounds per square inch?

22. The volume of a circular disc varies as its thickness and the square of the radius of its face. Two metallic discs having the thicknesses 3 and 2 and the radii 24 and 36 respectively are melted and recast in a single disc having the radius 48. What is its thickness?

Chapter 3

Powers and Roots

In this chapter, we study powers of numbers with integer and rational exponents.

3.1 Powers with a Natural Exponent

In Section 1.3.3, we discussed how to raise a rational number a to a power with a natural exponent n. We repeat the definition of the power a^n, but this time the base a may be any real number.

Definition 3.1.1. Let $a \in \mathbb{R}$ be a real number and let $n \in \mathbb{N}$ be a natural number. We define the **power** of a with a natural **exponent** n by

$$a^n = \underbrace{aa \cdots a}_{n \ times}.$$

We define $a^0 = 1$, for $a \neq 0$, $a \in \mathbb{R}$.

As in Theorem 1.3.13, we see that the following properties hold true for any power with a natural exponent:

Let a, b be real numbers and m, n natural numbers. Then

1. $a^n \cdot a^m = a^{n+m}$.

2. $(a^m)^n = a^{mn}$.

3. $(a \cdot b)^n = a^n \cdot b^n$.

4. $\frac{a^m}{a^n} = a^{m-n}$, where $a \neq 0$, and $m > n$.

5. $\left(\frac{a}{b}\right)^n = \frac{a^n}{b^n}$, where $b \neq 0$.

3.2   Powers with an Integer Exponent

Definition 3.2.1. Let $a \neq 0$ be a real number and let $n \in \mathbb{N}$ be a natural number. We define

$$a^{-n} = \frac{1}{a^n}.$$

Example 3.2.2.

$$2^{-3} = \frac{1}{2^3} = \frac{1}{8}, \quad x^{-2} = \frac{1}{x^2}, \quad \left(-\frac{4}{5}\right)^{-3} = \frac{1}{\left(-\frac{4}{5}\right)^3} = \frac{1}{-\frac{64}{125}} = -\frac{125}{64}.$$

We show below that the properties $(1) - (5)$ of powers with a natural exponent extend to a power with negative exponent, and consequently to powers with any integer exponent.

Let a, b be real numbers and m, n natural numbers. Then

1. $a^{-n} \cdot a^{-m} = a^{(-n)+(-m)}$. Indeed,

$$a^{-n} \cdot a^{-m} = \frac{1}{a^n} \cdot \frac{1}{a^n} = \frac{1}{a^n \cdot a^m} = \frac{1}{a^{n+m}} = a^{-(n+m)} = a^{(-n)+(-m)}.$$

2. $(a^{-m})^{-n} = a^{(-m)(-n)}$. Indeed,

$$(a^{-m})^{-n} = \frac{1}{(a^{-m})^n} = \frac{1}{\left(\frac{1}{a^m}\right)^n} = \frac{1}{\frac{1}{a^{mn}}} = a^{mn} = a^{(-m)(-n)}.$$

3. $(a \cdot b)^{-n} = a^{-n} \cdot b^{-n}$. Indeed,

$$(a \cdot b)^{-n} = \frac{1}{(a \cdot b)^n} = \frac{1}{a^n \cdot b^n} = \frac{1}{a^n} \cdot \frac{1}{b^n} = a^{-n} \cdot b^{-n}.$$

4. $\frac{a^{-m}}{a^{-n}} = a^{(-m)-(-n)}$, where $a \neq 0$. Indeed, if $n \geq m$, we have

$$\frac{a^{-m}}{a^{-n}} = \frac{\frac{1}{a^m}}{\frac{1}{a^n}} = \frac{a^n}{a^m} = a^{n-m} = a^{(-m)-(-n)}.$$

If $n < m$, then $m = n + k$, with $k \in \mathbb{N}$. Hence

$$\frac{a^{-m}}{a^{-n}} = \frac{a^n}{a^m} = \frac{\overbrace{aa\cdots a}^{n\ times}}{\underbrace{aa\cdots a}_{n+k\ times}} = \frac{1}{\underbrace{aa\cdots a}_{k\ times}} = \frac{1}{a^k} = a^{-k} = a^{-(m-n)} = a^{(-m)-(-n)}.$$

5. $\left(\frac{a}{b}\right)^{-n} = \frac{a^{-n}}{b^{-n}}$, where $b \neq 0$. We leave the proof of this property as an exercise to the reader.

Note that

$$\left(\frac{a}{b}\right)^{-n} = \left(\frac{b}{a}\right)^n.$$

Example 3.2.3. Compute

$$\frac{3^9 \cdot 18^{-4}}{2^{-5}}.$$

Solution.

$$\frac{3^9 \cdot 18^{-4}}{2^{-5}} = \frac{3^9 \cdot \frac{1}{18^4}}{\frac{1}{2^5}} = \frac{3^9 \cdot 2^5}{18^4} = \frac{3^9 \cdot 2^5}{(2 \cdot 3^2)^4} = \frac{3^9 \cdot 2^5}{2^4 \cdot 3^8} = \frac{2^5 \cdot 3^9}{2^4 \cdot 3^8} = 2 \cdot 3 = 6.$$

Example 3.2.4. Express each number in a power of 2 and then simplify.

$$\left[\left(\frac{1}{4}\right)^6 \cdot 64\right]^{-3} \cdot 32^{-2}.$$

Solution.

$$\left[\left(\frac{1}{2^2}\right)^6 \cdot 2^6\right]^{-3} \cdot (2^5)^{-2} = \left[\left(2^{-2} \cdot 2\right)^6\right]^{-3} \cdot 2^{-10} = \left[2^{-2+1}\right]^{(6)(-3)} \cdot 2^{-10} =$$

$$\left(2^{-1}\right)^{-18} \cdot 2^{-10} = (2)^{(-1)(-18)} \cdot 2^{-10} = 2^{18} \cdot 2^{-10} = 2^{18+(-10)} = 2^8 = 256.$$

Example 3.2.5. Simplify the expression. Write the answer without using negative exponents.

$$\left(\frac{3x^2y^{-4}}{2x^{-3}y^{-5}}\right)^{-3} \quad (x \neq 0,\ y \neq 0).$$

Solution. Using the properties of exponents, we have

$$\left(\frac{3x^2y^{-4}}{2x^{-3}y^{-5}}\right)^{-3} = \frac{(3x^2y^{-4})^{-3}}{(2x^{-3}y^{-5})^{-3}} = \frac{3^{-3}(x^2)^{-3}(y^{-4})^{-3}}{2^{-3}(x^{-3})^{-3}(y^{-5})^{-3}} = \left(\frac{3}{2}\right)^{-3} \cdot \frac{x^{-6}y^{12}}{x^9y^{15}} =$$

$$\left(\frac{2}{3}\right)^3 \cdot \frac{x^{-6}}{x^9} \cdot \frac{y^{12}}{y^{15}} = \frac{2^3}{3^3} \cdot x^{-6-9} \cdot y^{12-15} = \frac{8}{27} \cdot x^{-15} \cdot y^{-3} = \frac{8}{27x^{15}y^3}.$$

3.2.1   Scientific Notation

Scientists frequently work with numbers which are very large or very small. These numbers can be written compactly by expressing them in *scientific notation*.

Definition 3.2.6. A number is expressed in ***scientific notation*** when it is written in the form

$$N \times 10^k,$$

where $1 \leq |N| < 10$ and k is an integer.

Example 3.2.7. Write each number in scientific notation:

a. $-475,000$ b. 0.00000032.

Solution.

$$\textbf{a.} \quad -475,000 = -4.75 \times 10^5.$$

$$\textbf{b.} \quad 0.00000032 = 3.2 \times 10^{-7}.$$

Scientific notation has computational advantage.

Example 3.2.8. Use scientific notation to simplify

$$\frac{(45,000,000,000)(212,000)}{0.00018}.$$

Solution.

$$\frac{(45,000,000,000)(212,000)}{0.00018} = \frac{(4.5 \times 10^{10})(2.12 \times 10^5)}{1.8 \times 10^{-4}} =$$

$$\frac{(4.5)(2.12) \times 10^{15}}{1.8 \times 10^{-4}} = \frac{9.54}{1.8} \times 10^{15-(-4)} = 5.3 \times 10^{19}.$$

3.2.2 Exercises

Perform the operations.

1. $0^1 \cdot 1^0$

2. $(-2)^3 \cdot (-2)^2$

3. $-2^2 \cdot \left(-\frac{1}{2}\right)^4 \cdot \left(-\frac{1}{2}\right)^3$

4. $\left(-\frac{3}{5}\right)^5 \div \left(-\frac{3}{5}\right)^3$

5. $\left[\left(-\frac{1}{2}\right)^2\right]^3$

6. $\left[\left(-\frac{1}{2}\right)^2 \cdot \frac{3}{4}\right]^2$

7. $(-1.4)^3 \cdot \left(3\frac{4}{7}\right)^3$

8. $(a^4 \cdot a^{-4})^4$

9. $\left(3 \cdot 2^{20} + 7 \cdot 2^{19}\right) \cdot 52 \div \left(13 \cdot 8^4\right)^2$

10. $(3^2)^2 - [(-2)^3]^2 - (-5^2)^2$

11. $4^{-2} - 2^{-3} + [(-2)^3]^{-1}$

12. Simplify

$$\frac{(x^{-2})^3 \cdot x^5}{x^{-1}}$$

13. Compute

$$\frac{63^4 \cdot 35^3}{5^5 \cdot 21^7}$$

14. Compute

$$\frac{2^{-2} + 3^{-3}}{4^{-2} - 9^{-1}}$$

15. Simplify and write the answer without using negative exponents.

$$\frac{(a^4 a^{-6})^{-1}}{(a^5 a^{-3})^2}$$

16. Simplify and write the answer without using negative exponents.

$$\frac{(x^{-3}x^2)^2}{(x^2 x^{-5})^{-3}}$$

17. Simplify and write the answer without using negative exponents.

$$\left(\frac{3x^5 y^{-3}}{6x^{-5} y^3}\right)^{-2}$$

18. Simplify and write the answer without using negative exponents.

$$\frac{(ab^{-1})^4}{(a^{-2}b^3)^{-2}}$$

19. Simplify and write the answer without using negative exponents.

$$\frac{(x^3 y)^{-1}}{(5x^2 y^{-2})^3 (5xy^{-2})^{-1}}$$

20. Simplify and write the answer without using negative exponents.

$$\frac{(x^{-2}y^3z^4)^{-2}(xy^{-2}z^3)^4}{(xy^{-2}z^3)^{-4}(xy^2z)^{-1}}$$

21. Express each number in scientific notation:

   (a) $89,500$

   (b) $-25,489,000$

   (c) 0.00052

   (d) -0.000000089

22. Express each number in standard form:

   (a) 5.63×10^6

   (b) 7.81×10^{-5}

   (c) -3.2×10^{-3}

   (d) $5,300 \times 10^{-7}$

23. Use scientific notation to simplify

$$\frac{(65,000,000)(45,000)}{250000}$$

24. Use scientific notation to simplify

$$\frac{(0.00000045)(0.00000012)}{45,000,000}$$

25. Use scientific notation to simplify

$$\frac{(0.00000035)(17,000)}{0.0000085}$$

26. Use scientific notation to simplify

$$\frac{(0.00000000275)(4,750)}{500,000,000,000}.$$

3.3   Roots

In Chapter 1, we saw that taking the root of a number is an operation inverse to raising to a power. For instance, $\sqrt{16} = 4$, since $4^2 = 16$.

Here we shall study roots in more detail and we will derive their basic properties. We recall the definition.

Definition 3.3.1. An nth-**root** of a real number a is a number b, which when it is raised to the nth power gives the number a. The nth root of a is symbolized by $\sqrt[n]{a}$. Thus,

$$\sqrt[n]{a} = b \;\Leftrightarrow\; b^n = a.$$

Of course, $\sqrt[n]{0} = 0$, since $0^n = 0$. The definition also implies that

$$\left(\sqrt[n]{a}\right)^n = a.$$

For n even this is true only if $a \geq 0$.

As we have seen in Chapter 1, when n is even $(n = 2k)$ the root $\sqrt[2k]{a}$ makes sense only for $a \geq 0$ and $\sqrt[2k]{a} \geq 0$. In particular, when $n = 2$, for $a \geq 0$, we get the *square root* $\sqrt{a} \geq 0$. Thus,

$$\sqrt{a} = b \;\Leftrightarrow\; b^2 = a.$$

The expression $\sqrt[n]{x^n}$ makes sense for all $x \in \mathbb{R}$. In fact, if $n = 2k+1$ is odd then

$$\sqrt[2k+1]{x^{2k+1}} = x,$$

and if $n = 2k$ is even then

$$\sqrt[2k]{x^{2k}} = |x|.$$

For instance, $\sqrt[4]{(-2)^4} = \sqrt[4]{2^4} = 2 = |-2|$.

Theorem 3.3.2. *(**Existence of Roots**).For any real positive number $a > 0$ and a natural number n there is one and only one n^{th}-root of a. That is, for any $a > 0$ there exists a unique positive number b such that $b^n = a$.*

We are not going to prove the *existence* of $\sqrt[n]{a}$ (the proof of this statement is given in higher mathematics).

Let us prove the *uniqueness* of the nth-root.

Suppose $\sqrt[n]{a} = b$ and $\sqrt[n]{a} = c$, with $a > 0$, $b > 0$ and $c > 0$. Then, by definition, $b^n = c^n = a$. Hence

$$b^n - c^n = 0.$$

Factoring $b^n - c^n$ (see, Corollary 4.4.11) we obtain

$$b^n - c^n = (b - c)(b^{n-1} + b^{n-2}c + ... + c^{n-1}) = 0.$$

Since the factor $b^{n-1} + b^{n-2}c + ... + c^{n-1}$ is positive, as a sum of positive terms, we conclude that $b - c = 0$ or $b = c$. Thus, the $\sqrt[n]{a}$ is unique.

Properties of Roots

Roots satisfy the following basic properties.

Theorem 3.3.3. *Let $a \geq 0$, $b \geq 0$, and m, n, k be natural numbers. Then*

1. $\sqrt[n]{ab} = \sqrt[n]{a} \cdot \sqrt[n]{b}$.

2. $\sqrt[n]{\frac{a}{b}} = \frac{\sqrt[n]{a}}{\sqrt[n]{b}}$, $\quad b > 0$.

3. $\sqrt[n]{a^m} = \left(\sqrt[n]{a} \right)^m$.

4. $\sqrt[n]{a} = \sqrt[nk]{a^k}$.

5. $\sqrt[m]{\sqrt[n]{a}} = \sqrt[mn]{a}$.

Proof. 1. Using property (3) of the powers, and the fact $\left(\sqrt[n]{a} \right)^n = a$, we have

$$\left(\sqrt[n]{a} \cdot \sqrt[n]{b} \right)^n = \left(\sqrt[n]{a} \right)^n \left(\sqrt[n]{b} \right)^n = ab.$$

That is,

$$\left(\sqrt[n]{a} \cdot \sqrt[n]{b} \right)^n = ab \;\Rightarrow\; \sqrt[n]{ab} = \sqrt[n]{a} \cdot \sqrt[n]{b}.$$

2. Using property (5) of powers, this property is proved in the same way as the preceding property.

3. By property (1), we have

$$\sqrt[n]{a^m} = \sqrt[n]{\underbrace{aa\cdots a}_{m\ times}} = \underbrace{\sqrt[n]{a}\,\sqrt[n]{a}\cdots\sqrt[n]{a}}_{m\ times} = \left(\sqrt[n]{a}\right)^m.$$

4. Let $\sqrt[n]{a} = b$. This means $b^n = a$. Raising to the k power both sides we get $(b^n)^k = a^k$. Using property (2) of powers, we have $b^{nk} = a^k$. This implies $b = \sqrt[nk]{a^k}$ Thus,

$$\sqrt[n]{a} = \sqrt[nk]{a^k}.$$

5. By property (3), we have

$$\left(\sqrt[m]{\sqrt[n]{a}}\right)^{mn} = \sqrt[m]{\left(\sqrt[n]{a}\right)^{mn}} = \sqrt[m]{\left[\left(\sqrt[n]{a}\right)^n\right]^m} = \sqrt[m]{a^m} = a.$$

That is,

$$\left(\sqrt[m]{\sqrt[n]{a}}\right)^{mn} = a \;\Rightarrow\; \sqrt[mn]{a} = \sqrt[m]{\sqrt[n]{a}}.$$

$\square$

Example 3.3.4. Simplify the expression

$$2\sqrt{27} - \sqrt{48}.$$

Solution. We have

$$2\sqrt{27} - \sqrt{48} = 2\sqrt{9\cdot 3} - \sqrt{16\cdot 3} = 2\sqrt{9}\cdot\sqrt{3} - \sqrt{16}\cdot\sqrt{3} = 6\sqrt{3} - 4\sqrt{3} = 2\sqrt{3}.$$

Example 3.3.5. Simplify the numerical expression

$$\sqrt{8} - \sqrt{72} + \sqrt{50}.$$

Solution. We have

$$\sqrt{8} - \sqrt{72} + \sqrt{50} = \sqrt{4\cdot 2} - \sqrt{36\cdot 2} + \sqrt{25\cdot 2} =$$
$$\sqrt{4}\cdot\sqrt{2} - \sqrt{36}\cdot\sqrt{2} + \sqrt{25}\cdot\sqrt{2} = 2\sqrt{2} - 6\sqrt{2} + 5\sqrt{2} = \sqrt{2}.$$

Example 3.3.6. Simplify

$$\sqrt[6]{16} - \sqrt[4]{4} + \sqrt[3]{-4}.$$

Solution. We have

$$\sqrt[6]{16} - \sqrt[4]{4} + \sqrt[3]{-4} = \sqrt[6]{2^4} - \sqrt[4]{2^2} - \sqrt[3]{4} = \sqrt[3]{2^2} - \sqrt{2} - \sqrt[3]{2^2} = -\sqrt{2}.$$

Example 3.3.7. Simplify

$$1. \quad \sqrt{45a^5b^3}, \quad 2. \quad \sqrt[3]{-24a^6b^2}.$$

Solution. We have

1. $\sqrt{45a^5b^3} = \sqrt{9a^4b^2} \cdot \sqrt{5ab} = 3a^2|b|\sqrt{5ab}.$

2. $\sqrt[3]{-24a^6b^2} = \sqrt[3]{-8a^6} \cdot \sqrt[3]{3b^2} = -2a^2\sqrt[3]{3b^2}.$

3.3.1 Powers with a Rational Exponent

The notion and properties of powers with an integer exponent were considered in Section 3.2. Let us now introduce powers with a rational fractional exponent.

Definition 3.3.8. Let $a > 0$ and let $p = \frac{m}{n}$ be a rational number where m, n are natural numbers with $n \geq 2$. We define

$$a^p = a^{\frac{m}{n}} = \sqrt[n]{a^m}.$$

We also define

$$a^{-\frac{m}{n}} = \frac{1}{a^{\frac{m}{n}}} = \frac{1}{\sqrt[n]{a^m}}.$$

Note that

$$\sqrt[n]{a} = a^{\frac{1}{n}}, \quad \sqrt{a} = a^{\frac{1}{2}}$$

For instance,

$$a^{\frac{2}{3}} = \sqrt[3]{a^2}, \quad a^{-\frac{1}{2}} = \frac{1}{a^{\frac{1}{2}}} = \frac{1}{\sqrt{a}}.$$

$$8^{\frac{4}{3}} = \sqrt[3]{8^4} = \left(\sqrt[3]{8}\right)^4 = 2^4 = 16.$$

$$4^{-\frac{3}{2}} = \frac{1}{4^{\frac{3}{2}}} = \frac{1}{\sqrt{4^3}} = \frac{1}{\sqrt{64}} = \frac{1}{8}.$$

$$a^{1.6} = a^{\frac{16}{10}} = a^{\frac{8}{5}} = \sqrt[5]{a^8}.$$

Remark 1. Note that, if $\frac{m}{n}$ is an irreducible fraction, then for any fraction of the form $\frac{km}{kn} = \frac{m}{n}$, we have

$$a^{\frac{mk}{nk}} = \sqrt[nk]{a^{mk}} = \sqrt[n]{a^m} = a^{\frac{m}{n}}.$$

Remark 2. If $a < 0$, we must be careful using the notation $a^{\frac{m}{n}}$. For example,

$$(-8)^{\frac{1}{3}} = \sqrt[3]{-8} = -2, \text{ but } (-8)^{\frac{2}{6}} = \sqrt[6]{(-8)^2} = \sqrt[6]{8^2} = \sqrt[3]{8} = 2$$

a different answer.

Properties of Rational Exponents

The properties of powers with an integer exponent also extend to a power with any rational exponent and a positive base. Let a and b be positive numbers. Then

1. $a^{\frac{m}{n}} \cdot a^{\frac{k}{l}} = a^{\frac{m}{n} + \frac{k}{l}}$. Indeed,

$$a^{\frac{m}{n}} \cdot a^{\frac{k}{l}} = \sqrt[n]{a^m} \cdot \sqrt[l]{a^k} = \sqrt[nl]{a^{ml}} \cdot \sqrt[nl]{a^{kn}} = \sqrt[nl]{a^{ml} \cdot a^{kn}} = \sqrt[nl]{a^{ml+kn}} =$$

$$a^{\frac{ml+kn}{nl}} = a^{\frac{m}{n} + \frac{k}{l}}.$$

2. $\left(a^{\frac{m}{n}}\right)^{\frac{k}{l}} = a^{\frac{m}{n} \cdot \frac{k}{l}}$. Indeed,

$$\left(a^{\frac{m}{n}}\right)^{\frac{k}{l}} = \sqrt[l]{\left(a^{\frac{m}{n}}\right)^k} = \sqrt[l]{\left(\sqrt[n]{a^m}\right)^k} = \sqrt[l]{\sqrt[n]{a^{mk}}} = \sqrt[ln]{a^{mk}} = a^{\frac{mk}{nl}} = a^{\frac{m}{n} \cdot \frac{k}{l}}.$$

3. $(a \cdot b)^{\frac{m}{n}} = a^{\frac{m}{n}} \cdot b^{\frac{m}{n}}$. We have,

$$(ab)^{\frac{m}{n}} = \sqrt[n]{(ab)^m} = \sqrt[n]{a^m \cdot b^m} = \sqrt[n]{a^m} \cdot \sqrt[n]{b^m} = a^{\frac{m}{n}} \cdot b^{\frac{m}{n}}.$$

4. $\dfrac{a^{\frac{m}{n}}}{a^{\frac{k}{l}}} = a^{\frac{m}{n} - \frac{k}{l}}$.

5. $\left(\dfrac{a}{b}\right)^{\frac{m}{n}} = \dfrac{a^{\frac{m}{n}}}{b^{\frac{m}{n}}}$.

Properties (4) and (5) are proved similarly, and we leave them for the reader as an exercise.

Remark 3. From the above properties we see that calculations involving radicals become much easier if we replace roots by powers with rational exponents.

Example 3.3.9. Replace the radical signs by rational exponents and simplify.

$$\sqrt{a} \cdot \sqrt[5]{a^3} \cdot \sqrt[10]{a^4}.$$

Solution.

$$\sqrt{a} \cdot \sqrt[5]{a^3} \cdot \sqrt[10]{a^4} = a^{\frac{1}{2}} \cdot a^{\frac{3}{5}} \cdot a^{\frac{4}{10}} = a^{\frac{1}{2} + \frac{3}{5} + \frac{4}{10}} = a^{\frac{5}{10} + \frac{6}{10} + \frac{4}{10}} = a^{\frac{15}{10}} = a^{\frac{3}{2}} = \sqrt{a^3}.$$

Example 3.3.10. Replace the radical signs by rational exponents and simplify.

$$\sqrt{\sqrt{5}} \cdot \left(\frac{\sqrt[3]{\sqrt{5}}}{\sqrt[4]{\sqrt{5}}}\right)^2.$$

Solution.

$$\sqrt{\sqrt{5}} \cdot \left(\frac{\sqrt[3]{\sqrt{5}}}{\sqrt[4]{\sqrt{5}}}\right)^2 = \left(5^{\frac{1}{2}}\right)^{\frac{1}{2}} \cdot \left[\frac{\left(5^{\frac{1}{2}}\right)^{\frac{1}{3}}}{\left(5^{\frac{1}{2}}\right)^{\frac{1}{4}}}\right]^2 =$$

$$5^{\frac{1}{4}} \cdot \left(\frac{5^{\frac{1}{6}}}{5^{\frac{1}{8}}}\right)^2 = 5^{\frac{1}{4}} \cdot \frac{(5^{\frac{1}{6}})^2}{(5^{\frac{1}{8}})^2} = 5^{\frac{1}{4}} \cdot \frac{5^{\frac{1}{3}}}{5^{\frac{1}{4}}} = 5^{\frac{1}{3}}.$$

Example 3.3.11. Simplify

$$\left(\frac{-a^{\frac{2}{3}} \cdot a^{\frac{5}{3}}}{a^{-\frac{2}{3}}}\right)^{\frac{1}{3}}.$$

Solution.

$$\left(\frac{-a^{\frac{2}{3}}\cdot a^{\frac{5}{3}}}{a^{-\frac{2}{3}}}\right)^{\frac{1}{3}}=\left(-\frac{a^{\frac{2}{3}+\frac{5}{3}}}{a^{-\frac{2}{3}}}\right)^{\frac{1}{3}}=\left(-\frac{a^{\frac{7}{3}}}{a^{-\frac{2}{3}}}\right)^{\frac{1}{3}}=\left(-a^{\frac{7}{3}-\left(-\frac{2}{3}\right)}\right)^{\frac{1}{3}}=$$

$$\left(-a^{\frac{7}{3}+\frac{2}{3}}\right)^{\frac{1}{3}}=\left(-a^{\frac{9}{3}}\right)^{\frac{1}{3}}=\left(-a^{3}\right)^{\frac{1}{3}}=(-1)^{\frac{1}{3}}\left(-a^{3}\right)^{\frac{1}{3}}=-a.$$

3.3.2   Exercises

Simplify

1. $2\sqrt{3}+\sqrt{12}$

2. $\sqrt{7}+\sqrt{28}-\sqrt{63}$

3. $\sqrt{8}+\sqrt{50}-\sqrt{98}$

4. $\sqrt{12}-2\sqrt{27}+3\sqrt{75}-\sqrt{48}-\sqrt{80}+\sqrt{20}$

5. $\sqrt{6}\cdot\sqrt{10}\cdot\sqrt{15}$

6. $\sqrt{3a}\cdot\sqrt{12a^{3}}$

7. $\sqrt{16a^{2}b}-\sqrt{9a^{2}b},\ \ a\geq 0.$

8. $\sqrt{3}\div\sqrt{48}$

9. $\sqrt[3]{54}-\sqrt[3]{16}+\sqrt{12}-\sqrt{3}$

10. $4\sqrt[3]{24}+\sqrt[3]{-3}+\sqrt[3]{-81}$

11. $\sqrt[6]{16}-\sqrt[4]{4}+\sqrt[3]{-4}$

12. $\sqrt{64}-\sqrt[3]{64}-\sqrt[6]{64}$

13. $\sqrt{50}-\sqrt[4]{324}-\sqrt[6]{2916}+\sqrt[8]{256}$

14. $\left(\sqrt[3]{81}+2\sqrt[3]{24}-4\sqrt[3]{375}\right)\cdot\sqrt[3]{-3}$

15. $\left(\sqrt[3]{a^{5}b^{4}}\cdot a\cdot\sqrt[3]{b^{2}}\right)\div\sqrt[3]{a^{2}b^{12}}$

16. $\sqrt[3]{\dfrac{3}{\sqrt[4]{3}}}$

17. $\sqrt[n-1]{\dfrac{a}{\sqrt[n]{a}}}$

18. $\sqrt[3]{\sqrt[7]{-8a^3}}$

19. $(a^3)^{\frac{2}{3}} \cdot (a^{-\frac{1}{2}})^{\frac{4}{5}} \cdot \sqrt{a^{\frac{4}{5}}}$

20. $(a^{\frac{1}{2}} - b^{\frac{1}{2}})(a^{-\frac{1}{2}} + b^{-\frac{1}{2}}).$

21. $\left(-\sqrt[7]{-a\sqrt{3a}} \right)^{14}$

22. $\sqrt{2\sqrt{2\sqrt{2\sqrt{2}}}}.$

3.4 Review Exercises

1. Compute
$$24 \cdot (-\frac{1}{3})^3 \cdot (4\frac{1}{2})^2$$

2. Compute
$$\left(\frac{1}{2} \right)^3 + \left(\frac{1}{2} \right)^2 - \left[\left(-\frac{1}{2} \right)^3 \right]^2.$$

3. Compute
$$\left(\frac{2}{3} \right)^{-2} + \left(-\frac{5}{2} \right)^{-2} - \left(-2\frac{1}{2} \right)^{-1}.$$

4. Compute
$$(1\frac{1}{2})^{-4} \div (2^4 \div 3^2 - 2^3 \div 1\frac{1}{8}).$$

5. Compute
$$\frac{2 \cdot 4^{-2} + (81^{-\frac{1}{2}})^3 \cdot (\frac{1}{9})^{-3}}{125^{-\frac{1}{3}} \cdot (\frac{1}{5})^{-2} + (\frac{1}{2})^{-2}}$$

6. Use scientific notation to simplify

$$\frac{(45,000,000,000)(21,200)}{0.000018}$$

7. Simplify

$$5\sqrt{18} \cdot 3\sqrt{8}$$

8. Simplify

$$\sqrt[3]{6} \cdot \sqrt[3]{30} \cdot \sqrt[3]{150}$$

9. Simplify

$$\sqrt{a} \cdot \sqrt[5]{a^3} \cdot \sqrt[10]{a^4}$$

10. Simplify

$$3\sqrt[4]{a} \cdot 7\sqrt[6]{a^5 b} \cdot \sqrt[12]{a^3 b^{10}}$$

11. Compute

$$\left(27^{\frac{1}{3}} \cdot 8^{\frac{2}{3}} \cdot 32^{\frac{2}{5}} \cdot 81^{\frac{3}{4}}\right)^{\frac{1}{4}}.$$

12. Compute

$$3\sqrt{5\sqrt{48}} - 2\sqrt{90\sqrt{12}} + 2\sqrt{15\sqrt{27}}$$

13. Simplify

$$3\sqrt{\frac{2}{3}} - 2\sqrt{\frac{3}{2}} + \sqrt{6} + \sqrt{150}$$

14. Evaluate

$$\left(\frac{a^{\frac{5}{12}} \cdot a^{-\frac{3}{8}}}{a^{\frac{7}{24}}}\right)^{-\frac{4}{3}}$$

for $a = 125$.

15. Perform the operations $(a, b > 0)$

$$ab \cdot \sqrt[3]{\frac{b}{a^2}} - ab \cdot \sqrt[3]{\frac{a}{b^2}} + \frac{a}{b}\sqrt[3]{ab^4} - \frac{b}{a}\sqrt[3]{a^4 b}.$$

16. Use fractional exponents and simplify

$$\sqrt[3]{a^{-1}\sqrt[4]{a^3}}.$$

17. Use fractional exponents and simplify

$$\frac{\sqrt[3]{a^{\frac{2}{3}}\cdot\sqrt{a^{-3}}}}{\sqrt{\sqrt[3]{a^{-7}}\cdot\sqrt[3]{a}}}.$$

18. Use fractional exponents and simplify

$$\sqrt{\frac{a}{b}\cdot\sqrt{\frac{a}{b}\cdot\sqrt[3]{\frac{a}{b}}}}\cdot a^{-\frac{1}{3}}\cdot b^{\frac{1}{3}}.$$

19. Use fractional exponents and simplify

$$\sqrt{\frac{a}{b}\sqrt[3]{\frac{a^2}{b^2}\sqrt[4]{\frac{a^3}{b^3}}}}.$$

20. Simplify

$$\sqrt[3]{9a^2\sqrt{\frac{2b}{3a}}}\cdot\sqrt[3]{4b^2\sqrt{\frac{3a}{2b}}}.$$

21. Compute

$$\sqrt{\sqrt{5}}\cdot\left(\sqrt[3]{\sqrt{5}}\div\sqrt[4]{\sqrt{5}}\right)^2.$$

22. Compute

$$\sqrt{2\sqrt[3]{2}}\div\left(\sqrt[3]{2\sqrt{2}}\cdot\sqrt[4]{2\sqrt[3]{2}}\right)^{\frac{1}{2}}.$$

23. Find the value of the expression

$$\left\{\left[\left(3^{-\frac{1}{2}}\cdot 2^{-\frac{1}{3}}\right)\div\left(3^{-\frac{3}{4}}\cdot 2^{-\frac{5}{6}}\right)\right]\div\left(\frac{1}{864}\right)^{\frac{1}{4}}\right\}^{\frac{2}{7}}.$$

24. Compute

$$\left(\sqrt[3]{25\sqrt{5\sqrt{5\sqrt{\sqrt{5}}}}}\right)^{2}$$

25. Compute

$$\frac{\left(\sqrt[6]{5\sqrt{5}}\right)^{2}\cdot\left(\sqrt[6]{25\sqrt[3]{25}}\right)^{4}}{\left(\sqrt[3]{5\sqrt[4]{125}}\right)^{\frac{2}{3}}}.$$

Chapter 4

Algebraic Expressions

In this chapter we study the basic algebraic expressions. These are: monomials, polynomials, rational expressions and radical expressions.

4.1 Monomials

The simplest algebraic expressions are the monomials. A ***monomial*** is an algebraic expression obtained as a product of two or more factors each of which is either a number, a letter or a power of a letter with natural exponent. For example,

$$3x, \quad -2y, \quad 5x^2y^4, \quad -\frac{1}{2}xy^3z^4, \quad 2a^3b^2, \quad 3abc, \quad 4s^2t^5$$

are monomials.

A single number or a single letter can be regarded as a monomial. The numerical factor (say 3, in the expression, $3x^5y^2$) is called the *coefficient* of the monomial and the letter factors, x^5y^2, is called the *variable part* of the monomial.

Two monomials are said *like* monomials (or *similar* monomials) if they differ solely in the coefficient, ie, if they have the same variable part. For example, the monomials $5x^2y^6$ and $-x^2y^6$ are similar. So are the monomials ax^2y^2 and bx^2y^2, if the coefficients are taken to be a, b, respectively. The *degree* of a monomial is defined as the sum of the exponents of its literal factors. For example, $6x^2$ is a monomial of degree 2, $6xy^2z^5$ is a monomial of degree 8, and $6 = 6x^0$ is a monomial of degree 0 (any non-zero number regarded as a monomial has degree 0).

Operations with Monomials

Generally speaking, when we **add** or **subtract** monomials, the only simplification possible is in the case of similar monomials, in which case we add (or subtract) their coefficients. This procedure is called *collecting like terms*.

Example 4.1.1. Simplify

1. $2x + 8x - 3x = 7x$.

2. $4x^2 - 9x^2 + 6x^2 = x^2$.

3. $3xy^2z^3 + 6xy^2z^3 - xy^2z^3 = 8xy^2z^3$.

Example 4.1.2. Simplify

1. $ax^2y^2 + bx^2y^2 = (a+b)x^2y^2$.

2. $a - 2ab^2 + 5ab^2 = a + 3ab^2$.

3. $2x^4y^2 - 3x^4y^2 + \frac{1}{3}x^4y^2 = -\frac{2}{3}x^4y^2$.

Monomials are multiplied by means of the rules of multiplying numbers. To **multiply** two or more monomials, we multiply their coefficients separately and their variable parts separately. Such a product is simplified if the monomials include powers of the same letters. In this case we use the rules of powers; $a^na^m = a^{n+m}$, $(a^n)^k = a^{kn}$ and $(ab)^n = a^nb^n$.

Example 4.1.3. Multiply

1. $(2x)(3x^2) = (2)(3)xx^2 = 6x^3$.

2. $(-5ab^2)(8a^3b) = (-5)(8)aa^3b^2b = -40a^4b^3$.

3. $(3x^3y^5)(4x^2y^3z) = (3)(4)x^3x^2y^5y^3z = 12x^5y^8z$.

Example 4.1.4. Multiply

1. $(-5ab^2c^2)(-3a^3b^4c^6) = (-5)(-3)aa^3b^2b^4c^2c^6 = 15a^4b^6c^8$.

2. $(-2x^3y^2)^3 = (-2)^3(x^3)^3(y^2)^3 = -8x^9y^6$.

3. $(-\frac{2}{5}x^4z)(\frac{3}{2}x^3y^3)(-10y^2z) = (-\frac{2}{5})(\frac{3}{2})(-10)x^4 \cdot x^3 \cdot y^3 \cdot y^2 \cdot z \cdot z = 6x^7y^5z^2$.

The **division** of one monomial by another one can be simplified if the divident and the divisor contain powers of the same letters. To do so we use the rule; $\frac{a^n}{a^m} = a^{n-m}$.

Example 4.1.5. Simplify

1. $\frac{6x^4}{2x^3} = \frac{6}{2} \cdot \frac{x^4}{x^3} = 3x^{4-3} = 3x$.

2. $\frac{24a^2b}{-3a^2} = \frac{24}{-3} \cdot \frac{a^2}{a^2} \cdot \frac{b}{1} = -8b$.

3. $\frac{-42x^5}{-7x^3} = \frac{-42}{-7} \cdot \frac{x^5}{x^3} = 6x^2$.

Example 4.1.6. Simplify

$$\frac{15x^5y^4z^3}{3x^2y^3z} = \frac{15}{3} \cdot \frac{x^5}{x^2} \cdot \frac{y^4}{y^3}\frac{z^3}{z} = 5x^3yz^2.$$

Example 4.1.7. Simplify

$$\frac{-6x^3y^4z}{\frac{2}{3}x^2y^4} = -\frac{6}{\frac{2}{3}} \cdot \frac{x^3}{x^2} \cdot \frac{y^4}{y^4} \cdot \frac{z}{1} = -\frac{18}{2}xz = -9xz.$$

Example 4.1.8. Simplify

$$\frac{(2a^3b)^4(-3ab^2c^4)^2}{(-3a^4b^2c)^3}.$$

Solution.

$$\frac{(2a^3b)^4(-3ab^2c^4)^2}{(-3a^4b^2c)^3} = \frac{(16a^{12}b^4)(9a^2b^4c^8)}{-27a^{12}b^6c^3} = \frac{(16)(9)}{-27} \cdot \frac{a^{14}b^8c^8}{a^{12}b^6c^3} = -\frac{16}{3}a^2b^2c^5.$$

4.1.1   Exercises

Do the operations.

1. $-3x^2 + 5x - (-2x^2) - 5x$

2. $(8xy^2)(\frac{1}{4}x^3y)$

3. $(5abc)(abc^2)(2bc)$

4. $\left(-\frac{2}{5}x^4\right)\left(-\frac{3}{2}x^3\right)(10x^2)$

5. $(-2x^3)^2 \cdot (-x^2)^3$

6. $(-20x^5) \div (5x^2)$

7. $(-4x^4)^2 \div (-2x^2)^3$

8. $(-6x^4y^3z) \div (-2x^3y^2)$

9. $(7x^2y^2z) \cdot (-2x^2y^3) \div (-14x^4y^5z)$

10. $(2a^3b)^2 \cdot (-3ab^2c^2)^3 \cdot (-4a^4b^2c^2) \div (-3a^2b^3c^3)$

11. $(4x^7) \cdot (3xy^3z^4) \div (xyz)^2 \div (6yz)$

12. Simplify

$$x^7 \div \left[x^5 \div \left(x^4 \div x^2 \cdot x\right) \cdot \left(x^3 \cdot x \div x^2\right)\right].$$

4.2 Polynomials

Polynomials in one variable: A *polynomial in one variable x* is an algebraic expression which can be build up from the variable x and numbers or constants by means of the operations of addition, subtraction and multiplication alone, that is, a polynomial is a sum of two or more monomials in one variable x. For example,

$$5x - 2, \quad x^2 - 5x + 6, \quad x^3 - 1, \quad 3x^5 - 2x^4 + 5x^3 - 8x^2 - x + 1$$

are polynomials. Each monomial entering into a polynomial is said a *term* of the polynomial. Similar addends in a polynomial are called *like terms*. The greatest power of x is called the *degree* of the polynomial. For instance, the polynomial $3x^4 - 5x^3 + 6x^2 - 2x$ has degree 4.

Certain polynomials have special names according to their degrees:

1. *Linear polynomial* (degree 1): $ax + b$.

2. *Quadratic polynomial* (degree 2): $ax^2 + bx + c$.

3. *Cubic polynomial* (degree 3): $ax^3 + bx^2 + cx + d$.

4. *Quadric polynomial* (degree 4): $a_4x^4 + a_3x^3 + a_2x^2 + a_1x + a_0$.

More generally, we have.

Definition 4.2.1. An algebraic expression of the form

$$f(x) = a_nx^n + a_{n-1}x^{n-1} + ... + a_1x + a_0,$$

where n is a natural number and $a_0, a_1, ..., a_n$ are real numbers with $a_n \neq 0$, is called a ***polynomial*** in x of **degree** n.

The numbers $a_0, a_1, ..., a_n$ are called the *coefficients* of the polynomial and the terms $a_n x^n$ and a_0 are called the *leading term* and *constant term*, respectively. The polynomial $f(x)$ written as above is said to be written according to *descending powers* of x.

If $a_0 = a_1 = ... = a_n = 0$, then the polynomial $f(x) = 0$ is called the *zero polynomial*, that is, the zero polynomial is simply the number 0 regarded as a polynomial all whose coefficients are zero.

Two polynomials $f(x)$ and $g(x)$ of the same degree are called **equal** if their coefficients of equal powers of x are equal, ie, if their corresponding coefficients are equal. That is, if $f(x) = a_n x^n + ... + a_1 x + a_0$ and $g(x) = b_n x^n + ... + b_1 x + b_0$, then $f(x) = g(x)$ if and only if

$$a_n = b_n, \quad a_{n-1} = b_{n-1}, \quad .... \; , a_1 = b_1, \quad a_0 = b_0.$$

Polynomials in several variables: A polynomial in several variables, say x, y, z, is a sum of two or more monomials in the variables x, y, z. Here again, each monomial entering into a polynomial is said a *term* of the polynomial, and similar addends in a polynomial are called *like terms*. For example,

$$f(x, y) = 2x^4 y^3 + 6x^2 y - 7xy + 2y$$

is a polynomial in the two variables x, y. Its degree is the highest degree of its terms, that is, its degree is 7. Similarly,

$$f(x, y, z) = 3x^2 y^2 z^2 - 5x^2 yz + xyz - 3xy + 1$$

is a polynomial in the three variables x, y, z of degree 6.

A polynomial with two terms is also called a *binomial*, and a polynomial with three terms a *trinomial*.

4.2.1 Operations with Polynomials

1) Addition: To **add** two or more polynomials we simply combine like terms.

Example 4.2.2. Add the polynomials $x^4 + 8x^3 - 3x^2 + 6x - 7$ and $4x^3 - 2x^2 - 5x + 1$.

   Solution.

$$(x^4 + 8x^3 - 3x^2 + 6x - 7) + (4x^3 - 2x^2 - 5x + 1) = x^4 + 12x^3 - 5x^2 + x - 6.$$

Example 4.2.3. Add $2x^3y - 3xy + 4y^2$ and $5x^3y + xy - x^2 - y^2 + 8x - 1$.

   Solution.

$$(2x^3y - 3xy + 4y^2) + (5x^3y + xy - x^2 - y^2 + 8x - 1) = 7x^3y - 2xy + 3y^2 - x^2 + 8x - 1.$$

2) Subtraction: To **subtract** the polynomial $g(x)$ from the polynomial $f(x)$, we add to $f(x)$ the opposite of $g(x)$, that is,

$$f(x) - g(x) = f(x) + (-g(x)).$$

Example 4.2.4. Find $f(x) - g(x)$, where $f(x) = x^4 + 8x^3 - 3x^2 + 6x - 7$ and $g(x) = 4x^3 - 2x^2 - 5x + 1$.

   Solution. The opposite of $g(x)$ is $-g(x) = -4x^3 + 2x^2 + 5x - 1$. We have

$$f(x) - g(x) = (x^4 + 8x^3 - 3x^2 + 6x - 7) - (4x^3 - 2x^2 - 5x + 1)$$

$$= (x^4 + 8x^3 - 3x^2 + 6x - 7) + (-4x^3 + 2x^2 + 5x - 1) = x^4 + 4x^3 - x^2 + 11x - 8.$$

3) Multiplication: To **multiply a polynomial by a monomial**, we apply the distributive property $a(b + c) = ab + ac$ of multiplication over addition. That is, we multiply the monomial by each term of the polynomial and we add the resulting monomials. For example,

$$2x(4x - 3y) = 8x^2 - 6xy,$$

and

$$-3x^2(2x^3 - 5x^2 + 8x - 4) = -6x^5 + 15x^4 - 24x^3 + 12x^2.$$

To **multiply two polynomials**, we multiply each term of one polynomial by all the terms of the other and we add the resulting monomials.

Example 4.2.5. Multiply $(x + a)(x + b)$.
   Solution.

$$(x+a)(x+b) = x(x+b)+a(x+b) = x^2+bx+ax+ab = x^2+(a+b)x+ab.$$

Example 4.2.6. Multiply $(2x - 5)(3x + 4)$.
   Solution.

$$(2x-5)(3x+4) = 2x(3x+4)-5(3x+4) = 6x^2+8x-15x-20 = 6x^2-7x-20.$$

Example 4.2.7. Find the product $(2x - 3)(4x^2 + 5x - 6)$.
   Solution.

$$(2x - 3)(4x^2 + 5x - 6) = 2x(4x^2 + 5x - 6) - 3(4x^2 + 5x - 6)$$
$$= 8x^3 + 10x^2 - 12x - 12x^2 - 15x + 18 = 8x^3 - 2x^2 - 27x + 18$$

Example 4.2.8. Find the product $(3x^2+4x-2)(5x^4-6x^3-x+8)$.

   Solution.

$$(3x^2 + 4x - 2)(5x^4 - 6x^3 - x + 8)$$
$$= 3x^2(5x^4-6x^3-x+8)+4x(5x^4-6x^3-x+8)-2(5x^4-6x^3-x+8)$$
$$= 15x^6-18x^5-3x^3+24x^2+20x^5-24x^4-4x^2+32x-10x^4+12x^3+2x-16$$
$$= 15x^6 + 2x^5 - 34x^4 + 9x^3 + 20x^2 + 34x - 16.$$

Note that the degree of the product of two polynomials is the sum of the degrees of the two polynomials.

4.3 Identities

An *identity* is an equality of two expressions which is true for *all values* of the variables involved in the expressions. For instance

$$x + 5 = 5 + x, \quad 1 - x = -(x - 1), \quad xy = yx, \quad \frac{2}{3}x = \frac{2x}{3}$$

are some trivial identities.

Special Products-Identities

The following identities are important and are frequently used in Algebra.

1. $(a + b)^2 = (a + b)(a + b) = a^2 + ab + ba + b^2 = a^2 + 2ab + b^2$.
   Thus, we obtain the so-called *perfect square identity*

   $$(a + b)^2 = a^2 + 2ab + b^2.$$

2. $(a - b)^2 = (a - b)(a - b) = a^2 - ab - ba + b^2 = a^2 - 2ab + b^2$.
   Thus, we obtain the so-called *perfect square identity*

   $$(a - b)^2 = a^2 - 2ab + b^2.$$

3. $(a + b)(a - b) = a^2 - ab + ba - b^2 = a^2 - b^2$. Thus, we obtain the so-called *difference of squares identity*

   $$a^2 - b^2 = (a + b)(a - b).$$

4.

   $$(a+b+c)^2 = (a+b+c)(a+b+c) = a(a+b+c)+b(a+b+c)+c(a+b+c) =$$
   $$a^2 + b^2 + c^2 + 2ab + 2ac + 2bc.$$

   Thus,

   $$(a + b + c)^2 = a^2 + b^2 + c^2 + 2ab + 2ac + 2bc.$$

5. $(a+b)^3 = (a+b)(a+b)^2 = (a+b)(a^2+2ab+b^2) = a^3 + 3a^2b + 3ab^2 + b^3$. Thus,

$$(a+b)^3 = a^3 + 3a^2b + 3ab^2 + b^3.$$

6. $(a-b)^3 = (a-b)(a-b)^2 = (a+b)(a^2-2ab+b^2) = a^3 - 3a^2b + 3ab^2 - b^3$. Thus,

$$(a-b)^3 = a^3 - 3a^2b + 3ab^2 - b^3.$$

7. $(a+b)(a^2-ab+b^2) = a^3 - a^2b + ab^2 + ba^2 - ab^2 + b^3 = a^3 + b^3$. Thus,

$$a^3 + b^3 = (a+b)(a^2 - ab + b^2).$$

8. $(a-b)(a^2+ab+b^2) = a^3 + a^2b + ab^2 - ba^2 - ab^2 - b^3 = a^3 - b^3$. Thus,

$$a^3 - b^3 = (a-b)(a^2 + ab + b^2).$$

Similarly, by multiplying, we find the expansions of $(a \pm b)^4$, $(a \pm b)^5$, $(a \pm b)^6$ etc. For example,

$$(a+b)^4 = (a+b)(a+b)^3 = (a+b)(a^3+3a^2b+3ab^2+b^3) = a^4+4a^3b+6a^2b^2+4ab^3+b^4,$$

$$(a+b)^5 = (a+b)(a+b)^4 = a^5 + 5a^4b + 10a^3b^2 + 10a^2b^3 + 5ab^4 + b^5.$$

Replacing b by $-b$, in these formulas, we get

$$(a-b)^4 = (a+(-b))^4 = a^4 - 4a^3b + 6a^2b^2 - 4ab^3 + b^4,$$

$$(a-b)^5 = (a+(-b))^5 = a^5 - 5a^4b + 10a^3b^2 - 10a^2b^3 + 5ab^4 - b^5.$$

A more general formula called the **binomial formula** is given in the following theorem. Before we state it we introduce the *factorial notation*:

$$0! = 1, \quad 1! = 1, \quad 2! = 2 \cdot 1 = 2, \quad 3! = 3 \cdot 2 \cdot 1 = 6.$$

In general for $n \in \mathbb{N}$, $n!$ (reads n-**factorial**) is defined by

$$n! = n \cdot (n-1) \cdot (n-2) \cdot (n-3) \cdots 3 \cdot 2 \cdot 1.$$

Thus, for example, $5! = 5 \cdot 4 \cdot 3 \cdot 2 \cdot 1 = 120$ and $6! = 6 \cdot 5! = 720$.

Theorem 4.3.1. *(**Binomial theorem** (*)). Let a,b be real numbers and n a positive integer. Then*

$$(a+b)^n = \binom{n}{0} a^n + \binom{n}{1} a^{n-1}b + \binom{n}{2} a^{n-2}b^2 + \ldots + \binom{n}{k} a^{n-k}b^k + \ldots + \binom{n}{n} b^n,$$

where the binomial coefficients are computed by the formula

$$\binom{n}{k} = \frac{n!}{k!(n-k)!}$$

This number gives the number of **combinations** of n *objects taken k at a time*[1]

Example 4.3.2.

$$(a+b)^6 = \binom{6}{0} a^6 + \binom{6}{1} a^5b + \binom{6}{2} a^4b^2 + \binom{6}{3} a^3b^3 + \binom{6}{4} a^2b^4 + \binom{6}{5} ab^5 + \binom{6}{6} b^6$$

$$= a^6 + 6a^5b + \frac{6 \cdot 5}{1 \cdot 2} a^4b^2 + \frac{6 \cdot 5 \cdot 4}{1 \cdot 2 \cdot 3} a^3b^3 + \frac{6 \cdot 5}{1 \cdot 2} a^2b^4 + 6ab^5 + b^6.$$

That is,

$$(a+b)^6 = a^6 + 6a^5b + 15a^4b^2 + 20a^3b^3 + 15a^2b^4 + 6ab^5 + b^6.$$

---

[1]On a calculator the symbol $\binom{n}{k}$ is denoted by the key nCk.

Basic Identities

We collect the most important of these identities, which they should be carefully memorized.

1. $(a \pm b)^2 = a^2 \pm 2ab + b^2$ (**perfect square**)

2. $a^2 - b^2 = (a + b)(a - b)$ (**difference of squares**)

3. $(a \pm b)^3 = a^3 \pm 3a^2b + 3ab^2 \pm b^3$ (**perfect cube**)

4. $a^3 + b^3 = (a + b)(a^2 - ab + b^2)$ (**sum of cubes**)

5. $a^3 - b^3 = (a - b)(a^2 + ab + b^2)$ (**difference of cubes**)

4.3.1   Exercises

1. Add $(5x^2 + 6x - 5) + (-3x^2 - 2x + 9)$.

2. Subtract $(3x^2 - 4x + 2) - (8x^2 - 2x - 1)$

3. Add the polynomials $f(x) = 5x^3 - 4x^2 + 6x - 1$ and $g(x) = 2x^4 - x^3 + 10x^2 + 8x + 13$.

4. Add the polynomials $f(x) = 2x^5 - 3x^4 + +2x^3 + 7x - 6$ and $g(x) = -x^5 + 3x^4 - 6x + 2$.

5. Let $f(x) = 2x^4 - 5x^3 + 6x^2 - 14$ and $g(x) = -3x^3 + 5x^2 + 3x - 8$. Find $f(x) - g(x)$.

6. Let $f(x) = 3x - 5 + 6x^2 - 3x^3 + x^4$, $g(x) = -x^2 + 2x - x^3 - 6x^4 + 7$ and $h(x) = -x^4 + x^3 + 3x^2 - 2 - 2x$. Arrange the polynomials in descending order and find; $f(x) + g(x)$, $f(x) - g(x), f(x) + g(x) - h(x)$ and $f(x) - g(x) - h(x)$.

7. Let $f(x, y) = x^3y^3 - 3x^2y^2 - 2xy + y^4$ and $g(x, y) = 5x^3y^3 + 2x^2y^2 + 3xy - y^4$. Find $f(x, y) + g(x, y)$ and $f(x, y) - g(x, y)$.

8. Multiply $5x(3x - 2)$.

9. Multiply $2x^2(6x^3 - 4x)$.

10. Multiply $-4x(2x^5 - 3x^2 - x + 5)$.

11. Multiply $6x\left(-\frac{2}{3}x^4 + \frac{x^3}{3} - \frac{x}{2} + \frac{3}{2}\right)$.

12. Multiply $(-2xy^2)(5x^2y - 3xy - 2xy^2 + y^3)$.

13. Do the operation $3y(x^2 + 2y) + (-x)(3xy + y^2)$.

14. Multiply $(2x + 3)(3x^2 - 5x + 6)$.

15. Multiply $(3x^2 + 5x - 2)(4x^2 - 6x + 3)$.

16. Multiply $(x - 3)(x^3 - 7x^2 + 6x - 2)$.

17. Multiply $(2x^3 - 3x)(x^2 - 6x + 5)$.

18. Multiply $(x + 1)(x + 2)(x + 3)$.

19. Multiply $(x - 1)(x - 2)(x - 3)$.

20. Multiply $(x - y)(x^3 + xy^2 + x^2y + y^3)$.

21. Multiply $(2 - 2x^2)(x^3 + 2x^2 + 5x - 1)$.

22. Multiply $(x^3 - 3x^2 + x - 2)(x^3 - 2x^2 + 1)$.

23. Do the operation $(x + 5)(x - 1)(x - 3) - (x + 3)(x - 2)^2$.

24. Develope the square $(2a - 3b)^2$.

25. Develope the square $\left(\frac{3}{2}x^2 + 4xy\right)^2$.

26. Develope the square $(x - y + z)^2$.

27. Develope the square $(3x + 2y - 1)^2$.

28. Simplify $(a+b)^2 - (a-b)^2$.

29. Simplify $(a+b-c)^2 - (a-b+c)^2$.

30. Simplify $(a+b)^3 - (a-b)^3$.

31. Simplify $(a+b+c)[(a-b)^2 + (b-c)^2 + (c-a)^2]$.

32. Multiply and simplify
$$(2x+3)^2 + (2x-3)^2 + (2x+3)(2x-3) - 3(x-5)^2.$$

33. Multiply and simplify
$$(2x+5)^2 - (x-5)^2 + (3x-1)^2 - (2x+1)^2 - (2x+3)(2x-3).$$

34. Multiply and simplify
$$(3x^4 - 5x^2)^2 - (x^3 + 3x)^2 - (x^4 + 3x^2)(x^4 - 3x^2).$$

35. Prove the identity
$$\left(\frac{a+b}{2}\right)^2 - \left(\frac{a-b}{2}\right)^2 \equiv ab.$$

36. Prove the identity
$$(a^2 + b^2)(x^2 + y^2) - (ax + by)^2 \equiv (ay - bx)^2.$$

37. Prove the identity
$$(1-a)(1-a^2) - a(1+a) \equiv a^3 + 1.$$

38. Prove the identity
$$(a^3 + b^3)^2 - (a^2 + b^2)^3 + 3a^2b^2(a+b)^2 \equiv (2ab)^3.$$

39. Prove the identity
$$(x+a)(x+b)(x+c) \equiv x^3 + (a+b+c)x^2 + (ab+bc+ac)x + abc.$$

4.4　The Division Algorithm

When we divide a polynomial $f(x)$ (called the Divident) by another polynomial $g(x)$ (called the Divisor) of degree less or equal than the degree of $f(x)$, we obtain a polynomial $Q(x)$ (called the Quotient) and a Remainder $R(x)$, the remainder being either the zero polynomial or a polynomial whose degree is less than the degree of the divisor. To check our work, we verify (as we did, similarly, for numbers) that

$$Divident = (Divisor) \cdot (Quotient) + Remainder.$$

That is,

$$f(x) = g(x) \cdot Q(x) + R(x).$$

This checking is based on the so-called ***Division Algorithm*** for polynomials which we state next without proof.

Theorem 4.4.1. ***(Division algorithm for polynomials).*** *Given a polynomial $f(x)$ of degree n and a polynomial $g(x)$ of degree m with $1 \leq m \leq n$, then there are unique polynomials $Q(x)$ and $R(x)$ such that*

$$f(x) = g(x) \cdot Q(x) + R(x) \quad \Leftrightarrow \quad \frac{f(x)}{g(x)} = Q(x) + \frac{R(x)}{g(x)},$$

where $R(x)$ is either the zero polynomial or a polynomial of degree less than that of $g(x)$.

When the remainder $R(x) \equiv 0$, we say the division $f(x) \div g(x)$ is **exact**, and in this case we have

$$f(x) = g(x) \cdot Q(x) \quad \Leftrightarrow \quad \frac{f(x)}{g(x)} = Q(x).$$

To **divide a polynomial by a monomial**, divide each term of the polynomial by the monomial (whenever possible), and add the quotients so obtained.

Example 4.4.2. Divide $(3x^5 - 6x^4 + 15x^3) \div 3x^3$.

Solution.

$$(3x^5 - 6x^4 + 15x^3) \div 3x^3 = \frac{3x^5 - 6x^4 + 15x^3}{3x^3} = \frac{3x^5}{3x^3} - \frac{6x^4}{3x^3} + \frac{15x^3}{3x^3} = x^2 - 2x + 5.$$

Example 4.4.3. Divide $(ax^3 - 5a^2x^2) \div ax$.

Solution.

$$(ax^3 - 5a^2x^2) \div ax = \frac{ax^3 - 5a^2x^2}{ax} = \frac{ax^3}{ax} - \frac{5a^2x^2}{ax} = x^2 - 5ax.$$

Example 4.4.4. Divide $(4x^2 - 3x) \div x^2$.

Solution. The division $(4x^2 - 3x) \div x^2$ is not exact, because the term $-3x$ is not divisible by x^2. In this case the division leaves a nonzero remainder $-3x$. We can simplify to get

$$\frac{4x^2 - 3x}{x^2} = \frac{4x^2}{x^2} - \frac{3x}{x^2} = 4 - \frac{3x}{x^2} = 4 - \frac{3}{x}.$$

4.4.1 Division of a polynomial by a polynomial

To divide a polynomial $f(x)$ by a polynomial $g(x)$, where the degree of $g(x)$ is not more than the degree of $f(x)$, we follow a process called *long division*. It consists of the following steps:

1. Arrange both polynomials $f(x)$ and $g(x)$ according to descending powers of x.

2. Divide the leading term of $f(x)$ by the leading term of $g(x)$ to obtain the first term of the quotient $Q(x)$.

3. Multiply $g(x)$ by this first term of $Q(x)$, and subtract the product from $f(x)$.

4. Proceed in a similar manner with the remainder thus obtained, dividing its leading term by the leading term of $g(x)$, and so on.

5. Continue the process until a remainder is reached which is of lower degree than $g(x)$. We shall then have found all the terms of $Q(x)$, and the final remainder $R(x)$.

We illustrate in the following examples.

Example 4.4.5. Divide $(x^2 - 7x + 12) \div (x - 3)$.
   Solution.

$$
\begin{array}{r}
x - 4 \\
x - 3 \overline{)\ x^2 - 7x + 12} \\
\underline{-x^2 + 3x} \\
-4x + 12 \\
\underline{4x - 12} \\
0
\end{array}
\tag{4.1}
$$

Thus, $Q(x) = x - 4$, and $R(x) = 0$.

Example 4.4.6. Divide $(18x^3 + 9x^2 - 50x - 25) \div (3x - 5)$.
   Solution.

$$
\begin{array}{r}
6x^2 + 13x + 5 \\
3x - 5 \overline{)\ 18x^3 + 9x^2 - 50x - 25} \\
\underline{-18x^3 + 30x^2} \\
39x^2 - 50x - 25 \\
\underline{-39x^2 + 65x} \\
15x - 25 \\
\underline{-15x + 25} \\
0
\end{array}
\tag{4.2}
$$

Thus, $Q(x) = 6x^2 + 13x + 5$, and $R(x) = 0$.

Example 4.4.7. $(x^5 - x^4 - 3x^3 + 4x^2 + 5) \div (x^2 - 3)$.
  Solution.

$$
\begin{array}{r}
x^3 - x^2 + 1 \\
x^2 - 3 \overline{\smash{\big)}\ x^5 - x^4 - 3x^3 + 4x^2 + 5} \\
\underline{-x^5 \qquad\ + 3x^3} \\
-x^4 + 0 + 4x^2 + 5 \\
\underline{x^4 \qquad - 3x^2} \\
x^2 + 5 \\
\underline{-x^2 + 3} \\
8
\end{array}
$$

$$(4.3)$$

Thus, $Q(x) = x^3 - x^2 + 1$, and $R(x) = 8$.

4.4.2   Division of a polynomial by $(x - a)$.

If a polynomial $f(x)$ is divided by a first-degree binomial $x - a$, where a is a given number, then the remainder $R(x)$ is a zero degree polynomial, that is, $R(x) = R$, where R is a number.

Example 4.4.8. Divide $f(x) = x^3 - 5x^2 + 9x - 10$ by $x - 2$. The division gives quotient $Q(x) = x^2 - 3x + 3$ and remainder $R = -4$. By the division algorithm we also have

$$f(x) = (x - 2)(x^2 - 3x + 3) - 4,$$

and substituting $x = 2$, we see that $f(2) = -4$. On the other hand, substituting $x = 2$ into the given polynomial, we find $f(2) = 2^3 - 5(2^2) + 9(2) - 10 = -4$. That is, $R = f(2)$.

This property of the remainder is true in general and is called the *remainder theorem*.

Theorem 4.4.9. (Remainder theorem). *If the polynomial $f(x)$ is divided by $x - a$, then the remainder is $f(a)$.*

Proof. By the division algorithm, we have

$$f(x) = (x - a)Q(x) + R. \tag{4.4}$$

Substituting $x = a$ yieds

$$f(a) = (a - a)Q(a) + R = 0 \cdot Q(a) + R = R.$$

$\square$

In general the remainder of the division $f(x) \div (ax + b)$, where $a \neq 0$ is the number $f(-\frac{b}{a})$.

It may happen that $R = 0$. In this case we obtain the following important and useful consequence of the remainder theorem called the *factor theorem*.

Theorem 4.4.10. (Factor theorem). *Let $f(x)$ be a polynomial. Then $x - a$ is a factor of $f(x)$ if and only if $f(a) = 0$. That is,*

$$f(x) = (x - a)Q(x) \Leftrightarrow f(a) = 0.$$

Proof. Suppose $x - a$ is a factor of $f(x)$. This means that when $f(x)$ is divided by $x - a$ leaves remainder 0. That is, $f(x) = (x - a)Q(x)$. Therefore $f(a) = 0$. Conversely, if $f(a) = 0$, then $R = 0$ and (4.4), tells us that $f(x) = (x - a)Q(x)$. $\square$

Special Quotients

Here are some immediate consequences of the factor theorem.

Corollary 4.4.11. *The following hold true;*

1. *The polynomial $f(x) = x^n - a^n$ is exactly divisible by $(x - a)$ for any natural n, and*

$$x^n - a^n = (x - a)(x^{n-1} + x^{n-2}a + x^{n-2}a^2 + ... + xa^{n-2} + a^{n-1}).$$

2. *The polynomial* $f(x) = x^n - a^n$ *is exactly divisible by* $(x + a)$ *for any even* n, *i.e.,* $n = 2k$, *where* $k \in \mathbb{N}$,

$$x^n - a^n = (x + a)(x^{n-1} - x^{n-2}a + x^{n-2}a^2 - \ldots + xa^{n-2} - a^{n-1}).$$

3. *The polynomial* $f(x) = x^n + a^n$ *is exactly divisible by* $(x + a)$ *for any odd* n, *i.e.,* $n = 2k + 1$, *where* $k \in \mathbb{N}$,

$$x^n + a^n = (x + a)(x^{n-1} - x^{n-2}a + x^{n-2}a^2 - \ldots - xa^{n-2} + a^{n-1}).$$

Proof.   1. Indeed, $f(a) = a^n - a^n = 0$, and so the remainder of the division $(x^n - a^n) \div (x - a)$ is 0. Dividing we get the quotient

$$x^{n-1} + x^{n-2}a + x^{n-2}a^2 + \ldots + xa^{n-2} + a^{n-1}.$$

2. Indeed, $f(-a) = (-a)^{2k} - a^{2k} = a^{2k} - a^{2k} = 0$, and so the remainder of the division $(x^n - a^n) \div (x + a)$ is 0. Dividing we get the quotient $x^{n-1} - x^{n-2}a + x^{n-2}a^2 - \ldots + xa^{n-2} - a^{n-1}$.

3. Indeed, $f(-a) = (-a)^{2k+1} + a^{2k+1} = -a^{2k+1} + a^{2k+1} = 0$, and so the remainder of the division $(x^n + a^n) \div (x + a)$ is 0. Dividing we get the quotient $x^{n-1} - x^{n-2}a + x^{n-2}a^2 - \ldots - xa^{n-2} + a^{n-1}$.

$\square$

More-Identities

1. $a^4 - b^4 = (a - b)(a^3 + a^2b + ab^2 + b^3)$

2. $x^5 - y^5 = (x - y)(x^4 + x^3y + x^2y^2 + xy^3 + y^4)$.

3. $a^4 - b^4 = (a + b)(a^3 - a^2b + ab^2 - b^3)$.

4. $x^6 - y^6 = (x + y)(x^5 - x^4y + x^3y^2 - x^2y^3 + xy^4 - y^5)$.

5. $x^5 + y^5 = (x + y)(x^4 - x^3y + x^2y^2 - xy^3 + y^4)$.

The binomial $f(x) = x^n + a^n$ is not divisible by $(x - a)$, since the remainder in this case is $f(a) = 2a^n$. Similarly, $f(x) = x^{2k} + a^{2k}$ is not divisible by $(x + a)$, since the remainder is $f(-a) = (-a)^{2k} + a^{2k} = 2a^{2k}$. Note also that $f(x) = x^{2k+1} - a^{2k+1}$ is not divisible by $(x+a)$, since the remainder is $f(-a) = (-a)^{2k+1} - a^{2k+1} = -2a^{2k+1}$.

4.4.3   Exercises

A. Find the remainder of the division without performing the division.

1. $(3x^2 - 5x + 2) \div (x - 1)$.

2. $(7x^2 - 6x - 1) \div (x + 1)$.

3. $(3x^2 - 10x - 8) \div (3x + 2)$.

4. $(8x^3 + 125) \div (2x + 5)$.

5. $(4x^3 - 2x^2 + 6x + 3) \div (2x + 1)$.

6. $(2x^3 + 5x^2 - 2x + 9) \div (x + 3)$.

7. $(3x^5 - 7x^3 + 9x^2 - 10x + 20) \div (x + 2)$.

8. Find k so that the polynomial $f(x) = x^3 + 2x + k$ is divisible by $x - 1$. Then perform the division

$$f(x) \div (x - 1).$$

B. Perform the divisions.

1. $(15x^5 - 3x^4 + 6x^3) \div (-3x^3)$

2. $(-12ax^5 + 18ax^3 - 6ax^2) \div (-6ax^2)$.

3. $\left(\frac{12}{5}a^3b^2 - \frac{4}{5}a^3b^3 + \frac{8}{15}a^2b^2\right) \div \left(-\frac{4}{5}a^2b^2\right)$.

4. $(3x^2 - 5x + 2) \div (x - 1)$.

5. $(7x^2 - 6x - 1) \div (x + 1)$.

6. $(x^3 + 2x^2 + 3x + 1) \div (x - 1)$.

7. $(2x^3 + 5x^2 - 2x + 9) \div (x + 3)$.

8. $(18x^3 + 9x^2 - 50x - 25) \div (3x - 5)$.

9. $(2x^3 - 3x^2 - 17x - 12) \div (2x + 3)$.

10. $(2x^4 - 4x^3 - x + 1) \div (x + 2)$.

11. $(x^3 + 4x^2 - 11x - 30) \div (x^2 - x - 6)$.

12. $(x^4 - 2x^3 - 10x^2 - 4x + 30) \div (x^2 - 3)$.

13. $(y^4 - y^2 + 1) \div (y^2 + y + 1)$.

14. $(x^4 + 3x^3 - 6x^2 - 8) \div (2x^2 - 4x)$.

15. $(x^4 + 6x^3 + 3x^2 - 24x - 28) \div (x^2 + 4x - 3)$.

C. Find the quotient of the divisions.

1. $(x^3 - 8) \div (x - 2)$

2. $(x^4 - 1) \div (x - 1)$

3. $(x^4 - 1) \div (x + 1)$

4. $(x^4 + 1) \div (x - 1)$

5. Show that the numbers $3^{12} - 1$, $3^{40} - 1$, $3^{2n} - 1$ $(n \in \mathbb{N})$ are divisible by 8.

4.5 Factorization of Polynomials

Definition 4.5.1. To ***factor*** a polynomial means to express it as a product of two or several polynomials of lower degrees. This is called ***factorization of the polynomial***. If a polynomial can be factored, then it is called *reducible*; a polynomial is called *irreducible* or *prime* if it cannot be factored. To *factor completely* a polynomial means to factor the polynomial into its prime factors.

For instance, the polynomial $x^2 - 1$ is reducible, since

$$x^2 - 1 = (x + 1)(x - 1).$$

However, the polynomial $x^2 + 1$ is not factorable over the real numbers, ie, it is irreducible.[2]

The problem of factoring a polynomial is analogous to that of prime factorization of a whole number. Here irreducible polynomials play the role of prime numbers, and reducible polynomials the role of composite numbers. As we shall see in the following chapters, factorization of polynomials is important and it is used frequently in a number of places in Algebra (for instance, in simplifying rational expressions, in solving equations and inequalities, systems of equations, and so on).

Factorization Techniques

The main techniques of factoring polynomials, whenever the factorization is obtained by elementary means, are the following:

1. Taking out a common factor;

2. Factoring by grouping;

---

[2]However, $x^2 + 1 = (x + i)(x - i)$, where $i = \sqrt{-1}$ (see, Theorem 5.4.16).

3. Factoring using identities;

4. Factoring quadratic trinomials;

5. Facroring using the factor theorem;

6. Factoring by combining the above techniques.

Let us have a closer look at each of them.

1. Taking out a Common Factor

This is the simplest technique of factoring, and this should always be done first. It is based on the distributive property $ab + ac = a(b + c)$.

Example 4.5.2. Factor.

1. $2x - 10 = 2(x - 5)$

2. $3x^2 + 6x = 3x(x + 2)$.

3. $20x^3 - 15x^2 = 5x^2(4x - 3)$.

4. $10a^5 + 18a^4 - 6a^3 = 2a^3(5a^2 + 9a - 3)$.

Example 4.5.3. Factor.

1. $4x^2y - 6xy^2 = 2xy(2x - 3y)$.

2. $4a^3b + 6a^2b^2 - 2ab^3 = 2a^2b(2a + 3b - b^2)$.

3. $2x(a - b) - a + b = 2x(a - b) - (a - b) = (a - b)(2x - 1)$.

4. $2(x-1)^3 - 8(x-1)^2 = 2(x-1)^2[(x-1)-4] = 2(x-1)^2(x-5)$.

2. Factoring by Grouping

Factoring by grouping means to separate the terms of the polynomial in groups (of the same number of terms) so that in each group there is a common factor which after we factor out in each group, the same polynomial appears inside the parenthesis in all groups. Finally, factoring out the common parenthesis, we obtain the factorization of the given polynomial.

Example 4.5.4. Factor.

1. $ax + by + ay + bx = ax + ay + bx + by = a(x+y) + b(x+y) = (x+y)(a+b)$;
   or, grouping the terms differently,
   $ax + by + ay + bx = ax + bx + ay + by = x(a+b) + y(a+b) = (a+b)(x+y)$.

2. $ab - a - b + 1 = a(b-1) - (b-1) = (b-1)(a-1)$.

Example 4.5.5. Factor $3x^3 - 7x^2 + 3x - 7$.
   Solution. Grouping the terms, we have

$$3x^3 - 7x^2 + 3x - 7 = x^2(3x-7) + (3x-7) = (3x-7)(x^2+1).$$

Grouping the terms differently,

$$3x^3 - 7x^2 + 3x - 7 = 3x^3 + 3x - 7x^2 - 7 = 3x(x^2+1) - 7(x^2+1) = (x^2+1)(3x-7).$$

Example 4.5.6. Factor $6x^2 + 3a^2x + 8ax + 4a^3$.
   Solution. Grouping the terms, we have

$$6x^2 + 3a^2x + 8ax + 4a^3 = 3x(2x+a^2) + 4a(2x+a^2) = (2x+a^2)(3x+4a).$$

Example 4.5.7. Factor $x^5 + x^4 + x^3 + x^2 + x + 1$.

   Solution.

$$x^5 + x^4 + x^3 + x^2 + x + 1 = x^3(x^2+x+1) + (x^2+x+1) = (x^3+1)(x^2+x+1).$$

3. Factoring using Identities

Here we factor by applying known identities, such as the perfect square identity, the difference of two squares, the perfect cube identity, the sum and difference of cubes and other identities. We illustrate this next with several examples.

3.a. Factoring a Perfect Square. We use the identities

$$a^2 + 2ab + b^2 = (a+b)^2$$

$$a^2 - 2ab + b^2 = (a-b)^2.$$

Example 4.5.8. Factor.

1. $x^2 - 2x + 1 = x^2 - 2(x)(1) + 1^2 = (x-1)^2.$

2. $x^2 + 14x + 49 = x^2 + 2(x)(7) + 7^2 = (x+7)^2.$

3. $4x^2 + 4x + 1 = (2x)^2 + 2(2x)(1) + 1^2 = (2x+1)^2.$

4. $9x^2 - 12x + 4 = (3x)^2 - 2(3x)(2) + 2^2 = (3x-2)^2.$

Example 4.5.9. Factor $a^4x^2 + 2a^2b^3x + b^6$.

Solution.

$$a^4x^2 + 2a^2b^3x + b^6 = (a^2x)^2 + 2(a^2x)(b^3) + (b^3)^2 = (a^2x + b^3)^2.$$

Example 4.5.10. Factor $16b^2 + 49a^2b^4 - 56ab^3$.

Solution.

$$16b^2 + 49a^2b^4 - 56ab^3 = b^2(16 + 49a^2b^2 - 56ab) = b^2(4 - 7ab)^2.$$

Example 4.5.11. Factor $(x^2 + y^2)^2 + 4x^2y^2 + 4(x^2 + y^2)xy$.

Solution.

$$(x^2+y^2)^2+4x^2y^2+4(x^2+y^2)xy = [(x^2+y^2)+2xy]^2 = [(x+y)^2]^2 = (x+y)^4.$$

3.b. Factoring the Difference of two Squares. We use the identity

$$a^2 - b^2 = (a+b)(a-b).$$

Example 4.5.12. Factor.

1. $x^2 - 16 = x^2 - 4^2 = (x+4)(x-4)$.

2. $9x^2 - 1 = (3x)^2 - 1^2 = (3x+1)(3x-1)$.

3. $25x^2 - 64y^4 = (5x)^2 - (8y)^2) = (5x+8y^2)(5x-8y^2)$.

4. $x^5 - x = x(x^4-1) = x(x^2+1)(x^2-1) = x(x^2+1)(x+1)(x-1)$.

Example 4.5.13. Factor $x^2 - y^2 - z^2 + 2yz$

Solution.

$$x^2-y^2-z^2+2yz = x^2-(y^2-2yz+z^2) = x^2-(y-z)^2 = (x+y-z)(x-y+z).$$

Example 4.5.14. Factor $12a^3x^3 - 75axy^2$.

Solution.

$$12a^3x^3 - 75axy^2 = 3ax(4a^2x^2 - 25y^2) = 3ax(2ax+5y)(2ax-5y).$$

Example 4.5.15. Factor $a^{16} - b^8$,

Solution.

$$a^{16}-b^8 = (a^8)^2-(b^4)^2 = (a^8+b^4)(a^8-b^4) = (a^8+b^4)(a^4+b^2)(a^4-b^2)$$
$$= (a^8 + b^4)(a^4 + b^2)(a^2 + b)(a^2 - b).$$

3.c. Factoring the Difference or Sum of two Cubes. We use the identities

$$a^3 - b^3 = (a - b)(a^2 + ab + b^2)$$

$$a^3 + b^3 = (a + b)(a^2 - ab + b^2)$$

Example 4.5.16. Factor.

1. $x^3 - 1 = x^3 - 1^3 = (x - 1)(x^2 + x + 1)$.

2. $x^3 - 8 = x^3 - 2^3 = (x - 2)(x^2 + 2x + 4)$.

3. $8x^3 + 1 = (2x)^3 + 1^3 = (2x + 1)(4x^2 - 2x + 1)$.

Example 4.5.17. Factor.

1.

$$27x^3 + 125y^3 = (3x)^3 + (5y)^3 = (3x + 5y)(9x^2 - 15xy + 25y^2).$$

2.

$$(a-b)^3 - b^3 = [(a-b)-b][(a-b)^2 + (a-b)b + b^2] = (a-2b)(a^2 + b^2 - ab).$$

Example 4.5.18. Factor $(x + y)^3 + (x - y)^3$.

Solution.

$$(x+y)^3 + (x-y)^3 = [(x+y)+(x-y)][(x+y)^2 - (x+y)(x-y) + (x-y)^2] = (2x)(x^2 + 3y^2).$$

Example 4.5.19. Factor $(a + b)^3 - (a^3 + b^3)$.

Solution.

$$(a+b)^3 - (a^3+b^3) = (a+b)^3 - (a+b)(a^2 - ab + b^2) = (a+b)[(a+b)^2 - (a^2 - ab + b^2)] = (a+b)(3ab).$$

3.d. Factoring a perfect cube. We use the identities

$$a^3 + 3a^2b + 3ab^2 + b^3 = (a+b)^3$$

$$a^3 - 3a^2b + 3ab^2 - b^3 = (a-b)^3$$

Example 4.5.20. Factor.

1. $x^3 - 3x^2 + 3x - 1 = x^3 - 3(x^2)(1) + 3(x)(1^2) - 1^3 = (x-1)^3$.

2. $8x^3 + 12x^2 + 6x + 1 = (2x)^3 + 3(2x)^2(1) + 3(2x)(1^2) + 1^3 = (2x+1)^3$.

Example 4.5.21. Factor $27x^3 + 27x^2y + 9xy^2 + y^3$.

Solution. We have

$$27x^3 + 27x^2y + 9xy^2 + y^3 = (3x)^3 + 3(3x)^2y + 3(3x)y^2 + y^3 = (3x+y)^3.$$

Example 4.5.22. Factor $8x^6a^3 - 36x^5a^2 + 54x^4a - 27x^3$.

Solution. We have

$$8x^6a^3 - 36x^5a^2 + 54x^4a - 27x^3 = (2x^2a)^3 - 3(2x^2a)^2(3x) + 3(2x^2a)(3x)^2 - (3x)^3 = (2x^2a - 3x)^3.$$

4.5.1 Exercises

Factor completely.

1. $3x - 6$.

2. $ax^2 - a^3$

3. $3x^2y - 6xy^2 + 12xy$

4. $x^2 + ax + bx + ab$

5. $ax^2 - bx^2 - b^2x + a^2x + a - b$

6. $x^3 - 3x^2 - x + 3$

7. $6x^2 + 3a^2x + 2ax + a^3$

8. $x^3 - 4x^2 + 2x - 8$

9. $3x^4 + 3x^3 - 24x - 24$

10. $6xy + 15x - 4y - 10$

11. $3abx^2 + by - 6ax^2y - 2y^2$

12. $x^3 + 2ax^2 + 4a^2x + 8a^3$

13. $x^2 - 1$

14. $25 - 9x^2$

15. $4x^2 - 49y^2$

16. $3x^2 - 12.$

17. $64x^3 - 27$

18. $x^4 - 1$

19. $x^6 - 64$

20. $(a + b)^2 - c^2 + a + b + c$

21. $a^6 - a^9$

22. $a^6 - b^9$

23. $x^2 - 2xy + y^2 - 25.$

24. $8x^3 + 1 + 12x^2 + 6x$

25. $x^9 - 27x^6 - x^3 + 27$

26. $16a^2b^2 - 4b^4 - 4a^4 + a^2b^2$

27. $a^4b - a^2b^3 + a^3b^2 - ab^4$

28. $x^2 - y^2 - z^2 + 2yz + x + y - z$

29. $x^4 - 16a^4 + 9y^4 - 6x^2y^2$

30. $x^5 - x^3 - 8x^2 + 8$

31. $a^2 - b^2 + a^3 - b^3$

32. $a^3 - b^6 + b^2 - a$

33. $(x^2 + ab)^3 - 8a^3b^3 - x^2 + ab$

4. Factoring Quadratic Trinomials

4.a. Factoring the quadratic $x^2 + bx + c$ **by inspection.** This is sometimes possible when the coefficients b and c are integers. Note that, since

$$(x + p)(x + q) = x^2 + (p + q)x + pq,$$

we shall be able to factor $x^2 + bx + c$ by inspection, if we can find two numbers, p and q, such that $p + q = b$ and $pq = c$. To find p and q we list the factors of c and we select those two whose sum is equal to b. If we can not find such integers p and q, we say that the quadratic is not factorable[3] using integer numbers.

---

[3]However, we shall see in Theorem 5.4.16 that any quadratic polynomial factors as $ax^2 + bx + c = a(x - x_1)(x - x_2)$, where x_1, x_2 are the roots of the quadratic equation $ax^2 + bx + c = 0$.

Example 4.5.23. Factor $x^2 + 4x - 12$.

Solution. The coefficients are $a = 1$, $b = 4$, and $c = -12$. The two factors of -12 whose sum is 4 are 6 and -2. That is, here $p = 6$ and $q = -2$. Hence

$$x^2 + 4x - 12 = (x + 6)(x - 2).$$

Example 4.5.24. Factor.

1. $x^2 + 5x + 6 = x^2 + (3 + 2)x + (3)(2) = (x + 3)(x + 2)$.

2. $x^2 + 5x - 6 = x^2 + (-1 + 6)x + (-1)(6)$
   $= (x - 1)(x + 6)$.

Example 4.5.25. Factor.

1.

$$x^2 + 2x - 15 = x^2 + [5 + (-3)]x + (5)(-3) = (x + 5)(x - 3).$$

2.

$$x^2 - 16x + 15 = x^2 + [-15 + (-1)]x + (-15)(-1) = (x - 15)(x - 1).$$

4.b. Factoring the quadratic $ax^2 + bx + c$ by inspection. This is sometimes possible when the coefficients a, b and c are integers. Note that we can write

$$ax^2 + bx + c = \frac{1}{a}[(ax)^2 + b(ax) + ac].$$

Setting $t = ax$, the expression in the bracket becomes $t^2 + bt + ac$, which we can factor by the method 4.a. we discussed above, that is, by finding two integers p and q such that $pq = ac$ and $p + q = b$. Then, we may write

$$ax^2 + bx + c = \frac{1}{a}(t^2 + bt + ac) = \frac{1}{a}(t + p)(t + q) = \frac{1}{a}(ax + p)(ax + q).$$

In practice, we use the numbers p and q to factor directly ax^2+bx+c by grouping. Indeed, since $b = p + q$ and $c = \frac{pq}{a}$, we have

$$ax^2+bx+c = ax^2+(p+q)x+\frac{pq}{a} = ax^2+px+qx+\frac{pq}{a} = x(ax+p)+q(x+\frac{p}{a})$$

$$= x(ax + p) + \frac{q}{a}(ax + p) = (ax + p)(x + \frac{q}{a}) = \frac{1}{a}(ax + p)(ax + q).$$

We illustrate this method with the following example.

Example 4.5.26. Factor $2x^2 - x - 15$.

Solution. The coefficients are $a = 2$, $b = -1$, and $c = -15$. So that $ac = (2)(-15) = -30$. The two factors of -30 whose sum is -1 are 5 and -6. That is, here $p = -6$ and $q = 5$. Now, we write $b = -1 = -6 + 5$ to create four tems and we factor by grouping

$$2x^2 - x - 15 = 2x^2 + (-6 + 5)x - 15 = 2x^2 - 6x + 5x - 15$$

$$= 2x(x - 3) + 5(x - 3) = (x - 3)(2x + 5).$$

That is,
$$2x^2 - x - 15 = (2x + 5)(x - 3).$$

Note that $(2x + 5)(x - 3) = \frac{1}{2}(2x + 5)(2x - 6)$.

Example 4.5.27. Factor the trinomials.

1. $6x^2 - 5x - 1 = 6x^2 - 6x + x - 1 = 6x(x - 1) + (x - 1) = (x - 1)(6x + 1)$.

2. $5x^2 + 14x - 3 = 5x^2 + 15x - x - 3 = 5x(x + 3) - (x + 3) = (x + 3)(5x - 1)$.

Example 4.5.28. Factor the trinomials.

1. $8x^2 - 10x - 3 = 8x^2 + 2x - 12x - 3 = 2x(4x + 1) - 3(4x + 1) = (4x + 1)(2x - 3)$.

2. $15x^2 + 17x - 4 = 15x^2 + 20x - 3x - 4 = 5x(3x + 4) - (3x + 4) = (3x + 4)(5x - 1)$.

Example 4.5.29. Factor the trinomials.

1.
$$24x^2 + 2x - 15 = 24x^2 - 18x + 20x - 15 =$$
$$6x(4x - 3) + 5(4x - 3) = (4x - 3)(6x + 5).$$

2.
$$3x^2 - 5xy - 12y^2 = 3x^2 - 9xy + 4xy - 12y^2 =$$
$$3x(x - 3y) + 4y(x - 3y) = (x - 3y)(3x + 4y).$$

Example 4.5.30. Factor the trinomials.

1.
$$8m^2 - 10mn - 3n^2 = 8m^2 + 2mn - 12mn - 3n^2 =$$
$$2m(4m + n) - 3n(4m + n) = (4m + n)(2m - 3n).$$

2.
$$-6a^2 + 47ab - 35b^2 = -6a^2 + 42ab + 5ab - 36b^2 =$$
$$-6a(a - 7b) + 5b(a - 7b) = (a - 7b)(5b - 6a).$$

4.5.2 Integer and Rational Root Theorems

Definition 4.5.31. Let

$$f(x) = a_n x^n + a_{n-1} x^{n-1} + ... + a_1 x + a_0$$

be a polynomial. A number r is called a **root** or **zero** of the polynomial if

$$f(r) = 0.$$

The following two results in combination with the factor theorem provide another technique for factoring polynomials whose coefficients are all integers.

The first result gives a necessary condition for a polynomial with *integer coefficients* to have an integer root.

Theorem 4.5.32. *(Integer Root Theorem)*. *Let*

$$f(x) = a_n x^n + a_{n-1} x^{n-1} + \ldots + a_1 x + a_0$$

be a polynomial with integer coefficients ($a_n \neq 0$). If r is an integer root of the polynomial f, ie, $f(r) = 0$, then r is a factor of the constant term a_0 of the polynomial.

Proof. Since $f(r) = 0$, we have $a_n r^n + a_{n-1} r^{n-1} + \ldots + a_1 r + a_0 = 0$. This implies

$$r(a_n r^{n-1} + a_{n-1} r^{n-2} + \ldots + a_1) = -a_0.$$

Since the coefficients of the polynomial are all integers, the number $a_n r^{n-1} + a_{n-1} r^{n-2} + \ldots + a_1$ is also an integer, call it k. Then $rk = -a_0$. That is, r divides a_0. $\square$

The second result is known as the *Rational Root Theorem*.

Theorem 4.5.33. *(Rational Root Theorem)*. *Let*

$$f(x) = a_n x^n + a_{n-1} x^{n-1} + \ldots + a_1 x + a_0$$

be a polynomial with integer coefficients ($a_n \neq 0$, $a_0 \neq 0$). If $\frac{p}{q}$, in lowest terms, is a rational root of the polynomial f, then p is a factor of a_0, and q is a factor of a_n.

Proof. Since $f(\frac{p}{q}) = 0$, we have

$$a_n \frac{p^n}{q^n} + a_{n-1} \frac{p^{n-1}}{q^{n-1}} + \ldots + a_1 \frac{p}{q} + a_0 = 0.$$

Multiplying by q^n, we get

$$a_n p^n + a_{n-1} p^{n-1} q + \ldots + a_1 p q^{n-1} + a_0 q^n = 0.$$

From this we get

$$a_n p^n = -q(a_{n-1} p^{n-1} + \ldots + a_1 p q^{n-2} + a_0 p^{n-1}),$$

$$a_0 q^n = -p(a_n p^{n-1} + a_{n-1} p^{n-2} q + \ldots + a_1 q^{n-1}).$$

Since all the coefficients of the polynomial are integers, both numbers $-(a_{n-1} p^{n-1} + \ldots + a_1 p q^{n-2} + a_0 p^{n-1})$ and $-(a_n p^{n-1} + a_{n-1} p^{n-2} q + \ldots + a_1 q^{n-1})$ are also integers. Call them k_1 and k_2, respectively. Then we have

$$a_n p^n = q k_1 \text{ and } a_0 q^n = p k_2.$$

Now, $a_n p^n = q k_1$ tells us that q divides $a_n p^n$. Since q and p are relatively prime, and therefore so are q and p^n, we conclude that q divides a_n.

Similarly, from $a_0 q^n = p k_2$ we conclude that p divides a_0. $\square$

5. Factoring using the Factor Theorem

If a polynomial with integer coefficients has an integer or a rational root, we can use the Integer or Rational Root Theorem to find it. We can then apply the Factor Theorem to factor the polynomial. We illustrate this technique in the following examples.

Example 4.5.34. Factor $f(x) = x^3 - x^2 - 10x - 8$

Solution. The factors of the constant term $a_0 = -8$ are: ± 1, ± 2, ± 4, and ± 8. If f has an interger root, it will be found in this list. We test the potential root -1, by evaluating

$$f(-1) = (-1)^3 - (-1)^2 - 10(-1) - 8 = -1 - 1 + 10 - 8 = 0.$$

Since $f(-1) = 0$, the integer -1 is a root of f, and the factor theorem tells us that $x - (-1) = x + 1$ is a factor of f. Dividing f by $x + 1$, we get

$$f(x) = x^3 - x^2 - 10x - 8 = (x + 1)(x^2 - 2x - 8).$$

Factoring the quadratic we get $x^2 - 2x - 8 = (x - 4)(x + 2)$. Thus

$$f(x) = x^3 - x^2 - 10x - 8 = (x + 1)(x - 4)(x + 2).$$

Notice that f has three integer roots, -1, -2 and 4 and all three are among the list of potential roots.

Example 4.5.35. Factor $f(x) = 2x^3 + 11x^2 - 7x - 6$

Solution. We list all integers p that are factors of the constant term $a_0 = -6$ and all integers q that are factors of the leading coefficient $a_3 = 2$. The factors of -6 are p: $\pm 1, \pm 2, \pm 3, \pm 6$. The factors of 2 are q: $\pm 1, \pm 2$. If f has a rational root, it will be found in the list of all possible fractions

$$\frac{p}{q} : \pm 1, \pm 2, \pm 3, \pm 6, \pm \frac{1}{2}, \pm \frac{3}{2}.$$

By evaluating f at 1 we see $f(1) = 0$. The factor theorem then tells us that $x - 1$ is a factor of f. The division $f(x) \div (x - 1)$, yields

$$f(x) = 2x^3 + 11x^2 - 7x - 6 = (x - 1)(2x^2 + 13x + 6).$$

Factoring now the quadratic $2x^2 + 13x + 6 = 2x^2 + 12x + x + 6 = 2x(x + 6) + (x + 6) = (x + 6)(2x + 1)$, we get

$$f(x) = 2x^3 + 11x^2 - 7x - 6 = (x - 1)(x + 6)(2x + 1).$$

Notice that f has three rational roots, 1, -6 and $-\frac{1}{2}$ and all three are among the list of potential roots.

6. Factoring by Combining the above Techniques

Frequently to factor a polynomial, we may need to apply a combination of two or more of the techniques we studied above.

Here are several examples.

Example 4.5.36. Factor completely

$$6x^2y^3 + 3x^2y^2 + 18xy + 9x.$$

Solution. We have

$$6x^2y^3 + 3x^2y^2 + 18xy + 9x = 3x(2xy^3 + xy^2 + 6y + 3) =$$

$$3x\left[xy^2(2y+1) + 3(2y+1)\right] = 3x(2y+1)(xy^2+3).$$

Example 4.5.37. Factor completely

$$x^7 - x^5 - x^3 + x.$$

Solution. We have

$$x^7 - x^5 - x^3 + x = x\left[x^6 - x^4 - x^2 + 1\right] = x\left[x^4(x^2-1) - (x^2-1)\right] =$$

$$x(x^2-1)(x^4-1) = x(x^2-1)(x^2+1)(x^2-1) = x(x^2-1)^2(x^2+1) = x(x+1)^2(x-1)^2(x^2+1).$$

Example 4.5.38. Factor completely

$$x^9 - 27x^6 - x^3 + 27.$$

Solution. We have

$$x^9 - 27x^6 - x^3 + 27 = (x^9 - x^3) - (27x^6 - 27) = x^3(x^6-1) - 27(x^6-1) =$$

$$(x^6 - 1)(x^3 - 27) = (x^3 + 1)(x^3 - 1)(x^3 - 3^3) =$$

$$(x + 1)(x^2 - x + 1)(x - 1)(x^2 + x + 1)(x - 3)(x^2 + 3x + 9).$$

Example 4.5.39. Factor completely

$$(x^2 - 9)^2 - (x + 5)(x - 3)^2.$$

Solution. We have

$$(x^2 - 9)^2 - (x + 5)(x - 3)^2 = (x + 3)^3(x - 3)^2 - (x + 5)(x - 3)^2 =$$

$$(x-3)^2 \left[(x + 3)^2 - (x + 5)\right] = (x-3)^2(x^2+5x+4) = (x-3)^2(x+4)(x+1).$$

Sometimes, when the polynomial has an odd number of terms, we may need to split one (or more) terms as a sum of other like terms, in order to be able to factor the polynomial (see the examples below).

Example 4.5.40. Factor completely

$$a^4 + a^2b^2 + b^4.$$

Solution. We have

$$a^4 + a^2b^2 + b^4 = a^4 + 2a^2b^2 - a^2b^2 + b^4 = a^4 + 2a^2b^2 + b^4 - a^2b^2 =$$

$$(a^2 + b^2)^2 - (ab)^2 = (a^2 + b^2 + ab)(a^2 + b^2 - ab).$$

Example 4.5.41. Factor completely

$$x^3 - 3x + 2.$$

Solution. We have

$$x^3-3x+2 = x^3-x-2x+2 = x(x^2-1)-2(x-1) = x(x+1)(x-1)-2(x-1) =$$

$$(x-1)\left[x(x + 1) - 2\right] = (x-1)(x^2+x-2) = (x-1)(x-1)(x+2) = (x-1)^2(x+2).$$

Example 4.5.42. Factor completely

$$x^3 + 2x^2 - 3.$$

Solution. We have

$$x^3+2x^2-3 = x^3+2x^2-2-1 = x^3-1+2x^2-2 = (x^3-1)+2(x^2-1) =$$

$$(x-1)(x^2+x+1)+2(x+1)(x-1) = (x-1)\left[(x^2 + x + 1) + 2(x + 1)\right] = (x-1)(x^2+3x+3).$$

Example 4.5.43. Factor completely

$$x^4 + x^3 - 3x^2 - 5x - 2.$$

Solution.

$$x^4+x^3-3x^2-5x-2 = x^4+x^3-3x^2-3x-2x-2 = x^3(x+1)-3x(x+1)-2(x+1) =$$

$$(x+1)(x^3-3x-2) = (x+1)(x^3-x-2x-2) = (x+1)[x(x^2-1)-2(x+1)] =$$

$$(x+1)[x(x+1)(x-1)-2(x+1)] = (x+1)^2(x^2-x-2) = (x+1)^2(x-2)(x+1) = (x+1)^3(x-2).$$

4.5.3   Exercises

Factor completely.

1. $x^2 + 7x + 10$

2. $x^2 + 3x - 10$

3. $x^2 - x - 6$

4. $x^2 - 8x - 20$

5. $3x^2 + 7x - 6$

6. $3x^2 + 5x - 2$

7. $2x^2 - x - 6$

8. $6y^2 - y - 1$

9. $3x^2 - 12x - 36$

10. $10x^2 - 17xy + 6y^2$

11. $x^2 - 4xy + 3y^2$

12. $4x^3 - 10x^2 - 6x$

13. $2 - a^3 - a$

14. $x^3 + 2x^2 - 3$

15. $x^3 + x^2 - 2$

16. $x^3 + 3x^2 - 10x - 24$

17. $x^3 + 6x^2 + 11x + 6$

18. $4x^3 - 3x - 1$

19. $15x^3 + 29x^2 - 8x - 12$

20. $8x^3 + 1 + 12x^2 + 6x$

21. $6x^3 - 13x - 14x - 3$

22. $x^4 + 3x^2 + 4$

23. $x^4 + x^2 + 1$

24. $x^4 + 4$

25. $x^4 - 2x^2 + 3x - 2$

26. $2x^4 - x^3 - 9x^2 + 13x$

27. $6x^4 + 5x^3 + 3x^2 - 3x - 2$

28. $6x^5 + 19x^4 + 22x^3 + 23x^2 + 16x + 4$

29. $x^8 + x^4 + 1$

## 4.6	Greatest Common Factor

Given two polynomials $f(x)$ and $g(x)$, we call a *common factor* of these polynomials every polynomial $p(x)$ which divides exactly both $f(x)$ and $g(x)$. For example, $p(x) = x - 1$ is a common factor of $f(x) = x^3 - 1$ and $g(x) = x^2 - 1$. So is $c(x - 1)$, where c is a constant $c \neq 0$.

The **greatest common factor** or **GCF** of two or more polynomials is the polynomial of greatest degree which divides exactly each of the given polynomials.

*To find the **GCF** of two or more polynomials, we factor each polynomial, and we form the product of the common factors taken each with the smallest exponent.*

Example 4.6.1. Find the GCF of the monomials $18x^4y^2z^3w$, $30x^2y^5zw^2$, $-24x^3y^3z^2$.

Solution. The prime factorization of the coefficients is:
$$18 = 2 \cdot 3^2 \quad 30 = 2 \cdot 3 \cdot 5, \quad 24 = 2^3 \cdot 3.$$
So the GCF$(18, 30, 24) = 2 \cdot 3 = 6$. The common factor of the variable parts is x^2y^2z.

Hence the GCF of the monomials is $6x^2y^2z$.

Example 4.6.2. Find the GCF of the polynomials
$(x - 1)^3(x + 2)^2$, $5x(x - 1)^2(x + 2)^3$, $(x + 1)(x + 2)(x - 1)^2$.

Solution. The GCF is $(x - 1)^2(x + 2)$.

4.6.1 Exercises

A. Find the GCF of the monomials.

1. $18x$, $6x^2$, $2x^3$.

2. $12abx$, $6ax^2y$, $3abxy^2$.

3. $45a^2bxy^2$, $-15a^2b^3x^3y^2$, $5a^3bx^2y^4$.

4. $10x^3y^2z^5$, $4x^5yz^3$, $6x^4y^3z^5$, and $8x^4y^4z^4w$.

B. Find the GCF of the polynomials.

1. $(x+y)^2(x-y)$, $(x+y)(x-y)^2$, $x^3y - xy^3$.

2. $a^2 - b^2$, $a^3 - b^3$, $a^4 - b^4$.

3. $x^2 - 1$, $x^2 + 2x + 1$, $x^3 + 1$.

4. $x^2 + 5x + 6$, $x^2 + x - 2$, $x^2 - 14x - 32$.

5. $x^2 - 1$, $x^2 + 1$, $x^4 - 1$, and $x^8 - 1$.

6. $(x^2 - 1)^2(x + 3)$, $(x^2 + 3x)(x + 1)^2$, $(x^2 + 6x + 9)(x - 1)^2$.

7. $(x^2 - 1)^2(x - 2)^2$, $(x^2 + 3x + 2)(2x^3 - 5x^2 + 5x - 6)$.

4.7 Least Common Multiple

The **least common multiple** or **LCM** of two or more polynomials is the polynomial of smallest degree which is divided exactly by each of the given polynomials.

*To find the **LCM** of two or more polynomials, we factor each polynomial, and we form the product of the common and not common factors taken each with the greatest exponent.*

Example 4.7.1. Find the LCM of the monomials $18x^4y^2z^3w$, $30x^2y^5zw^2$, $-24x^3y^3z^2$.

   Solution. The LCM of the coefficients is $\text{LCM}(18, 30, 24) = 2^3 \cdot 3^2 \cdot 5 = 360$.

The LCM of the monomials is $360x^4y^5z^3w^2$.

Example 4.7.2. Find the LCM of the polynomials
$(x-1)^3(x+2)^2$, $5x(x-1)^2(x+2)^3$, $(x+1)(x+2)(x-1)^2$.

   Solution. The LCM of the polynomials is

$$5x(x+1)(x-1)^3(x+2)^3.$$

4.7.1   Exercises

A. Find the LCM of the monomials.

1. 6, $3x$, $8x^2$, $4x^5$.

2. $18x$, $6x^2$, $2x^3$.

3. $12abx$, $6ax^2y$, $3abxy^2$.

4. $45a^2bxy^2$, $-15a^2b^3x^4y^3$, $5a^3bx^2y$.

5. $10x^3y^2$, $4y^2z$, $6x^4y^3z^2$, and $8ax^4y^2$.

B. Find the LCM of the polynomials.

1. x, $x^3 - x$, $x - 1$.

2. $x - 1$, $x^2 - 1$, $x^2 + 1$.

3. $x^2 - 3x + 2$, $x^2 - 5x + 6$, $x^2 - 4x + 3$.

4. $x^2 + 5x + 6$, $x^2 + x - 2$, $x^2 - 14x - 32$.

5. $x^3 + x^2 + x$, $x^5 - x^3$, $x^6 - x^3$.

6. $a^2 - b^2$, $a^3 - b^3$, $a^4 - b^4$.

7. $(x^2 + 3x + 2)(x^2 + 7x + 12)$, and $(x^2 + 5x + 6)(2x^2 - 3x - 5)$.

8. $(x^2 - 1)^2(x + 3)$, $(x^2 + 3x)(x + 1)^2$, $(x^2 + 6x + 9)(x - 1)^2$.

9. $(x^2 - 1)^2(x - 2)^2$, $(x^2 + 3x + 2)(2x^3 - 5x^2 + 5x - 6)$, $(x - 1)(x - 2)^2$.

10. $x^3 - 6x^2 + 11x - 6$, $2x^3 - 7x^2 + 7x - 2$, $x^3 + x^2 - 13x + 6$.

11. $8x^3 - 18xy^2$, $8x^2 - 2xy - 15y^2$, $8x^3 + 8x^2y - 6xy^2$.

4.8 Rational Expressions

A *rational expression* is a fractional expressions of the form

$$\frac{P(x)}{Q(x)} = \frac{a_nx^n + a_{n-1}x^{n-1} + ... + a_1x + a_0}{b_mx^m + b_{m-1}x^{m-1} + ... + b_1x + b_0},$$

where polynomials $P(x)$ and $Q(x)$ are polynomials of degree n and m respectively. The polynomial $P(x)$ is the numerator and $Q(x)$ the denominator. The fraction $\frac{P(x)}{Q(x)}$ is considered only for *permissible values* of x for which $Q(x) \neq 0$.

In general, any algebraic fraction $\frac{P}{Q}$, where P and Q are polynomials in one or many variables is also called a rational expression. The following are some examples of rational expressions:

Example 4.8.1. $\frac{x^2+3}{x-5}$, $x \neq 5$; $\quad \frac{2x^3-6x}{x^2-1}$, $x \neq -1, x \neq 1$; $\quad \frac{1}{x^2+1}$;

$\frac{3x-1}{x^2-3x-10}$, $x \neq 2, x \neq -5$; $\quad \frac{a^2+b^2}{a-b}$, $a \neq b$; $\quad \frac{xy+5y}{3x-6y}$, $x \neq 2y$.

In the sequel we shall assume only permissible values for the variable(s), that is, we assume the denominators are not 0.

Two rational expressions $\frac{P}{Q}$, $\frac{A}{B}$ are said to be **equal**

$$\frac{P}{Q} = \frac{A}{B} \quad \text{if} \quad PB = QA.$$

For instance, $\frac{x+1}{x^2-1} = \frac{1}{x-1}$, since $(x+1)(x-1) = 1(x^2-1)$.

Like any numerical fraction, an algebraic fraction $\frac{P}{Q}$ is *simplified*, if P and Q have common factor. When a fraction is simplified, we say the fraction is in *lowest terms*.

Example 4.8.2. Simplify

$$\frac{x^2 - x}{x^2 - 1}.$$

Solution. We factor both the numerator and denominator and we cancel the common factor $x - 1$, that is,

$$\frac{x^2 - x}{x^2 - 1} = \frac{x(x - 1)}{(x + 1)(x - 1)} = \frac{x}{x + 1}.$$

Example 4.8.3. Simplify

$$\frac{x^2 - 4}{x^2 + 5x + 6}.$$

Solution. We factor both the numerator and denominator and we cancel the common factor $x + 2$, that is,

$$\frac{x^2 - 4}{x^2 + 5x + 6} = \frac{(x + 2)(x - 2)}{(x + 3)(x + 2)} = \frac{x - 2}{x + 3}.$$

4.8.1 Operations with Rational Expressions

The rules of operations with rational expressions are the same with those of numerical fractions.

Addition and Subraction of rational expressions

To **add** or **subtract** algebraic fractions with *like* denominators, we add or subtract the numerators and keep the common denominator. For example,

$$\frac{x+10}{x+3} + \frac{5x+8}{x+3} = \frac{6x+18}{x+3} = \frac{6(x+3)}{x+3} = \frac{6}{1} = 6.$$

To add or subtract algebraic fractions with *unlike* denominators, we must find the **least common denominator (LCD)**. That is, the LCM of the denominators.

Example 4.8.4. Perform the operations

$$\frac{7}{3a^2b} + \frac{4b}{a} - \frac{5}{ab}.$$

Solution. Since the least common denominator is $3a^2b$, we have

$$\frac{7}{3a^2b} + \frac{4b}{a} - \frac{5}{ab} = \frac{7}{3a^2b} + \frac{12ab^2}{3a^2b} - \frac{15a}{3a^2b} = \frac{7 + 12ab^2 - 15a}{3a^2b}.$$

Example 4.8.5. Perform the operations and simplify

$$\frac{2x-1}{2x} + \frac{2x}{1-2x} - \frac{1}{2x-4x^2}.$$

Solution. We factor $2x - 4x^2 = 2x(1-2x)$, and we see that the least common denominator is $2x(1-2x)$. Making equivalent fractions having this common denominator, we have

$$\frac{(2x-1)(1-2x)}{2x(1-2x)} + \frac{(2x)(2x)}{2x(1-2x)} - \frac{1}{2x(1-2x)} =$$

$$= \frac{(2x-1)(1-2x) + 4x^2 - 1}{2x(1-2x)} =$$

$$= \frac{4x-2}{2x(1-2x)} = \frac{-2(1-2x)}{2x(1-2x)} = -\frac{1}{x}.$$

Example 4.8.6. Perform the operations and simplify

$$\frac{1}{x^2 + x} + \frac{1}{x^2 + 3x + 2} + \frac{1}{x^2 + 5x + 6} - \frac{2}{x^2 + 3x}.$$

Solution. We factor all denominators to find the least common denominator. We have $x^2 + x = x(x+1)$, $x^2 + 3x + 2 = (x+1)(x+2)$, $x^2 + 5x + 6 = (x+2)(x+3)$ and $x^2 + 3x = x(x+3)$. The LCD is $x(x+1)(x+2)(x+3)$. Now we have

$$\frac{1}{x^2 + x} + \frac{1}{x^2 + 3x + 2} + \frac{1}{x^2 + 5x + 6} - \frac{2}{x^2 + 3x} =$$

$$\frac{1}{x(x+1)} + \frac{1}{(x+1)(x+2)} + \frac{1}{(x+2)(x+3)} - \frac{2}{x(x+3)} =$$

$$\frac{(x+2)(x+3) + x(x+3) + x(x+1) - 2(x+1)(x+2)}{x(x+1)(x+2)(x+3)} =$$

$$\frac{x^2 + 3x + 2}{x(x+1)(x+2)(x+3)} = \frac{(x+1)(x+2)}{x(x+1)(x+2)(x+3)} =$$

$$\frac{1}{x(x+3)}.$$

Multiplication and Division of rational expressions

To **multiply** rational expressions we use the same rule as when we multiply numerical fractions. That is, we multiply the numerators, the denominators and we cancel out the common factors.

Example 4.8.7. Multiply and simplify

$$\frac{x^2 - x}{3x^2 + x} \cdot \frac{3x^2 + 7x + 2}{x^2 + 2x}.$$

Solution. We have

$$\frac{x^2 - x}{3x^2 + x} \cdot \frac{3x^2 + 7x + 2}{x^2 + 2x} = \frac{x(x-1)}{x(3x+1)} \cdot \frac{(x+2)(3x+1)}{x(x+2)} = \frac{x-1}{x}.$$

Example 4.8.8. Multiply and simplify

$$\frac{x^2 + x}{2x^2 + 3x} \cdot \frac{2x^2 + x - 3}{x^2 - 1}.$$

Solution. We have

$$\frac{x^2 + x}{2x^2 + 3x} \cdot \frac{2x^2 + x - 3}{x^2 - 1} = \frac{x(x+1)}{x(2x+3)} \cdot \frac{(x-1)(2x+3)}{(x+1)(x-1)} = 1.$$

As with numerical fractions, to **divide** rational expressions, we convert division to multiplication by the reciprocal of the divisor. That is,

$$\frac{P}{Q} \div \frac{A}{B} = \frac{P}{Q} \cdot \frac{B}{A}.$$

Example 4.8.9. Divide

$$\frac{x^2 + 3x - 10}{x^2 + 2x - 15} \div \frac{x^2 - 4}{x^2 - 2x - 3}.$$

Solution. We have

$$\frac{x^2 + 3x - 10}{x^2 + 2x - 15} \div \frac{x^2 - 4}{x^2 - 2x - 3} = \frac{x^2 + 3x - 10}{x^2 + 2x - 15} \cdot \frac{x^2 - 2x - 3}{x^2 - 4} =$$

$$= \frac{(x+5)(x-2)}{(x+5)(x-3)} \cdot \frac{(x-3)(x+1)}{(x+2)(x-2)} = \frac{x+1}{x+2}.$$

Complex Fractions

A ***complex fraction*** is a fraction that has a fractional numerator or a fractional denominator, or both. A complex fraction is reduced to a simple fraction if we divide its numerator by its denominator.

Example 4.8.10. Simplify the complex fraction

$$\frac{\frac{3a}{b}}{\frac{6ac}{b^2}}.$$

Solution. We have

$$\frac{\frac{3a}{b}}{\frac{6ac}{b^2}} = \frac{3a}{b} \div \frac{6ac}{b^2} = \frac{3a}{b} \cdot \frac{b^2}{6ac} = \frac{b}{2c}.$$

Example 4.8.11. Simplify the complex fraction

$$\frac{\frac{x}{x+1} + \frac{x-1}{x}}{\frac{x}{x+1} - \frac{x-1}{x}}.$$

Solution. The numerator becomes

$$\frac{x}{x+1} + \frac{x-1}{x} = \frac{x^2 + x^2 - 1}{x(x+1)} = \frac{2x^2 - 1}{x(x+1)}.$$

The denominator becomes

$$\frac{x}{x+1} - \frac{x-1}{x} = \frac{x^2 - x^2 + 1}{x(x+1)} = \frac{1}{x(x+1)}.$$

Now, we have

$$\frac{\frac{x}{x+1} + \frac{x-1}{x}}{\frac{x}{x+1} - \frac{x-1}{x}} = \frac{\frac{2x^2-1}{x(x+1)}}{\frac{1}{x(x+1)}} = \frac{2x^2 - 1}{x(x+1)} \div \frac{1}{x(x+1)} =$$

$$= \frac{2x^2 - 1}{x(x+1)} \cdot \frac{x(x+1)}{1} = 2x^2 - 1.$$

4.8.2   Exercises

A. Simplify the following rational expressions.

1. $\frac{12x^3y^2}{4x^2y^2}$

2. $\frac{27a^3b^2x}{18a^4bx^2}$

3. $\frac{x-1}{1-x}$

4. $\frac{3x^2+3x}{2x^3-2x}$

5. $\frac{x^2-6x+9}{x^2-4x+3}$

6. $\frac{(x^2-4)^2-(x+2)^2}{x^2-4x+3}$

7. $\frac{(1+xy)^2-(x+y)^2}{1-x^2}$

8. $\frac{a^2-b^2}{a^2-a-b-b^2}$

9. $\frac{x^4-81}{x^2-9}$

10. $\frac{x^3-x^2-4}{x^2-5x+6}$

11. $\frac{x^3-8x^2+19x-12}{2x^3-13x^2+17x+12}$

12. $\frac{(a-b)^3+(b-c)^3+(c-a)^3}{(a-b)(b-c)(c-a)}$

B. Perform the additions and subtractions and simplify.

1. $\frac{1}{a}+\frac{1}{b-a}$

2. $\frac{3}{5x^2y}+\frac{5}{4xy^2}$

3. $\frac{a}{a-b}+\frac{b}{a+b}$

4. $\frac{a}{a-b}+\frac{ab}{b^2-a^2}$

5. $\frac{7}{3x+5} - \frac{2}{x-1}$

6. $7 + \frac{2a}{a+b} - \frac{3b}{a-b}$

7. $2x - 1 + \frac{3-5x^2}{x+3}$

8. $\frac{x^2}{x-y} + \frac{y^2}{y-x}$

9. $\frac{13}{2x-6} + \frac{2x}{x^2-9}$

10. $\frac{x-1}{x+3} - \frac{x-3}{x+1}$

11. $\frac{1}{x+1} - \frac{2}{x+2} + \frac{1}{x-3}$

12. $\frac{1}{x+1} + \frac{2}{x-1} - \frac{3}{x^2-1}$

13. $\frac{5-x}{x-y} + \frac{6-x}{x^2-y^2} - \frac{4-x}{x+y}$

14. $\frac{1}{(a-b)(a-c)} + \frac{1}{(b-c)(b-a)} + \frac{1}{(c-a)(c-b)}$

15. $\frac{1}{x^2-3x+2} + \frac{1}{x^2-5x+6} + \frac{1}{x^2-4x+3}$

16. $\frac{1+x^2+x}{1-x^3} + \frac{x-x^2}{(1-x)^3}$

17. $x + \frac{1}{3x-2} - \frac{8x^4-33x}{8x^3-27} - \frac{2x+6}{4x^2+6x+9}.$

C. Perform the multiplications and divisions and simplify.

1. $\frac{4ab}{15dcx} \cdot \frac{5xy}{8ab^2} \cdot \frac{2bx}{3ab}$

2. $\frac{2x^2y}{3yz} \cdot \frac{5xz^2}{7xy} \div \frac{21x^2y^3z^2}{40xy^2z}$

3. $\frac{a+b}{a-b} \cdot \frac{a^2-ab}{2a^2-2b^2}$

4. $\frac{3x+2}{5x^2} \cdot \frac{10x}{9x^2-4}$

5. $\frac{x^2+xy}{5x^2-5y^2} \cdot \frac{3x^3-3y^3}{x^2-xy}$

6. $\frac{a}{a^3-b^3} \div \frac{ab^2}{a^2-b^2}$

7. $\frac{x^2-1}{a+b} \div \frac{x+1}{a^2-b^2}$

8. $\left(\frac{m^2-n^2}{m^2n^2} \div \frac{am+an}{mn} \right) \cdot \frac{mn}{a(m-n)}$

9. $\frac{x^2-5x+6}{x^2+3x-4} \cdot \frac{x^2+7x+12}{x^2-8x+15} \div \frac{x^2+x-6}{x^2-4x-6}$

10. $\frac{x^2-5x+6}{x^2+5x+4} \div \frac{x^2-4x+3}{2x^2+3x+1} \cdot \frac{x^2+3x-4}{2x^2-3x-2}$

D. Perform the operations and simplify.

1.
$$\frac{x-1}{x-2} + \frac{x+1}{x+2} - \frac{4}{4-x^2} + \frac{2}{2-x}$$

2.
$$\left(\frac{1}{x^3} - \frac{1}{x^2} + \frac{1}{x} \right) (x^4 + x^3)$$

3.
$$\left(\frac{2a}{x^2-a^2} + \frac{1}{a-x} \right) \cdot \frac{x^2-a^2}{a+x}$$

4.
$$\frac{1-a^2}{1+b} \cdot \frac{1-b^2}{a+a^2} \cdot \left(1 + \frac{a}{1-a} \right)$$

5.
$$\left(\frac{a+b}{a-b} - \frac{a^2+b^2}{a^2-b^2} \right) \cdot \left(\frac{a+b}{a-b} + \frac{a^2+b^2}{a^2-b^2} \right)$$

6.
$$\left(\frac{a+b}{a-b} + \frac{a-b}{a+b} \right) \div \left(\frac{1}{(a+b)^2} + \frac{1}{(a-b)^2} \right)$$

7.
$$\left(\frac{x}{x+1} + \frac{x-1}{x}\right) \div \left(\frac{x}{x+1} - \frac{x-1}{x}\right)$$

8.
$$\left(\frac{a+3b}{(a-b)^2} + \frac{a-3b}{a^2-b^2}\right) \div \frac{a^2+3b^2}{(a-b)^2}$$

9.
$$\left[\frac{1}{(a+b)^2}\left(\frac{1}{a^2} + \frac{1}{b^2}\right) + \frac{2}{(a+b)^3}\left(\frac{1}{a} + \frac{1}{b}\right)\right] a^2 b^2.$$

10.
$$\frac{1}{x} - \left\{1 - \left[\frac{x-1}{x} + \frac{1}{2}\left(\frac{x-1}{x+1} - \frac{(x-2)(x-3)}{x(x+1)}\right)\right]\right\}$$

11.
$$\frac{\frac{x^2}{4y}}{\frac{x}{12y}}$$

12.
$$\frac{4x^2 + 2x}{1 + \frac{1}{1+\frac{1}{x}}}.$$

13.
$$\frac{\frac{x}{y} + \frac{y}{x}}{\frac{x}{y} - \frac{y}{x}} \div \frac{\frac{1}{x^4} - \frac{1}{y^4}}{\left(\frac{1}{x} + \frac{1}{y}\right)^2}$$

4.9 Radical Expressions

Let $A(x)$ be an algebraic expression. Any expression containing a radical $\sqrt[n]{A(x)}$ is called a **radical expression**; n is the *index*, and $A(x)$ is the *radicand*. For n even, the radicand is assumed to have positive values. Here are some example of radical expressions

$$2 + \sqrt{x}, \quad \sqrt{3x - 2}, \quad \frac{x}{\sqrt{x^2 + 1}}, \quad \frac{2x}{5 + \sqrt{x - 1}}, \quad 8x + \sqrt[3]{x^2 - 4x}.$$

As with numerical radicals (roots), to ***simplify*** a radical expression means to write the radical in the form in which the radicand has the simplest expression possible. Radicals which when reduced to their simplest forms, differ in their coefficients only are called ***similar***.

The rules of simplifying radical expressions are based on the properties of roots, which we discussed in Section 3.3. We restate these properties, for the convenience of the reader.

Let $a \geq 0$, $b \geq 0$, and m, n, k be natural numbers. Then

1. $\sqrt[n]{ab} = \sqrt[n]{a} \cdot \sqrt[n]{b}$.

2. $\sqrt[n]{\frac{a}{b}} = \frac{\sqrt[n]{a}}{\sqrt[n]{b}}$.

3. $\sqrt[n]{a^m} = \left(\sqrt[n]{a}\right)^m$.

4. $\sqrt[n]{a} = \sqrt[nk]{a^k}$.

5. $\sqrt[m]{\sqrt[n]{a}} = \sqrt[mn]{a}$.

Example 4.9.1. Simplify

1.

$$\sqrt[4]{32x^8y^5} = \sqrt[4]{16x^8y^4} \cdot \sqrt[4]{2y} = 2x^2|y|\sqrt[4]{2y}.$$

2.

$$\sqrt{\frac{b^3c^5}{25a^4}} = \frac{\sqrt{b^3c^5}}{\sqrt{25a^4}} = \frac{\sqrt{b^2c^4} \cdot \sqrt{bc}}{5a^2} = \frac{|b|c^2\sqrt{bc}}{5a^2}.$$

Example 4.9.2. Simplify

1.

$$\left(\sqrt[3]{2x^3y^2}\right)^2 = \sqrt[3]{(2x^3y^2)^2} = \sqrt[3]{4x^6y^4} = \sqrt[3]{x^6y^3} \cdot \sqrt[3]{4y} = x^2y\sqrt[3]{4y}.$$

2.

$$\sqrt[5]{\sqrt[3]{-x^5}} = \sqrt[5]{\left(\sqrt[3]{-x}\right)^5} = \sqrt[3]{-x} = \sqrt[3]{(-1)x} = \sqrt[3]{-1} \cdot \sqrt[3]{x} = -\sqrt[3]{x}.$$

Alternatively,

$$\left[(-x^5)^{\frac{1}{3}}\right]^{\frac{1}{5}} = (-x^5)^{\frac{1}{15}} = (-x)^{\frac{1}{3}} = -x^{\frac{1}{3}} = -\sqrt[3]{x}.$$

4.9.1   Operations with Radical Expressions

To **add** or **subtract** radical expressions, we simplify them first and then we combine the similar radicals (if any).

Example 4.9.3. Do the operations:

1.

$$\sqrt{64x^5y} - \sqrt{36x^5y} + 3\sqrt{xy} = 8x^2\sqrt{xy} - 6x^2\sqrt{xy} + 3\sqrt{xy} =$$
$$= 2x^2\sqrt{xy} + 3\sqrt{xy} = (2x^2 + 3)\sqrt{xy}.$$

2. For $x \geq 0$,

$$\sqrt{9a^2 + 9} + \sqrt{4a^2 + 4} + \sqrt{x^2 + a^2x^2} - 5\sqrt{1 + a^2} =$$
$$3\sqrt{a^2 + 1} + 2\sqrt{a^2 + 1} + x\sqrt{a^2 + 1} - 5\sqrt{a^2 + 1} = x\sqrt{a^2 + 1}.$$

Example 4.9.4. Do the operations:

1. For $x \geq 3$,

$$\sqrt{ax^3 + 6ax^2 + 9ax} - \sqrt{ax^3 - 4ax^2 + 4ax} =$$

$$= \sqrt{ax(x^2 + 6x + 9)} - \sqrt{ax(x^2 - 4x + 4)} = \sqrt{ax(x-3)^2} - \sqrt{ax(x+2)^2} =$$

$$= (x-3)\sqrt{ax} - (x+2)\sqrt{ax} = [(x-3) - (x+2)]\sqrt{ax} = -5\sqrt{ax}.$$

2.

$$9\sqrt[3]{2a^6x} - 3\sqrt[3]{16a^3x} + \sqrt[3]{2x} = 9a^2\sqrt[3]{2x} - 6a\sqrt[3]{2x} + \sqrt[3]{2x} =$$

$$= (9a^2 - 6a + 1)\sqrt[3]{2x} = (3a+1)^2\sqrt[3]{2x}.$$

To **multiply** radical expressions we use the property

$$\sqrt[n]{a} \cdot \sqrt[n]{b} = \sqrt[n]{ab}.$$

Example 4.9.5. Multiply and simplify:

1. $\sqrt[3]{x^2yz^2} \cdot \sqrt[3]{xy^2z^4} = \sqrt[3]{x^3y^3z^6} = \sqrt[3]{(xyz^2)^3} = xyz^2.$

2. $3\sqrt{ab} \cdot 5\sqrt[3]{a^2b^2} = 15\sqrt[6]{a^3b^3} \cdot \sqrt[6]{a^4b^4} = 15\sqrt[6]{a^7b^7} = 15ab\sqrt[6]{ab}.$

3. $\sqrt[2n]{a} \cdot \sqrt[4n]{a} = \sqrt[4n]{a^2} \cdot \sqrt[4n]{a} = \sqrt[4n]{a^3}.$

To **divide** radical expressions we use the property

$$\frac{\sqrt[n]{a}}{\sqrt[n]{b}} = \sqrt[n]{\frac{a}{b}}.$$

Example 4.9.6. Divide and simplify:

1. $\sqrt{a^3} \div \sqrt[3]{a^4} = \sqrt[6]{a^9} \div \sqrt[6]{a^8} = \frac{\sqrt[6]{a^9}}{\sqrt[6]{a^8}} = \sqrt[6]{\frac{a^9}{a^8}} = \sqrt[6]{a}.$

2.

$$\frac{\sqrt[10]{a^7b^9} \cdot \sqrt[15]{a^{12}b^{14}}}{\sqrt[3]{ab^2} \cdot \sqrt[6]{ab^5}} = \frac{\sqrt[30]{a^{21}b^{27}} \cdot \sqrt[30]{a^{24}b^{28}}}{\sqrt[6]{a^2b^4} \cdot \sqrt[6]{ab^5}} = \frac{\sqrt[30]{a^{45}b^{55}}}{\sqrt[6]{a^3b^9}} =$$

$$= \frac{\sqrt[30]{a^{45}b^{55}}}{\sqrt[30]{a^{15}b^{45}}} = \sqrt[30]{\frac{a^{45}b^{55}}{a^{15}b^{45}}} = \sqrt[30]{a^{30}b^{10}} = a\sqrt[30]{b^{10}} = a\sqrt[3]{b}.$$

4.9.2   Rationalizing Factors

When the product of two given radical expressions is rational, each of the expression is called a ***rationalizing factor*** for the other. The binomial radical expressions $a+\sqrt{b}$ and $a-\sqrt{b}$ are rationalizing factor of each other, because

$$(a + \sqrt{b})(a - \sqrt{b}) = a^2 - (\sqrt{b})^2 = a^2 - b.$$

Similarly, since

$$(\sqrt{a} + \sqrt{b})(\sqrt{a} - \sqrt{b}) = (\sqrt{a})^2 - (\sqrt{b})^2 = a - b$$

each expression $\sqrt{a} + \sqrt{b}$ and $\sqrt{a} - \sqrt{b}$ is a rationalizing factor of each other.

We call $a - \sqrt{b}$ the ***conjugate*** of $a + \sqrt{b}$ and vice versa. Similarly, $\sqrt{a} - \sqrt{b}$ is the ***conjugate*** of $\sqrt{a} + \sqrt{b}$ and vice versa.

Rationalizing the Denominator

Computations with algebraic fractions involving radicals are often simplified by what is termed ***rationalizing the denominator***. This means elimininating the radicals in the denominator[4].

The most common fractions of these form are the following:

1. Fractions of the form

$$\frac{1}{\sqrt{a}}, \quad a > 0.$$

Here we multiply the numerator and denominator by $\sqrt{a}$. That is,

$$\frac{1}{\sqrt{a}} = \frac{1}{\sqrt{a}} \cdot \frac{\sqrt{a}}{\sqrt{a}} = \frac{\sqrt{a}}{\sqrt{a^2}} = \frac{\sqrt{a}}{a}.$$

---

[4] In certain occasions if the radical appears only in the numerator, to simplify, we may need similarly, to rationalize the numerator.

Example 4.9.7. Rationalize the denominator

$$\frac{2}{\sqrt{3}} = \frac{2}{\sqrt{3}} \cdot \frac{\sqrt{3}}{\sqrt{3}} = \frac{2\sqrt{3}}{\sqrt{3^2}} = \frac{2\sqrt{3}}{3}.$$

2. Fractions of the form

$$\frac{1}{\sqrt[n]{a^m}}, \quad a > 0, \ n > m.$$

Here we multiply the numerator and denominator by $\sqrt[n]{a^{n-m}}$. That is,

$$\frac{1}{\sqrt[n]{a^m}} = \frac{1}{\sqrt[n]{a^m}} \cdot \frac{\sqrt[n]{a^{n-m}}}{\sqrt[n]{a^{n-m}}} = \frac{\sqrt[n]{a^{n-m}}}{\sqrt[n]{a^n}} = \frac{\sqrt[n]{a^{n-m}}}{a}.$$

Example 4.9.8. Rationalize the denominator $\frac{1}{\sqrt[3]{5}}$.
Solution.

$$\frac{1}{\sqrt[3]{5}} = \frac{1}{\sqrt[3]{5}} \cdot \frac{\sqrt[3]{5^2}}{\sqrt[3]{5^2}} = \frac{\sqrt[3]{25}}{\sqrt[3]{5^3}} = \frac{\sqrt[3]{25}}{5}.$$

3. Fractions of the form

$$\frac{1}{a + \sqrt{b}}, \quad \frac{1}{\sqrt{a} + \sqrt{b}}, \quad a > 0, \ b > 0.$$

Here we multiply the numerator and denominator by the conjugate of the denominator, that is, by $a - \sqrt{b}$ or $\sqrt{a} - \sqrt{b}$ respectively. (When the denominator is of the form $a - \sqrt{b}$ or $\sqrt{a} - \sqrt{b}$, the conjugate of the denominator is, of course, $a + \sqrt{b}$ or $\sqrt{a} + \sqrt{b}$, respectively).

$$\frac{1}{a + \sqrt{b}} = \frac{1}{a + \sqrt{b}} \cdot \frac{a - \sqrt{b}}{a - \sqrt{b}} = \frac{a - \sqrt{b}}{(a + \sqrt{b})(a - \sqrt{b})} = \frac{a - \sqrt{b}}{a^2 - b}$$

or

$$\frac{1}{\sqrt{a}+\sqrt{b}} = \frac{1}{\sqrt{a}+\sqrt{b}}\cdot\frac{\sqrt{a}-\sqrt{b}}{\sqrt{a}-\sqrt{b}} = \frac{\sqrt{a}-\sqrt{b}}{(\sqrt{a}+\sqrt{b})(\sqrt{a}-\sqrt{b})} = \frac{\sqrt{a}-\sqrt{b}}{a-b}.$$

Example 4.9.9. Rationalize the denominator

$$\frac{1}{2-\sqrt{3}}.$$

Solution. We have

$$\frac{1}{2-\sqrt{3}} = \frac{1}{2-\sqrt{3}}\cdot\frac{2+\sqrt{3}}{2+\sqrt{3}} = \frac{2+\sqrt{3}}{(2-\sqrt{3})(2+\sqrt{3})} = \frac{2+\sqrt{3}}{2^2-3} = 2+\sqrt{3}.$$

Example 4.9.10. Rationalize the denominator

$$\frac{1}{\sqrt{5}+\sqrt{3}}.$$

Solution. We have

$$\frac{1}{\sqrt{5}+\sqrt{3}} = \frac{1}{\sqrt{5}+\sqrt{3}}\cdot\frac{\sqrt{5}-\sqrt{3}}{\sqrt{5}-\sqrt{3}} = \frac{\sqrt{5}-\sqrt{3}}{(\sqrt{5}+\sqrt{3})(\sqrt{5}-\sqrt{3})} =$$

$$\frac{\sqrt{5}-\sqrt{3}}{5-3} = \frac{\sqrt{5}-\sqrt{3}}{2}.$$

4. Fractions of the form

$$\frac{1}{\sqrt[n]{a}+\sqrt[n]{b}}.$$

a). If $n = 2k+1$, we write

$$\frac{1}{\sqrt[n]{a}+\sqrt[n]{b}} = \frac{1}{a+b}\cdot\frac{a+b}{\sqrt[n]{a}+\sqrt[n]{b}} = \frac{1}{a+b}\cdot\frac{(\sqrt[n]{a})^n + (\sqrt[n]{b})^n}{\sqrt[n]{a}+\sqrt[n]{b}}$$

$$= \frac{1}{a+b}\cdot(\sqrt[n]{a^{n-1}} - \sqrt[n]{a^{n-2}b} + \sqrt[n]{a^{n-3}b^2} - ... + \sqrt[n]{b^{n-1}}).$$

Example 4.9.11. Rationalize the denominator
$$\frac{1}{\sqrt[3]{2} + \sqrt[3]{3}}.$$

Solution.
$$\frac{1}{\sqrt[3]{2} + \sqrt[3]{3}} = \frac{1}{2+3} \cdot \frac{2+3}{\sqrt[3]{2} + \sqrt[3]{3}} = \frac{1}{5} \cdot \frac{(\sqrt[3]{2})^3 + (\sqrt[3]{3})^3}{\sqrt[3]{2} + \sqrt[3]{3}} = \frac{1}{5} \cdot (\sqrt[3]{4} - \sqrt[3]{6} + \sqrt[3]{9}).$$

b). If $n = 2k$, we write
$$\frac{1}{\sqrt[n]{a} + \sqrt[n]{b}} = \frac{1}{a-b} \cdot \frac{a-b}{\sqrt[n]{a} + \sqrt[n]{b}} = \frac{1}{a-b} \cdot \frac{(\sqrt[n]{a})^n - (\sqrt[n]{b})^n}{\sqrt[n]{a} + \sqrt[n]{b}}$$
$$= \frac{1}{a-b} \cdot (\sqrt[n]{a^{n-1}} - \sqrt[n]{a^{n-2}b} + \sqrt[n]{a^{n-3}b^2} - \dots - \sqrt[n]{b^{n-1}}).$$

Example 4.9.12. Rationalize the denominator
$$\frac{1}{\sqrt[3]{2} + \sqrt{3}}.$$

Solution.
$$\frac{1}{\sqrt[3]{2} + \sqrt{3}} = \frac{1}{\sqrt[6]{4} + \sqrt[6]{27}} = \frac{1}{4-27} \cdot \frac{4-27}{\sqrt[6]{4} + \sqrt[6]{27}} = -\frac{1}{23} \cdot \frac{(\sqrt[6]{4})^6 - (\sqrt[6]{27})^6}{\sqrt[6]{4} - \sqrt[6]{27}} =$$
$$= -\frac{1}{23}(\sqrt[6]{4^5} - \sqrt[6]{4^4 \cdot 27} + \sqrt[6]{4^3 \cdot 27^2} - \sqrt[6]{4^2 \cdot 27^3} + \sqrt[6]{4 \cdot 27^4} - \sqrt[6]{27^5}).$$

5. Fractions of the form
$$\frac{1}{\sqrt[n]{a} - \sqrt[n]{b}}.$$

We write
$$\frac{1}{\sqrt[n]{a} - \sqrt[n]{b}} = \frac{1}{a-b} \cdot \frac{a-b}{\sqrt[n]{a} - \sqrt[n]{b}} = \frac{1}{a-b} \cdot \frac{(\sqrt[n]{a})^n - (\sqrt[n]{b})^n}{\sqrt[n]{a} - \sqrt[n]{b}}$$
$$= \frac{1}{a-b} \cdot (\sqrt[n]{a^{n-1}} + \sqrt[n]{a^{n-2}b} + \sqrt[n]{a^{n-3}b^2} + \dots + \sqrt[n]{b^{n-1}}).$$

Example 4.9.13. Rationalize the denominator

$$\frac{1}{\sqrt[4]{3} - \sqrt[4]{2}}.$$

Solution.

$$\frac{1}{\sqrt[4]{3} - \sqrt[4]{2}} = \frac{1}{3 - 2} \cdot \frac{3 - 2}{\sqrt[4]{3} + \sqrt[4]{2}} = 1 \cdot \frac{(\sqrt[4]{3})^4 + (\sqrt[4]{2})^4}{\sqrt[4]{3} - \sqrt[4]{2}} = \sqrt[4]{27} + \sqrt[4]{18} + \sqrt[4]{12} + \sqrt[4]{8}.$$

4.9.3 Exercises

A. Simplify.

   1. a. $\sqrt{128a^4b^6c^8}$ b. $\sqrt[3]{16x^3y^9z^{12}}$.

   2. a. $\sqrt{a^2b^2 - a^2c^2}$ b. $\sqrt{(x^2 - y^2)(x + y)}$.

   3. a. $\sqrt{\frac{a+b}{a-b}}$ b. $\sqrt{\frac{a^2x^2}{b^3} - \frac{2ax}{b^2} + \frac{1}{b}}$.

   4. a. $(5 - \sqrt{2})(5 + \sqrt{2})$ b. $(\sqrt{2} - 3)(2 - \sqrt{2})$

   5. a. $(\sqrt{2} - \sqrt{3})^2$ b. $(\sqrt{2} - 1)^3$

   6. a. $\sqrt[3]{\sqrt{8}}$ b. $(\sqrt{3} + \sqrt{2})(\sqrt{3} - \sqrt{2})$

   7. a. $\left(\sqrt[3]{a^2}\right)^6$ b. $\sqrt{x^2y^3} \div \sqrt[6]{x^5y^5} = \sqrt[3]{x^2y^2}$.

B. Perform the operations and simplify.

   1.

$$(\sqrt[3]{a} + \sqrt[3]{b})(\sqrt[3]{a^2} - \sqrt[3]{ab} + \sqrt[3]{b^2}).$$

   2.

$$\sqrt{ax^3 + 6ax^2 + 9ax} - \sqrt{ax^3 - 4a^2x^2 + 4a^3x}.$$

3.

$$\frac{\sqrt{1+a^2}}{\sqrt{1-a^2}} + \frac{\sqrt{1-a^2}}{\sqrt{1+a^2}} - \frac{2}{\sqrt{1-a^2}}.$$

4.

$$\frac{x+\sqrt{x^2-1}}{x-\sqrt{x^2-1}} - \frac{-+\sqrt{x^2-1}}{x+\sqrt{x^2-1}}.$$

5.

$$5\sqrt{\frac{a^3+a^2}{x^3-x^2}} - \frac{1}{x}\sqrt{\frac{4a^2+4a^3}{x-1}} - \frac{3a}{x}\sqrt{\frac{a+1}{x-1}}.$$

C. Rationalize the denominator.

1. a. $\frac{1}{\sqrt{2}}$ b. $\frac{3}{\sqrt{6}}$.

2. a. $\frac{10}{\sqrt{5}}$ b. $\frac{7}{\sqrt{7}}$.

3. a. $\frac{1}{1+\sqrt{2}}$ b. $\frac{1}{3-\sqrt{5}}$.

4. a. $\frac{a+b}{\sqrt{a+b}}$ b. $\frac{a+\sqrt{b}}{a-\sqrt{b}}$.

5. a. $\frac{1-\sqrt{2}}{1-\sqrt{3}}$ b. $\frac{\sqrt{3}-\sqrt{2}}{1-\sqrt{2}}$.

6. a. $\frac{\sqrt{3}-\sqrt{2}}{2\sqrt{3}+3\sqrt{3}}$ b. $\frac{1+\sqrt{2}+\sqrt{3}}{1-\sqrt{2}+\sqrt{3}}$.

7. a. $\frac{5}{\sqrt[3]{2}+\sqrt[3]{3}}$ b. $\frac{1}{\sqrt[4]{3}-\sqrt[4]{2}}$.

4.10   Review Exercises

A. Perform the operation.

1. $(2x^n - 5)(3x^n - 2)$, $n \in \mathbb{N}$.

2. $(x - 1)^3(x + 1)^3$

3. $(2x^2 - 1)^2 - (3x + 2)^4$

4. $(x^3 + 4x^2 - 18x + 2) \div (x^2 + 1)$.

5. $(x^4 - 2x^3 - 19x^2 - 8x + 60) \div (x^2 - 5x + 6)$.

B. Factor completely.

1. $(a + b)^3 - (a - b)^3 - 3a$.

2. $(x^3 - 9)^2 - (x + 5)(x - 3)^2$.

3. $ab^2 + bc^2 + ca^2 - a^2b - b^2c - ac^2$.

4. $(x - a)^2 + 12a^2(x - a) + 36a^4$.

5. $(x + y)^2 - 1 - (x + y + 1)xy$.

6. $\frac{x^3}{27} - \frac{x^2}{3} + x - 1$.

7. $x^6 - (a^3 - 1)x^3 - a^3$.

8. $x^3y^{4n+5} - y^5x^{4n+3}$, $n \in \mathbb{N}$.

9. $3x^5 - 3x^3 - 11x^2 - 10x - 6$

10. $x^7 + y^7$.

C. Prove the following identities.

1.

$$a^2b + a^2c + b^2a + b^2c + c^2a + c^2b + 2abc \equiv (a+b)((b+c)(a+c)).$$

2.
$$(a + b + c)^3 \equiv a^3 + b^3 + c^3 + 3(a + b)(b + c)(c + a).$$

3.

$$a^3 + b^3 + c^3 - 3abc \equiv (a + b + c)(a^2 + b^2 + c^2 - ab - bc - ac).$$

4.
$$(a + b)^3 - a^3 - b^3 \equiv 3ab(a + b).$$

5.
$$(a + b)^5 - a^5 - b^5 \equiv 5ab(a + b)(a^2 + ab + b^2).$$

6.

$$a^4 + b^4 + c^4 - 2a^2b^2 - 2b^2c^2 - 2a^2c^2 \equiv (a+b+c)(a+b-c)(a-b+c)(a-b-c).$$

7. Prove that if $a \neq b \neq c$, then

$$a^3 + b^3 + c^3 = 3abc \iff a + b + c = 0.$$

8. Prove that if $a + b + c = 0$, then

$$\frac{a^2}{2a^2 + bc} + \frac{b^2}{2b^2 + ac} + \frac{c^2}{2c^2 + ab} = 1$$

9. Prove that if $a + b + c = 0$, then

$$\frac{a^2 - b^2 - 2bc}{a + b} + \frac{b^2 - c^2 - 2ac}{b + c} + \frac{c^2 - a^2 - 2ab}{a + c} = 0$$

D. Find the LCM of the polynomials.

1. $x^2 - 1$, $x^3 - 1$, $x^2 + 2x + 1$.

2. $x^2 - 1$, $x^4 - 1$, $3x^3 - 5x^2 - 3x + 5$.

3. $x^3 + x^2 + x + 1$, $x^3 - x^2 + x - 1$.

E. Perform the operations and simplify.

1.
$$\frac{1 - x^2}{(1 + ax)^2 - (a + x)^2}.$$

2.
$$\frac{3}{x + 2} + \frac{4}{(x + 2)^2} + \frac{5}{(x - 1)^2} - \frac{3}{x - 1}$$

3.
$$\frac{a^2 - b^2}{a - b} - \frac{a^3 - b^3}{a^2 - b^2}$$

4.
$$\left(\frac{x^2}{y^3} + \frac{1}{x} \right) \div \left(\frac{x}{y^2} - \frac{1}{y} + \frac{1}{x} \right)$$

5.
$$\frac{x^3 + y^3}{x + y} \div (x^2 - y^2) + \frac{2y}{x + y} - \frac{xy}{x^2 - y^2}$$

6.
$$\left(\frac{x}{x^2 - 4} - \frac{8}{x^2 + 2x} \right) \frac{x^2 - 2x}{4 - x} + \frac{x + 8}{x + 2}$$

7.
$$\left(\frac{x^2 - 1}{x^4 - 2x^2 + x^2} \div \frac{x^2 + 2x + 1}{x^3 - 1} \right) \frac{x^2}{(x + 1)^2 - x}$$

8.

$$\frac{a^3 - b^3}{a^2 - b^2 + \frac{2b^2}{1 + \frac{a+b}{a-b}}}$$

9.

$$\left(\frac{1}{a - \sqrt{2}} - \frac{a^2 + 4}{a^3 - \sqrt{8}}\right) \div \left(\frac{a}{\sqrt{2}} + 1 = \frac{\sqrt{2}}{a}\right)^{-1}$$

10.

$$\left(\frac{1}{2 + 2\sqrt{a}} + \frac{1}{2 - 2\sqrt{a}} - \frac{a^2 + 1}{1 - a^2}\right)\left(1 + \frac{1}{a}\right)$$

F. Perform the operations and simplify.

1.

$$\left(x + \sqrt{x^2 - 1}\right)^2 + \left(x + \sqrt{x^2 - 1}\right)^{-2} + 2(1 - 2x^2)$$

2.

$$\frac{x^{\frac{1}{2}} + x^{-\frac{1}{2}}}{1 - x} + \frac{1 - x^{-\frac{1}{2}}}{1 + \sqrt{x}}$$

3.

$$\frac{x - x^{-2}}{x^{\frac{1}{2}} - x^{-\frac{1}{2}}} - \frac{2}{x^{\frac{3}{2}}} - \frac{1 - x^{-2}}{x^{\frac{1}{2}} - x^{-\frac{1}{2}}}$$

4.

$$\left(\frac{4a - 9a^{-1}}{2a^{\frac{1}{2}} - 3a^{-\frac{1}{2}}} + \frac{a - 4 + \frac{3}{a}}{a^{\frac{1}{2}} - a^{-\frac{1}{2}}}\right)^2$$

5.

$$\left(\frac{3x^{-\frac{1}{3}}}{x^{\frac{2}{3}} - 2x^{-\frac{1}{3}}} - \frac{x^{\frac{1}{3}}}{x^{\frac{4}{3}} - x^{\frac{1}{3}}}\right)^{-1} - \left(\frac{1 - 2x}{3x - 2}\right)^{-1}$$

6.

$$\sqrt{x} + \frac{1}{\sqrt{x}} + \frac{(1 - x)\left(1 - \frac{1}{\sqrt{x}}\right)}{1 + \sqrt{x}}.$$

7.

$$\left(\frac{a\sqrt{a} + b\sqrt{b}}{\sqrt{a} + \sqrt{b}} - \sqrt{ab}\right) \cdot \left(\frac{\sqrt{a} + \sqrt{b}}{a - b}\right)^2.$$

8.

$$\left(\frac{3}{\sqrt{1 + x}} + \sqrt{1 - x}\right) \div \left(\frac{3}{\sqrt{1 - x^2}} + 1\right).$$

9.

$$\frac{\sqrt{x} + 1}{1 + \sqrt{x} + x} \div \frac{1}{x^2 - \sqrt{x}}.$$

10.

$$\left(\frac{1 + \sqrt{1 - x}}{1 - x + \sqrt{1 - x}} + \frac{1 + \sqrt{1 + x}}{1 + x - \sqrt{1 + x}}\right)^2 \cdot \frac{x^2 - 1}{2} + 1.$$

11.

$$\frac{\sqrt{x} + 1}{x + \sqrt{x} + 1} \div \frac{1}{\sqrt{x^3 - 1}}.$$

12.

$$\left(\frac{\sqrt{x}}{2} - \frac{1}{2\sqrt{x}}\right)^2 \cdot \left(\frac{\sqrt{x} - 1}{\sqrt{x} + 1} - \frac{\sqrt{x} + 1}{\sqrt{x} - 1}\right).$$

13.

$$a\left(\frac{\sqrt{a} + \sqrt{b}}{2b\sqrt{a}}\right)^{-1} + b\left(\frac{\sqrt{a} + \sqrt{b}}{2a\sqrt{b}}\right)^{-1}.$$

14.

$$\frac{\frac{a-b}{\sqrt{a} - \sqrt{b}} - \frac{a-b}{\sqrt{a} + \sqrt{b}}}{\frac{\sqrt{a} - \sqrt{b}}{a - b} + \frac{\sqrt{a} + \sqrt{b}}{a - b}} \cdot \frac{2\sqrt{ab}}{b - a}.$$

15.

$$\left(\frac{2x + \sqrt{xy}}{3x}\right)^{-1} \cdot \left(\frac{\sqrt{x^3} - \sqrt{y^3}}{x - \sqrt{xy}} - \frac{x - y}{\sqrt{x} + \sqrt{y}}\right).$$

Chapter 5

Equations I

In this chapter, we revisit the first degree equations, we introduce the set of complex numbers and we study second degree equations.

Equations and the methods for solving them occupy a central part of algebra. We recall that an **equation** in one unknown x is an equality $A(x) = B(x)$ of two algebraic expressions $A(x)$ and $B(x)$. A **solution** or **root** of an equation is a value of x which when substituted into the equation yields a true statement (or as we say this value of x satisfies the equation). To **solve** an equation means to find all its solutions. An equation may have solution(s) or may be impossible (in this case, we say the equation has no solution). We also recall that two equations are said to be **equivalent** if every solution of one is also a solution of the other, and vice versa. That is, equivalent equations have the *same solutions*[1]. For instance, the equation $x^2 - 9 = 0$ is equivalent to $(x+3)(x-3) = 0$. Clearly the solutions of both equations are $x = -3$ and $x = 3$.

We shall see in Chapter 6, that the first degree and second degree equations are the basic building blocks of all polynomial equations. For this reason the *first and second degree equations play an important role in the study of equations.*

---

[1]If both equations have no solution, then they are also regarded as equivalent.

5.1   First Degree Equations; revisited

The simplest equations are the ***first degree equations***

$$ax + b = 0,$$

where $a \neq 0$. In Section 2.2, we saw that a first degree equation has one solution $x = -\frac{b}{a}$. These equations are also called ***linear equations***.

As a general rule, before we solve an equation, we perform the operations and we *simplify* (whenever possible) the expressions $A(x)$ and $B(x)$ involved in the equation $A(x) = B(x)$. Let us solve a some examples of equations which, after simplification lead to a first degree equation.

Example 5.1.1. Solve the equation

$$(x + 3)^2 = x(x - 5).$$

Solution. We simplify both sides and we have

$$x^2 + 6x + 9 = x^2 - 5x \iff x^2 + 6x - x^2 + 5x + 9 = 0$$

$$\iff 11x + 9 = 0 \iff 11x = -9 \iff x = -\frac{9}{11}.$$

Example 5.1.2. Solve the equation

$$(3x - 1)(x + 5) - 7x = 3(x + 2)^2 + 5(2 - x).$$

Solution. We perform the operations in both sides and we have

$$3x^2 + 15x - x - 5 - 7x = 3x^2 + 12x + 12 + 10 - 5x$$

$$\iff 3x^2 + 7x - 5 = 3x^2 + 7x + 22$$

$$\iff 3x^2 - 3x^2 + 7x - 7x = 22 + 5$$

$$\iff 0x = 27 \iff 0 = 27.$$

This is impossible. Hence, the equation has no solution

Example 5.1.3. Solve the equation

$$\frac{3x - 5}{2(x + 2)} = \frac{3x - 1}{2x + 5} - \frac{1}{x + 2}.$$

Solution. We multiply both sides by $2(x + 2)(2x + 5)$ so as to clear the equation of fractions, and we have

$$(3x - 5)(2x + 5) = 2(3x - 1)(x + 2) - 2(2x + 5)$$

$$\Leftrightarrow\ 6x^2 + 5x - 25 = 6x^2 + 6x - 14$$

$$\Leftrightarrow\ 6x^2 + 5x - 6x^2 - 6x = 14 + 25$$

$$\Leftrightarrow\ -x = 11\ \Leftrightarrow\ x = -11.$$

5.1.1   Exercises

Solve each equation.

1. $\frac{5}{3}x - 8 = 7$.

2. $(3x + 5) - (x - 2) = 2(x - 1) + 3$.

3. $(x - 2)(x - 3) = (x + 3)(x + 4)$.

4. $(x + 1)(x - 1) - (x + 2)(x - 3) = 4$.

5. $(2x - 1) - (3x + 7) = 5 - [(x - 3) - 4x]$.

6. $3(x - 2) - (5 - 12x) + x(x - 4) = (x + 2)^2 + 7x - 15$.

7. $2\{3\,[4(5x - 1) - 8] - 20\} - 7 = 1$.

8. $(2x - \frac{3}{5})(5x + \frac{2}{3}) = 10(x - 1)(x + 1) - \frac{2}{5}$.

9. $(5x - 2)^2 - 2(4x - 3)^2 = (7x + 2)(1 - x) + 14$.

10.
$$x + \frac{2x - 7}{3} - \frac{x - 5}{2} = 1.$$

11.
$$3x - \frac{x-2}{3} - \frac{2x-1}{2} - 1 = \frac{3(x-1)}{2} + \frac{x-1}{6}.$$

12.
$$\frac{4x}{7} - \frac{2(3x-2)}{21} - \frac{x-5}{3} = \frac{5(3-4x)}{7} + \frac{1}{3}.$$

13.
$$\frac{x + \frac{1}{3}}{\frac{2}{5}} - \frac{2x - \frac{1}{2}}{\frac{3}{4}} = \frac{1}{6} - \frac{3x}{4}.$$

14.
$$\frac{2}{x+1} + \frac{1}{3} = \frac{1}{x+1}.$$

15.
$$\frac{3}{x-2} + \frac{1}{x} = \frac{3}{x-2}.$$

16.
$$\frac{2}{x-2} + \frac{1}{x+1} = \frac{1}{x^2 - x - 2}.$$

17.
$$\frac{x}{x+2} - 1 = \frac{3x+2}{x^2 + 4x + 4}.$$

18.
$$\frac{x+1}{a+b} + \frac{x-1}{a-b} = \frac{2a}{a^2 - b^2}.$$

5.2 Equations Involving Absolute Value

We shall consider equations involving the absolute value. The simplest such equation is of the form

$$|x| = a, \text{ where } a \geq 0.$$

To solve it, we consider two cases: if $x \geq 0$, then $|x| = x$. Hence $x = a$. If $x < 0$, then $|x| = -x$. Hence $-x = a$ or equvalently $x = -a$. Thus,

$$|x| = a \iff x = a \text{ or } x = -a.$$

In general, let $A(x)$ and $B(x)$ be two algebraic expressions. Then the equation

$$|A(x)| = |B(x)|$$

is equivalent to two equations

$$A(x) = B(x) \text{ or } A(x) = -B(x).$$

Example 5.2.1. Solve the equation

$$|2x - 3| = 7.$$

Solution. The equation is equivalent to two equations

$$2x - 3 = 7 \text{ or } 2x - 3 = -7.$$

We solve the first equation

$$2x - 3 = 7 \iff 2x = 10 \iff x = 5.$$

We solve the second equation

$$2x - 3 = -7 \iff 2x = -4 \iff x = -2.$$

Thus, the solutions of the given equation are 5 and -2.

Example 5.2.2. Solve the equation.

$$|3x + 8| = |x - 2|.$$

Solution. The equation is equivalent to two equations

$$3x + 8 = x - 2 \quad \text{or} \quad 3x + 8 - (x - 2).$$

We solve the first equation

$$3x + 8 = x - 2 \iff 2x = -10 \iff x = -5.$$

We solve the second equation

$$3x + 8 = -(x - 2) \iff 3x + 8 = -x + 2, \iff 4x = -6 \iff x = -\frac{3}{2}.$$

Thus, the solutions of the given equation are -5 and $-\frac{3}{2}$.

Example 5.2.3. Solve the equation.

$$|x| - |x - 1| = 5 - 3x.$$

Solution. We consider the cases $x \leq 0$, $0 < x \leq 1$ and $1 < x$ separately.

If $x \leq 0$, then $|x| = -x$ and $|x - 1| = -(x - 1)$. Therefore, the equation becomes

$$-x + x - 1 = 5 - 3x \iff -1 = 5 - 3x \iff 3x = 6 \iff x = 2.$$

Since $2 > 0$ we reject it.

If $0 < x \leq 1$, then $|x| = x$ and $|x - 1| = -(x - 1)$. Therefore, the equation becomes

$$x + x - 1 = 5 - 3x \iff 2x - 1 = 5 - 3x \iff 5x = 6 \iff x = \frac{6}{5}.$$

Since $\frac{6}{5} > 1$, we reject it.

Finally, if $1 < x$, then $|x| = x$ and $|x - 1| = x - 1$. Therefore, the equation becomes

$$x - (x - 1) = 5 - 3x \iff 1 = 5 - 3x \iff 3x = 4 \iff x = \frac{4}{3}.$$

Since $\frac{4}{3} > 1$ we accept it. Thus the solution of the equation is $\frac{4}{3}$.

5.2.1   Exercises

1. Solve the equations.

   (a) $|x - 1| = 3$.

   (b) $2|x| - 3 = 0$.

2. Solve the equations.

   (a) $|2x + 5| = 3$.

   (b) $|3x - 1| = 5$.

3. Solve the equations.

   (a) $|x - 5| = x + 1$.

   (b) $|3x + 8| = x - 2$.

4. Solve the equations.

   (a) $\left|\frac{2x-4}{5}\right| = 2$.

   (b) $\left|\frac{x-2}{3}\right| = |6 - x|$.

5. Solve the equations.

   (a) $|x + 3| = |x|$.

   (b) $|x - 1| = |x + 2|$.

6. Solve the equations.

   (a) $|3x| - 2x = |x| + 8$.

   (b) $|x| - |x - 2| = 2$.

   (c) $|x - 2| - |x - 1| = 1$.

5.3   Complex Numbers

The developemnt of algebra called for numbers of a new kind besides the positive and negative real numbers we studied in Chapter 1. For example, since $x^2 \geq 0$ for any real number $x \in \mathbb{R}$, there is no real number satisfying the equation

$$x^2 + 1 = 0 \quad \Leftrightarrow \quad x^2 = -1.$$

The purpose of the imaginary numbers is to remidy this deficiency. Thus it was introduced a new number defined by

$$i = \sqrt{-1} \quad \Leftrightarrow \quad i^2 = -1$$

called the **imaginary unit**. Since $i^2 = -1$ and $(-i)^2 = (-1)^2 i^2 = -1$, we see that the numbers i and $-i$ are the solutions of the equation $x^2 = -1$.

5.3.1   Imaginary Numbers

Any number of the form bi, where $b \in \mathbb{R}$ with $b \neq 0$, is called an **imaginary number**. Thus, the numbers

$$2i, \quad -3i \ \tfrac{1}{2}i, \quad -\tfrac{5}{6}i, \quad 0.25i, \quad \text{and} \quad \pm i\sqrt{3} \text{ are imaginary.}$$

The square root of any negative real number is an imaginary number. Indeed, for example

$$\sqrt{-25} = \sqrt{25(-1)} = \sqrt{25} \cdot \sqrt{-1} = 5i$$

and

$$\sqrt{-3} = \sqrt{(-1)3} = \sqrt{-1} \cdot \sqrt{3} = i\sqrt{3}.$$

We remark that if a and b are both negative, then $\sqrt{ab} \neq \sqrt{a} \cdot \sqrt{b}$. For example the *correct* simplification of $\sqrt{-4} \cdot \sqrt{-9}$ is

$$\sqrt{-4} \cdot \cdot \sqrt{-9} = (2i)(3i) = 6i^2 = 6(-1) = -6.$$

The incorrect calculation

$$\sqrt{-4}\sqrt{-9} = \sqrt{(-4)(-9)} = \sqrt{36} = 6$$

gives a different result!

Powers of i

We have:

$$i^0 = 1, \ \ i^1 = i, \ \ i^2 = -1, \ \ i^3 = i^2 \cdot i = (-1)i = -i, \ \ i^4 = i^2 \cdot i^2 = (-1)(-1) = 1.$$

$$i^5 = i^4 \cdot i = i, \qquad i^6 = i^4 \cdot i^2 = 1(-1) = -1,$$
$$i^7 = i^4 \cdot i^3 = 1(-i) = -i, \qquad i^8 = i^4 \cdot i^4 = 1(1) = 1.$$

In general, for $n = 0, 1, 2, ...$, we have

$$i^{4n} = (i^4)^n = 1^n = 1, \qquad i^{4n+1} = i^{4n} \cdot i = 1 \cdot i = i,$$

$$i^{4n+2} = i^{4n} \cdot i^2 = 1(-1) = -1, \qquad i^{4n+3} = i^{4n} \cdot i^3 = 1(-i) = -i.$$

Thus, to find any positive integer power i^p, with $p > 4$ we divide p by 4 to get $p = 4n + r$. Then

$$i^p = i^{4n+r} = i^{4n} \cdot i^r = i^r$$

and according to the values of the remainder $r = 0, 1, 2, 3$, the power $i^p = 1, i, -1, -i$ respectively.

Note that even powers of i are ± 1, and odd powers of i are $\pm i$.

Example 5.3.1. Find i^{65} and i^{487}.

Solution. We divide 65 by 4. We get $65 = (4)(16) + 1$. Hence

$$i^{65} = i^{(4)(16)+1} = i^{(4)(16)} \cdot i^1 = (i^4)^{16} \cdot i = 1^{16} \cdot (i) = 1(i) = i.$$

Similarly, $487 = (4)(121) + 3$. Hence

$$i^{487} = i^{(4)(121)+3} = i^{(4)(121)} \cdot i^3 = (i^4)^{121} \cdot i^3 = (1)^{121}(-i) = 1(-i) = -i.$$

Example 5.3.2. Show that $i^7 + i^8 + i^9 + i^{10} = 0$.

Solution. We have

$$i^7 + i^8 + i^9 + i^{10} = i^7(1 + i + i^2 + i^3) = i^7(1 + i - 1 - i) = i^7(0) = 0.$$

Example 5.3.3. Find i^{-1}, i^{-2}, i^{-3}, i^{-4}.

Solution. We have

$$i^{-1} - \frac{1}{i} = \frac{1}{i} \cdot \frac{i}{i} = \frac{i}{i^2} = \frac{i}{-1} = -i.$$

$$i^{-2} = \frac{1}{i^2} = \frac{1}{-1} = -1.$$

$$i^{-3} = \frac{1}{i^3} = \frac{1}{-i} = -\frac{1}{i} = -(-i) = i.$$

$$i^{-4} = \frac{1}{i^4} = \frac{1}{1} = 1.$$

5.3.2   Complex Numbers

A **complex number** is a sum of the form $a + bi$, where a, b are real numbers. The number a is called the **real part**, and b is called the **imaginary part**. The set of complex numbers is denoted by $\mathbb{C}$.

That is,

$$\mathbb{C} = \{z = a + bi : a, b \in \mathbb{R}\}.$$

Any real number a can be written as $a = a + 0i$. In particular, $0 = 0 + 0i$ and $1 = 1 + 0i$. Similarly, any imaginary number bi can be written as $bi = 0 + bi$. Hence, we see that *the set of complex numbers $\mathbb{C}$ contains the set of real numbers $\mathbb{R}$ and the set of imaginary numbers $\mathbb{R}i$.* Thus, for the sets of numbers we have the following inclusions

$$\mathbb{N} \subset \mathbb{Z} \subset \mathbb{Q} \subset \mathbb{R} \subset \mathbb{C}.$$

Equality of Complex Numbers

Two complex numbers $z_1 = a_1 + b_1 i$ and $z_2 = a_2 + b_2 i$ are called **equal** if they have equal real parts and imaginary parts. That is,

$$z_1 = z_2 \iff a_1 = a_2 \text{ and } b_1 = b_2$$

Operations with Complex Numbers

Addition and Subtraction. Complex numbers can be added and subtracted as if they were algebraic binomials. Specifically, let $z = a + bi$ and $w = c + di$ be in $\mathbb{C}$. Then the **sum** of z and w is defined by

$$z + w = (a + bi) + (c + di) = (a + c) + (b + d)i.$$

Example 5.3.4. $(-2 + 5i) + (4 - 8i) = (-2 + 4) + (5 - 8)i = 2 - 3i$

The **difference** of z and w is defined by

$$z - w = (a + bi) - (c + di) = (a - c) + (b - d)i.$$

Example 5.3.5. $(7 + 2i) - (4 - 3i) = (7 - 4) + (2 - (-3))i = 3 + 5i$

Multiplication. The **product** of the complex numbers $z = a + bi$ and $w = c + di$ is defined by

$$zw = (ac - bd) + (ad + bc)i.$$

Note that

$$zw = (a+bi)(c+di) = ac+adi+bci+bdi^2 = ac+adi+bci+bd(-1) = (ac-bd)+(ad+bc)i,$$

that is, $a + bi$ and $c + di$ are multiplied as algebraic binomials.

Example 5.3.6. Multiply $(3 + 2i)(1 - 2i)$.
   Solution.

$$(3 + 2i)(1 - 2i) = 3 - 6i + 2i - 4i^2 = 3 - 6i + 2i + 4 = 7 - 8i.$$

These operations have the following properties whose easy verification we leave to the reader. Let $z, w, r \in \mathbb{C}$. Then

1. $z + (w + r) = (z + w) + r$, $z(wr) = (zw)r$.

2. $z + w = w + z$, $zw = wz$.

3. $z + 0 = z$, $z1 = z$

4. $z + (-z) = 0$,

5. $z(w + r) = zw + zr$

The ***complex conjugate*** $\bar{z}$ of $z = a + bi$ is defined by

$$\bar{z} = \overline{a + bi} = a - bi.$$

For example, if $z = 3 + 5i$, then $\bar{z} = 3 - 5i$.

Properties of Conjugation

Conjugation of complex numbers has the following properties;

$$\bar{\bar{z}} = z$$

$$z + \bar{z} = (a + bi) + (a - bi) = 2a$$

$$z - \bar{z} = (a + bi) - (a - bi) = 2bi$$

$$z\bar{z} = (a + bi)(a - bi) = a^2 - (bi)^2 = a^2 + b^2.$$

Moreover, for $z, w \in \mathbb{C}$, we have

$$\overline{z + w} = \bar{z} + \bar{w},$$

$$\overline{zw} = \bar{z} \cdot \bar{w}.$$

Division of Complex Numbers

In accordance with the definition of the division of real numbers we give the following definition. To **divide** a complex number $z = a + bi$ (dividend) by a complex number $w = c + di$ (divisor) means to find a number $q = x + yi$ (quotient) such that, $wq = z$. That is, $\frac{z}{w} = q$ or

$$\frac{z}{w} = \frac{a + bi}{c + di} = x + yi = p.$$

If $w \neq 0$, then division is always possible and the quotient is unique.

A convinient practical way to find the quotient is to ***multiply the numerator and denominator of the fraction*** $\frac{a+bi}{c+di}$ ***by the complex conjugate*** $c - di$ ***of the denominator***.

Applying this, we have

$$\frac{a + bi}{c + di} \cdot \frac{c - di}{c - di} = \frac{(a + bi)(c - di)}{(c + di)(c - di)} = \frac{(ac + bd) + (bc - ad)i}{c^2 + d^2} =$$

$$\frac{ac + bd}{c^2 + d^2} + \frac{bc - ad}{c^2 + d^2}i = x + yi,$$

where $x = \frac{ac+bd}{c^2+d^2}$ and $y = \frac{bc-ad}{c^2+d^2}$.

Example 5.3.7. Find the quotient

$$\frac{7 - 4i}{3 + 2i}.$$

Solution. We have

$$\frac{7 - 4i}{3 + 2i} \cdot \frac{3 - 2i}{3 - 2i} = \frac{(7 - 4i)(3 - 2i)}{(3 + 2i)(3 - 2i)} = \frac{13 - 26i}{3^2 + 2^2} = \frac{13 - 26i}{13} = 1 - 2i.$$

Example 5.3.8. Simplify and write the result in the form $a + bi$.

$$\frac{-2 + 5i}{4 - 3i} + \frac{(1 + 2i)^2}{2 + i}$$

Solution. We have

$$\frac{-2+5i}{4-3i} \cdot \frac{4+3i}{4+3i} + \frac{(1+2i)^2}{2+i} \cdot \frac{2-i}{2-i} =$$

$$\frac{(-2+5i)(4+3i)}{(4-3i)(4+3i)} + \frac{(1-4+4i)(2-i)}{(2+i)(2-i)} = \frac{-23+14i}{25} + \frac{-2+11i}{5} =$$

$$\frac{-23+14i-10+55i}{25} = \frac{-33+69i}{25} = -\frac{33}{25} + \frac{69}{25}i.$$

Absolute Value of a Complex Number

We define the *absolute value* of $z = a + bi$ by

$$|z| = \sqrt{a^2 + b^2}.$$

For example

$$|i| = |0 + 1i| = \sqrt{0^2 + 1^2} = \sqrt{1} = 1 \text{ and } |1 + 3i| = \sqrt{1^2 + 3^2} = \sqrt{10}.$$

Note that

$$|\bar{z}| = |z|.$$

Moreover,

$$z\bar{z} = a^2 + b^2 = |z|^2.$$

5.3.3   Exercises

1. Show that

   (a) $i^{42} = i^{-14} = -1$.

   (b) $i^{-(4n+1)} = i^{4n+3} = -i$, where $n \in \mathbb{N}$.

   (c) $i^{4m+1} \div i^{4n-1} = -1$, where $m, n \in \mathbb{N}$.

2. Perform the operations.

(a) $-5i^3(-i^7)$

(b) $-5i^2 + i(2i - i^4)$

(c) $i^{-1} + i^{-2} + i^{-3} + i^{-4}$

(d) $i^n + i^{n+1} + i^{n+2} + i^{n+3}$.

3. Perform the operations. Give all answers in $a + bi$ form.

    (a) $(1 - 7i) + (3 + 2i)$

    (b) $(5 - 8i) - (2 - 4i)$

    (c) $-2i(-1 + i) - (-3 + 2i)$

    (d) $(3 + 2i)(3 - 2i)$.

4. Perform the operations. Give all answers in $a + bi$ form.

    (a) $(5 + 3i)(4 - 6i)$

    (b) $3i(1 - i)(5 + 7i) + i(8i)$

    (c) $(2 + 3i)^2$

    (d) $(1 + i)^3$.

5. Perform the operations. Give all answers in $a + bi$ form.

    (a) $\frac{2}{1+i}$

    (b) $\frac{5}{3-i}$

    (c) $\frac{3-i}{1+i}$

    (d) $\frac{34+2i}{2-4i}$

6. Simplify
$$\frac{(1+2i)^2 - (1-i)^3}{(3+2i)^2 - (2+i)^3}.$$

7. Let $z_1 = 2 + i$ and $z_2 = 1 - 2i$. Find
$$z = z_1 + z_2 + z_1 z_2 + \frac{z_1}{z_2} = -1 + 3i.$$

8. Let $z = a + bi$ and $w = c + di$. Show that

   (a) $\overline{z + w} = \overline{z} + \overline{w}$

   (b) $\overline{z - w} = \overline{z} - \overline{w}$

   (c) $\overline{zw} = \overline{z} \cdot \overline{w}$

   (d) $\overline{\frac{z}{w}} = \frac{\overline{z}}{\overline{w}}$

9. Find the absolute value of the numbers.
$$|-i|, \quad |1 - i|, \quad |1 + i\sqrt{3}|, \quad |3 - 2i|, \quad \left|\frac{1 + 2i}{1 - 2i}\right|.$$

10. Let $z = a + bi$ and $w = c + di$. Show that

    (a) $|z| \geq 0$ and $|z| = 0$ if and only if $z = 0$.

    (b) $|zw| = |z||w|$

    (c) $|z + w| \leq |z| + |w|$ (triangle inequality).

    (d) $\left|\frac{1}{z}\right| = \frac{1}{|z|}$

    (e) $\left|\frac{z}{w}\right| = \frac{|z|}{|w|}$.

11. Let $z = a + bi$ and $w = c + di$. Show that

    (a) $|z + \overline{w}| = |\overline{z} + w|$.

    (b) If $|z|^2 + |w|^2 = |w - z|^2$, then $ac - bd = 0$.

    (c) If $|2z - 1| = |z - 2|$, then $a^2 + b^2 = 1$.

5.4 Quadratic Equations

A *quadratic equation* or *second degree equation* is an equation of the form

$$ax^2 + bx + c = 0, \tag{5.1}$$

where a, b and c are given real numbers with $a \neq 0$, and x is the unknown. The numbers a, b and c are called the *coefficients* of the quadratic equation.

The quadratic equation $ax^2 + bx + c = 0$ ($a \neq 0$) is said *incomplete* if $b = 0$ or $c = 0$. An incomplete quadratic equation is an equation of the following form:

1. $ax^2 = 0$, when $b = c = 0$.

2. $ax^2 + bx = 0$, when $b \neq 0$ and $c = 0$.

3. $ax^2 + c = 0$, when $b = 0$ and $c \neq 0$.

Let us first solve the incomplete quadratic equations.

1. The quadratic equation

$$ax^2 = 0 \iff x = 0 \text{ (double solution)}.$$

2. Factoring x we have

$$ax^2 + bx = 0 \iff x(ax + b) = 0 \iff x = 0, \ x = -\frac{b}{a}.$$

3. The quadratic equation

$$ax^2 + c = 0 \iff ax^2 = -c \iff x^2 = -\frac{c}{a}.$$

Two cases are possible.

If $\frac{c}{a} < 0$, then $-\frac{c}{a} > 0$ and the equation $x^2 = -\frac{c}{a}$ has two real solutions

$$x_1 = \sqrt{-\tfrac{c}{a}}, \quad x_1 = -\sqrt{-\tfrac{c}{a}}.$$

This is frequently written in the form

$$x_{1,2} = \pm\sqrt{-\tfrac{c}{a}} \quad (-\tfrac{c}{a} > 0).$$

If $\frac{c}{a} > 0$, then $-\frac{c}{a} < 0$ and the equation $x^2 = -\frac{c}{a}$ has two complex solutions

$$x_1 = \sqrt{-\tfrac{c}{a}} = i\sqrt{\tfrac{c}{a}}, \quad x_1 = -\sqrt{-\tfrac{c}{a}} = -i\sqrt{\tfrac{c}{a}},$$

where $i = \sqrt{-1}$ is the imaginary unit. This is written in the form

$$x_{1,2} = \pm i\sqrt{\tfrac{c}{a}} \quad (\tfrac{c}{a} > 0).$$

Example 5.4.1. Solve the quadratic equation

$$6x^2 + 5x = 0.$$

Solution. Factoring x, we have

$$6x^2 + 5x = 0 \iff x(6x+5) = 0 \iff x = 0,\ 6x+5 = 0 \iff x = 0,\ x = -\frac{5}{6}.$$

Example 5.4.2. Solve the quadratic equation

$$2x^2 - 18 = 0.$$

Solution. We have

$$2x^2 - 18 = 0 \iff 2x^2 = 18 \iff x^2 = \frac{18}{2} \iff x^2 = 9 \iff x = \pm\sqrt{9} = \pm 3.$$

Altenativelly, we have $2x^2 - 18 = 0 \iff 2(x^2 - 9) = 0 \iff x^2 - 9 = 0$ $\iff (x+3)(x-3) = 0 \iff x+3 = 0,\ x-3 = 0 \iff x = -3,\ x = 3.$

Example 5.4.3. Solve the quadratic equation

$$2x^2 + 18 = 0.$$

Solution. We have $2x^2 + 18 = 0 \iff 2x^2 = -18 \iff x^2 = -\frac{18}{2} \iff$ $x^2 = -9 \iff x = \pm\sqrt{-9} = \pm i\sqrt{-9} = \pm 3i.$

Completing the Square

The quadratic equations $ax^2 + bx + c = 0$, where $a \neq 0$, $b \neq 0$, $c \neq 0$, are solved applying the method of **completing the square**. The method of completing the square in a quadratic polynomial is important and is used in several occasions in mathematics. Before we solve the general quadratic equation $ax^2 + bx + c = 0$, we clarify this method with the following example.

Example 5.4.4. Solve the quadratic equation

$$3x^2 + 2x - 5 = 0.$$

Solution. We transform the quadratic trinomial $3x^2 + 2x - 5$, as follows

$$3x^2 + 2x - 5 = 3 \left[x^2 + \frac{2}{3}x - \frac{5}{3} \right] =$$

$$3 \left[x^2 + 2 \cdot x \cdot \frac{1}{3} + \left(\frac{1}{3} \right)^2 - \left(\frac{1}{3} \right)^2 - \frac{5}{3} \right] = 3 \left[\left(x + \frac{1}{3} \right)^2 - \frac{16}{9} \right].$$

Thus we completed the square, that is, we wrote

$$3x^2 + 2x - 5 = 3 \left[\left(x + \frac{1}{3} \right)^2 - \frac{16}{9} \right].$$

Now we solve

$$3 \left[\left(x + \frac{1}{3} \right)^2 - \frac{16}{9} \right] = 0 \Leftrightarrow \left(x + \frac{1}{3} \right)^2 - \frac{16}{9} = 0 \Leftrightarrow \left(x + \frac{1}{3} \right)^2 = \frac{16}{9}.$$

Hence,

$$x + \frac{1}{3} = \pm \sqrt{\frac{16}{9}} = \pm \frac{4}{3}.$$

Consequently,

$$x_1 = -\frac{1}{3} + \frac{4}{3} = 1, \; x_2 = -\frac{1}{3} - \frac{4}{3} = -\frac{5}{3}.$$

The Quadratic Formula

Next we solve the general quadratic equation $ax^2 + bx + c = 0$.

Theorem 5.4.5. *The quadratic equation*

$$ax^2 + bx + c = 0,$$

with $a \neq 0$, $b \neq 0$, $c \neq 0$, has two solutions x_1, x_2 given by the quadratic formula

$$x_{1,2} = \frac{-b \pm \sqrt{b^2 - 4ac}}{2a}.$$

Proof. First, we complete the square in $ax^2 + bx + c$. We have,

$$ax^2 + bx + c = a\left[x^2 + \frac{b}{a}x + \frac{c}{a}\right] =$$

$$a\left[x^2 + 2 \cdot x \cdot \frac{b}{2a} + \left(\frac{b}{2a}\right)^2 - \left(\frac{b}{2a}\right)^2 - \frac{c}{a}\right] = a\left[\left(x + \frac{b}{2a}\right)^2 - \frac{b^2 - 4ac}{4a^2}\right].$$

Thus we completed the square, that is, we wrote

$$ax^2 + bx + c = a\left[\left(x + \frac{b}{2a}\right)^2 - \frac{b^2 - 4ac}{4a^2}\right] \qquad (5.2)$$

Now we solve

$$a\left[\left(x + \frac{b}{2a}\right)^2 - \frac{b^2 - 4ac}{4a^2}\right] = 0 \Leftrightarrow \left(x + \frac{b}{2a}\right)^2 - \frac{b^2 - 4ac}{4a^2} = 0 \Leftrightarrow$$

$$\left(x + \frac{b}{2a}\right)^2 = \frac{b^2 - 4ac}{4a^2}.$$

Hence,

$$x + \frac{b}{2a} = \pm\sqrt{\frac{b^2 - 4ac}{4a^2}} = \pm\frac{\sqrt{b^2 - 4ac}}{2a}.$$

Consequently,

$$x_1 = -\frac{b}{2a} + \frac{\sqrt{b^2 - 4ac}}{2a}, \quad x_2 = -\frac{b}{2a} - \frac{\sqrt{b^2 - 4ac}}{2a}.$$

$\square$

The expression

$$D = b^2 - 4ac$$

is called the **discriminant** of the quadratic equation. The quadratic formula now can be written as

$$x_{1,2} = \frac{-b \pm \sqrt{D}}{2a}.$$

The discriminat $D = b^2 - 4ac$ determines the type of solutions (roots) of the quadratic equation $ax^2 + bx + c = 0$. There are three possibilities:

1. If $D > 0$, then $x_1 \neq x_2$ are two real and distinct solutions.

2. If $D = 0$, then $x_1 = x_2 = -\frac{b}{2a}$ is one double solution.

3. If $D < 0$, then the equation has two complex conjugate solutions of the form $x_1 = \lambda + i\mu$ and $x_2 = \lambda - i\mu$, where $\lambda, \mu \in \mathbb{R}$.

Example 5.4.6. Solve the quadratic equation.

$$4x^2 + 7x - 2 = 0.$$

Solution. Here $a = 4$, $b = 7$ and $c = -2$. The discriminant is $D = b^2 - 4ac = (7)^2 - 4(4)(-2) = 49 + 32 = 81$. Now the quadratic formula gives

$$x_{1,2} = \frac{-b \pm \sqrt{D}}{2a} = \frac{-7 \pm \sqrt{81}}{8} = \frac{-7 \pm 9}{8}.$$

Hence $x_1 = \frac{-7+9}{8} = \frac{2}{8} = \frac{1}{4}$, $x_2 = \frac{-7-9}{8} = \frac{-16}{8} = -2$.

Example 5.4.7. Solve the quadratic equation.

$$3x^2 - 5x + 1 = 0.$$

Solution. Here $a = 3$, $b = -5$ and $c = 1$. The discriminant is $D = b^2 - 4ac = (-5)^2 - 4(3)(1) = 25 - 12 = 13$. Now the quadratic formula gives

$$x_{1,2} = \frac{-b \pm \sqrt{D}}{2a} = \frac{-(-5) \pm \sqrt{13}}{6} = \frac{5 \pm \sqrt{13}}{6}.$$

Hence $x_1 = \frac{5+\sqrt{13}}{6}$, $x_2 = \frac{5-\sqrt{13}}{6}$.

Example 5.4.8. Solve the quadratic equation.

$$x^2 - 6x + 13 = 0.$$

Solution. Here $a = 1$, $b = -6$ and $c = 13$. The discriminant is $D = b^2 - 4ac = (-6)^2 - 4(1)(13) = 36 - 52 = -16$. Now the quadratic formula gives

$$x_{1,2} = \frac{-b \pm \sqrt{D}}{2a} = \frac{-(-6) \pm \sqrt{-16}}{2} = \frac{6 \pm 4i}{2} = 3 \pm 2i.$$

Hence $x_1 = 3 + 2i$, $x_2 = 3 - 2i$.

Remark. Note that in the case $D > 0$, and $a, b, c \in \mathbb{Q}$ there are two possibilities:

1. If D is a perfect square, then x_1, x_2 are rational numbers[2].

2. If D is not a perfecr square, then x_1, x_2 are irrational (conjugate) numbers of the form $x_1 = p + q\sqrt{d}$ and $x_2 = p - q\sqrt{d}$, where $p, q, d \in \mathbb{Q}$.

Solving Equations by Factoring

The method of solving equations by factoring is based on the property of **zero products**, that is,

$$A \cdot B = 0 \quad \Rightarrow \quad A = 0 \ or \ B = 0.$$

Example 5.4.9. Solve the quadratic equation by factoring

$$x^2 - 3x - 15 = 0.$$

Solution. We have

$$x^2 - 3x - 15 = 0 \quad \Leftrightarrow \quad (x + 3)(x - 5) = 0.$$

The property of zero products implies $x + 3 = 0$, $\quad x - 5 = 0$. Hence $x = -3$ and $x = 5$.

---

[2]When D is a perfect square the quadratic trinomial can be factored by inspection by the methods we studied in Section 4.5. In this case the quadratic equation can be solved also by factoring.

Example 5.4.10. Solve the quadratic equation

$$8x^2 - 10x - 3 = 0.$$

Solution. Factoring the quadratic trinomial we have

$$8x^2 - 10x - 3 = 0 \iff (4x+1)(2x-3) = 0.$$

This implies $4x + 1 = 0$, $\quad 2x - 3 = 0$. Hence $x = -\frac{1}{4}$ and $x = \frac{3}{2}$.

Sum and Product of Roots

Theorem 5.4.11. *If x_1, x_2 are the roots of $ax^2 + bx + c = 0$, then*

$$x_1 + x_2 = -\frac{b}{a}, \quad x_1 x_2 = \frac{c}{a}.$$

Proof. Let $x_1 = \frac{-b+\sqrt{D}}{2a}$ and $x_2 = \frac{-b-\sqrt{D}}{2a}$ be the roots of the equation. Then we have

$$x_1 + x_2 = \frac{-b+\sqrt{D}-b-\sqrt{D}}{2a} = \frac{-2b}{2a} = -\frac{b}{a},$$

$$x_1 x_2 = \frac{(-b+\sqrt{D})(-b-\sqrt{D})}{4a^2} = \frac{(-b)^2 - D}{4a^2} = \frac{b^2 - (b^2 - 4ac)}{4a^2} = \frac{c}{a}.$$

$\square$

In particular, for the equation

$$x^2 + px + q = 0,$$

the discriminant is $D = p^2 - 4q$, and

$$x_1 + x_2 = -p \quad \text{and} \quad x_1 x_2 = q.$$

Example 5.4.12. Set up the quadratic equation having the roots $x_1 = 2$ and $x_2 = 3$.
Solution. We have

$$p = -(x_1 + x_2) = -(2 + 3) = -5 \quad \text{and} \quad q = x_1 x_2 = (2)(3) = 6.$$

Thus, we obtain the equation $x^2 - 5x + 6 = 0$.

Example 5.4.13. Set up the quadratic equation having the roots $x_1 = \frac{1}{2}$ and $x_2 = 4$.

Solution. We have

$$p = -(x_1 + x_2) = -(\tfrac{1}{2} + 4) = -\tfrac{9}{2} \quad \text{and} \quad q = x_1 x_2 = (\tfrac{1}{2})(4) = 2.$$

Thus, we obtain the equation $x^2 - \frac{9}{2}x + 2 = 0 \quad \Leftrightarrow \quad 2x^2 - 9x + 4 = 0$.

Example 5.4.14. If x_1, x_2 are the roots of the equation $3x^2 - 2x + 6 = 0$, find

$$x_1^2 + x_2^2, \quad \frac{1}{x_1} + \frac{1}{x_2}, \quad \frac{1}{x_1^2} + \frac{1}{x_2^2}, \quad \frac{1}{x_1^3} + \frac{1}{x_2^3}.$$

Solution. Since $x_1 + x_2 = \frac{2}{3}$, and $x_1 x_2 = \frac{6}{3} = 2$, we have

$$x_1^2 + x_2^2 = (x_1 + x_2)^2 - 2x_1 x_2 = \left(\frac{2}{3}\right)^2 - 2(2) = \frac{4}{9} - 4 = -\frac{32}{9}.$$

$$\frac{1}{x_1} + \frac{1}{x_2} = \frac{x_1 + x_2}{x_1 x_2} = \frac{\frac{2}{3}}{2} = \frac{2}{6} = \frac{1}{3}.$$

$$\frac{1}{x_1^2} + \frac{1}{x_2^2} = \frac{x_1^2 + x_2^2}{x_1^2 x_2^2} = \frac{-\frac{32}{9}}{2^2} = \frac{-\frac{32}{9}}{4} = -\frac{8}{9}.$$

$$\frac{1}{x_1^3} + \frac{1}{x_2^3} = \frac{x_1^3 + x_2^3}{x_1^3 x_2^3} = \frac{(x_1 + x_2)^3 - 3x_1 x_2(x_1 + x_2)}{x_1^3 x_2^3} = \frac{\left(\frac{2}{3}\right)^3 - 3(2)\left(\frac{2}{3}\right)}{2^3} = -\frac{25}{54}.$$

Example 5.4.15. For which values of $k \in \mathbb{R}$ the sum of the squares of the roots of the equation

$$(2k - 1)x^2 - (2k - 2)x + 3k = 0$$

is equal to 4?

Solution. The coefficients of the quadratic equation are $a = 2k - 1$, $b = 2k - 2$ and $c = 3k$, with $k \neq \frac{1}{2}$.. Let x_1, x_2 be the roots of the equation. Then $x_1 + x_2 = -\frac{2k-2}{2k-1}$ and $x_1 x_2 = \frac{3k}{2k-1}$. We want to find k so that $x_1^2 + x_2^2 = 4$. Since $x_1^2 + x_2^2 = (x_1 + x_2)^2 - 2x_1 x_2$, we have

$$\left(-\frac{2k-2}{2k-1} \right)^2 - 2 \cdot \frac{3k}{2k-1} = 4 \Leftrightarrow \frac{(2k-2)^2}{(2k-1)^2} - \frac{6k}{2k-1} = 4 \Leftrightarrow$$

$$(2k-2)^2 - 6k(2k-1) - 4(2k-1)^2 = 0 \Leftrightarrow -24k^2 + 14k = 0 \Leftrightarrow$$

$$-2k(12k - 7) = 0 \Leftrightarrow k_1 = 0, k_2 = \frac{7}{12}.$$

Remark. If the quadratic equation $ax^2 + bx + c = 0$ has real roots x_1, x_2, then the formulas $x_1 + x_2 = -\frac{b}{a}$ and $x_1 x_2 = \frac{c}{a}$ make it possible to determine the signs of the roots by the sign of the coefficients of the equation. For instance,

1. If $x_1 x_2 = \frac{c}{a} > 0$, then the roots have the same sign. In this case, if $x_1 + x_2 = -\frac{b}{a} > 0$, then both roots are positive, and if $x_1 + x_2 = -\frac{b}{a} < 0$, then both roots are negative.

2. If $x_1 x_2 = \frac{c}{a} < 0$, then the roots are of opposite signs. In this case, if $x_1 + x_2 = -\frac{b}{a} > 0$, then the greater in absolute value root is the positive, and if $x_1 + x_2 = -\frac{b}{a} < 0$, then the greater in absolute value root is the negative.

Factorization of a Quadratic Trinomial

Theorem 5.4.16. *Let $f(x) = ax^2 + bx + c$ be a quadratic trinomial. If x_1 and x_2 are the roots of the quadratic equation $ax^2 + bx + c = 0$, then*

$$f(x) = ax^2 + bx + c = a(x - x_1)(x - x_2).$$

Proof. Since $x_1 + x_2 = -\frac{b}{a}$ and $x_1 x_2 = \frac{c}{a}$, we have

$$f(x) = ax^2 + bx + c = a\left(x^2 + \frac{b}{a}x + \frac{c}{a} \right) = a[x^2 - (x_1 + x_2)x + x_1 x_2] =$$

$$a[x^2 - x_1 x - x_2 x + x_1 x_2] = a[x(x - x_1) - x_2(x - x_1)] = a(x - x_1)(x - x_2).$$

$\square$

Example 5.4.17. Factor the quadratic trinomials of Examples 5.4.6., 5.4.7 and 5.4.8

$$\textbf{(1)}\ \ 4x^2 + 7x - 2 \quad \textbf{(2)}\ \ 3x^2 - 5x + 1 \quad \textbf{(3)}\ \ x^2 - 6x + 13.$$

Solution. 1). The roots of $4x^2 + 7x - 2 = 0$ are $x_1 = \frac{1}{4}, \ \ x_2 = -2$ Hence,

$$4x^2 + 7x - 2 = 4(x - \frac{1}{4})(x + 2) = (4x - 1)(x + 2).$$

2). The roots of $3x^2 - 5x + 1 = 0$ are $x_1 = \frac{5+\sqrt{13}}{6}, \ \ x_2 = \frac{5-\sqrt{13}}{6}$. Hence,

$$3x^2 - 5x + 1 = 3\left(x - \frac{5 + \sqrt{13}}{6}\right)\left(x - \frac{5 - \sqrt{13}}{6}\right).$$

3). The roots of $x^2 - 6x + 13 = 0$ are $x_1 = 3 + 2i, \ \ x_2 = 3 - 2i$. Hence,

$$x^2 - 6x + 13 = [x - (3 + 2i)][x - (3 - 2i)] = (x - 3 - 2i)(x - 3 + 2i).$$

5.4.1   Exercises

1. Solve the equations.

   (a) $x^2 - 3x = 0.$

   (b) $4x^2 + 6x = 0.$

   (c) $2x^2 - 10x = 0.$

   (d) $3x^2 - 48 = 0.$

   (e) $2x^2 - 72 = 0.$

2. Solve the equations.

   (a) $x^2 - 6x + 5 = 0.$

    (b) $x^2 - 9x + 14 = 0.$

    (c) $3x^2 + 4x - 4 = 0.$

    (d) $4x^2 - 4x + 1 = 0.$

    (e) $2x^2 + 2x + 5 = 0.$

    (f) $9x^2 - 6x + 4 = 0.$

3. Solve the equations.

    (a) $5x^2 + \sqrt{3}x - 1 = 0.$

    (b) $x^2 - 3x - 1 + \sqrt{3} = 0.$

    (c) $(4x - 1)^2 + 3(4x - 1) = 0.$

    (d) $(x + 1)^2 - (x - 1)(x + 2) = -2x(x - 3).$

    (e) $(3x + 2)(5x - 1) + (3x + 7)(1 - 5x) = (1 - 5x)(2 + 15x).$

4. Set up an equation having the roots:

    (a) $-\frac{3}{4}$ and $-\frac{1}{2}$.

    (b) $5 + \sqrt{3}$ and $5 - \sqrt{3}$.

    (c) $\frac{2}{2} + \frac{3}{2}i$ and $\frac{2}{2} - \frac{3}{2}i$.

5. For what value of k are the roots of the equation $kx^2 + 3x + 2 = 0$ equal?

6. For what values of k the equation

$$x^2 - 2(k + 2)x + 2k^2 - 17 = 0$$

has a double root?

7. Let

$$x^2 - (k + 1)x + 21k = 0.$$

Find the value of k for which the equation has: (1) opposite roots. (2) reciprocal roots.

8. For what value of k is one root of the equation

$$x^2 - (2k - 1)x + k^2 + 2 = 0$$

twice the value of the other root?

5.5   Applications. Word Problems

Example 5.5.1. Find two consecutive positive integers the sum of whose squares is 481.

   Solution. Let $x, x + 1$ be the integers with $x > 0$. Then we have

$$x^2 + (x+1)^2 = 481 \Leftrightarrow 2x^2 + 2x - 480 = 0 \Leftrightarrow (x+16)(x-15) = 0$$

The solutions are, $x = -16$ which is negative and we reject it, and $x = 15$. Thus, the integres are 15, 16.

Example 5.5.2. The length of a rectangle exceeds its width by 7 feet. If its area is 60 square feet, find its dimensions.

   Solution. Let $x > 0$ represent the width of the rectangle. Then its length is $x + 7$. The area A of a rectangle is $A = (x+7)x$. Since $A = 60$, we get

$$60 = (x+7)x \Leftrightarrow x^2 + 7x - 60 = 0 \Leftrightarrow (x-5)(x+12) = 0.$$

The solutions are $x = 5$ and $x = -12$. The solution $x = -12$ is rejected. Thus, the width is 5 and the length is 12.

Example 5.5.3. A train was delayed by a signal for 16 minutes and made up for the delay on a section of 80 miles travelling with the speed 10 miles per hour higher than that which accorded the schedule. Find the speed of the train which accorded the schedule.

   Solution. Recall that that the distance travelled by an object moving at a constant speed for a certain time is given by the formula $d = v \cdot t$. Solving for the time, we get $t = \frac{d}{v}$.

   Let v miles per hour be the speed of the train before it stops. Then $v + 10$ miles per hour is its speed for the next 80 miles. According to the time-table, the train should have covered this distance in $\frac{80}{v}$ hours but it actually covered it in $\frac{80}{v+10}$ hours. Since the delay is 16 minutes or $\frac{16}{60}$ of the hour, it follows that

$$\frac{80}{v} = \frac{16}{60} + \frac{80}{v+10}.$$

Multiplying both sides by $v(v + 10)$ to eliminate the denominators, and simplying the resulting equation, we obtain the quadratice equation

$$v^2 + 10v - 3000 = 0 \iff (v - 50)(v + 60) = 0 \iff v = 50, \ v = -60.$$

We reject $v = -60$ because it is negative. Hence $v = 50$ miles per hour is the required speed of the train.

Example 5.5.4. A swimming pool is fitted with three pipes. The first two pipes, operating simultaneously, fill the pool within the same time during which the pool is filled by the third pipe alone. Operating alone, the second pipe fills the pool 5 hours faster than the first pipe and 4 hours slower than the third pipe. In what time will the pool be filled by each pipe operating separately?

Solution. Let x hours be the time in which the second pipe fills the pool. Then $(x + 5)$ hours is the time during which the pool is filled by the first pipe, and $(x - 4)$ hours by the third pipe. If V is the volume of the pool, then the capacity of each of the pipes is respectively equal to $\frac{V}{x+5}$, $\frac{V}{x}$ and $\frac{V}{x-4}$.

By the hypothesis of the problem, we get

$$\frac{V}{x + 5} + \frac{V}{x} = \frac{V}{x - 4}.$$

Multiplying both sides by $\frac{1}{V}[x(x + 5)(x - 4)]$, and simplifying we obatin the quadratic equation

$$x^2 - 8x - 20 = 0.$$

Solving it, we find $x = 10$, $x = -2$. Rejecting the negative solution, we get $x = 10$, and so the required times are 15 hours, 10 hours, and 6 hours.

5.6   Review Exercises

1. Solve the equations

    (a) $4x - 6 = x - 4$

    (b) $0.1x + 0.4 = 0.9x$

    (c) $2(x + 4) - (x + 6) = 12 - x.$

    (d) $\frac{x}{2} - x = 2x + \frac{3}{2}.$

    (e) $2x + \frac{x}{6} + \frac{5}{3} = 2(1 + x).$

2. Solve the equations

    (a) $2x^2 - 5x + 3 = 0.$

    (b) $3x^2 + 2x - 5 = 0.$

    (c) $x^2 + x - 1 = 0.$

    (d) $3x^2 + 4x + 2 = 0.$

3. Solve the equations

    (a) $x^2 - 7|x| + 12 = 0.$

    (b) $x^2 + 2|x| - 35 = 0.$

    (c) $(x - 1)^2 + 4|x - 1| - 5 = 0.$

    (d) $x^2 + |x - 1| = 1.$

4. Set up a quadratic equation having roots the numbers

    (a) 3 and -7

    (b) 1 and $\frac{1}{2}.$

    (c) $\frac{1}{10 + 6\sqrt{2}}$ and $\frac{1}{10 - \sqrt{72}}.$

    (d) $2 + 3i$ and $2 - 3i.$

5. Show that the equation

$$x^2 - (5 - \sqrt{2})x + 6 - 3\sqrt{2} = 0$$

has the solutions 3 and $2 - \sqrt{2}$.

6. Find two numbers whose sum is 2 and their product is -15.

7. Find two numbers whose sum is -5 and their product is -36.

8. Find the values of k for which the equation

$$(k - 1)x^2 + (k + 4)x + k + 7 = 0$$

has equal roots.

9. Show that the equation $x^2 + 2kx - 8 = 0$ has real roots for all $k \in \mathbb{R}$.

10. If one root of the equation

$$x^2 + 2kx - 8 = 0$$

is equal to the square of the other root, then find the roots and the value of k.

11. Without solving the equation

$$x^2 - 3x - 10 = 0,$$

compute the sum of the cubes $x_1^3 + x_2^3$ of its roots x_1, x_2.

12. The roots x_1 and x_2 of the euation

$$x^2 + kx + 12 = 0$$

satisfy $x_2 - x_1 = 1$. Find k.

13. Find the values of λ for which the sum of the roots of the equation

$$x^2 - 2\lambda(x - 1) - 1 = 0$$

is equal to the sum of the squares of its roots.

14. Find $\frac{1}{x_1^3} + \frac{1}{x_2^3}$, where x_1 and x_2 are the roots of the equation

$$2x^2 - 3ax - 2 = 0.$$

15. Find the coefficients of the equation

$$x^2 + bx + c = 0$$

such that its roots are equal to b and c.

16. A rectangle is 4 feet longer than it is wide. If its area is 32 square feet, find its dimensions.

17. The hypotenuse of a right triangle is 26 feet. If one of its perpendicular sides is 14 feet shorter than the other side, find the lenght of the two sides.

18. Find the dimensions of a rectangle whose diagonal is 20 feet and its area is 192 feet.

19. The base of a triangle is one-third as long as its height. Given that its area is 24 square feet, find its base.

20. Find a number of two digits for which the product of the digits is 48, and if the digits be interchanged the number is diminished by 18.

21. A student sold a book for $48, and his gain percent was equal one half the cost of the book in dollars. What was the cost of the book?

22. A train had to cover a distance of 54 miles. Having covered 14 miles, it was delayed at a signal for 10 minutes and then continued on its way with a new speed that was 10 miles per hour faster than the initial speed. The train arrived at it place of destination 2 minutes late. Find the initial speed of the train.

23. A cyclist rides a distance of 40 miles. his return trip takes 2 hours longer, because his speed decreased by 10 miles per hour. How fast does her rides each way?

24. A cyclist covers the distance from city A to city B during 3 hours. To cover the distance from city A to city C for the same time, he must ride each kilometer one minute faster since the distance from A to C is 30 km longer than that from A to B. Find the distance from A to B.

25. A man travels 50 miles by the train A and then after a wait of 5 minutes returns by the train B, which runs 5 miles per hour faster than the train A. The entire journey took $2\frac{4}{9}$ hours. What are the speeds of the two trains?

26. A man walked 12 miles with a certain speed and then 6 miles farther with a speed $\frac{1}{2}$ miles per hour greater. Had he walked the entire distance with a greater speed, his time would have been 20 minutes less. How long did it take him to walk the 18 miles?

27. By increasing his speed by 25 miles per hour a taxi-driver decreased the time on a 25 miles trip by 10 minutes. What was his speed?

28. If two pipes operate simultaneously, the tank will be filled in 12 hours. Through one pipe the tank is filled up 10 hours faster than through the pother. How many hours does it take the second pipe to fill the tank?

29. A bus company has $3,000$ passangers daily, paying a 25 cents fare. For each nickel increase in fare, the company projects that it will lose 8 passangers. What fare increase will produce \$994 in daily revenue?

30. (**Ballistics**). The height of a projectile fired upward withan initial velocity of 400 feet per second is given by the formula $h = 400t - 16t^2$, where h is the height in feet and t the time in seconds. Find the required time for the projectile to return to earth.

31. (**Ballistics**). If from a certain height of a feet an object be thrown *vertically upward* with an initial velocity of b feet per second, its height h at the end of t seconds is given by the formula

$$h = a + bt - 16t^2.$$

The corresponding formula when the object is thrown *vertically downward* is

$$h = a - bt - 16t^2.$$

(a) if an object be thrown vertically upward from the ground with initial velocity of 32 feet per second, when will it be at a height of 7 feet? When will it be at a height ofof 16 feet? Will it ever reach a height of 17 feet?

(b) An object is thrown from a height of 64 feet vertically downward with an initial velocity of 48 feet per second. When will it reach the height of 36 feet? When will it reach the ground?

(c) If an object be dropped from a height of 36 feet, when will it reach the ground?

Chapter 6

Equations II

In this chapter, we study polynomial equations, rational and radical equations. We also study the Fundamental Theorem of Algebra and some of its immediate consequences.

6.1   Polynomial Equations

Equations involving polynomial expressions, are called **polynomial equations**. Thus we have: first-degree, second-degree, third-degree, fourth-degree equations and so on. We have shown earlier how to solve first-degree equations and quadratic equations and derived formulas for finding their solutions. Concerning polynomial equations whose degree is higher that two, there are formulas for third-degree and fourth-degree, but they are very cumbersome and are seldom used, and there are no formulas of this kind for fifth-degree equations[1]. The existence of solutions of polynomial equations is quaranteed by the **Fundamental Theorem of Algebra**, which states: *Any polynomial of degree $n \geq 1$ has at least one complex root* (see, Section 6.4 for more details).

---

[1]For polynomial equations of degree $n = 5$ this was proved by the Norwegian mathematician Neils Abel (1802-1829), and for equations of degree $n > 5$, by the French mathematician Everiste Galois (1811-1832).

Here we solve several special cases of higher-degree polynomial equations. The main techniques we use are either by making a substitution or/and by factoring the polynomial and thus reducing the problem to solve linear and/or quadratic equations.

6.1.1   Biquadratic Equations

An equation of the form

$$ax^4 + bx^2 + c = 0$$

is called a ***biquadratic equation***.

By the substitution $y = x^2$, the biquadratic equation is reduced to the (auxiliary) quadratic equation $ay^2 + by + c = 0$, which has two roots y_1, y_2. Hence the biquadratic equation has four roots

$$x_{1,2} = \pm\sqrt{y_1}, \quad x_{3,4} = \pm\sqrt{y_2}.$$

Example 6.1.1. Solve the equations

$$x^4 - 10x^2 + 9 = 0.$$

Solution. We set $y = x^2$. Then the equation becomes

$$y^2 - 10y + 9 = 0.$$

The solutions of this equation are $y_1 = 9$ and $y_2 = 1$. Solving the equation $x^2 = 9$ we obtain $x_{1,2} = \pm\sqrt{9} = \pm 3$. Solving the equation $x^2 = 1$, we obtain $x_{3,4} = \pm\sqrt{1} = \pm 1$.

Example 6.1.2. Solve the equations

$$9x^4 - 5x^2 - 4 = 0.$$

Solution. Setting $y = x^2$, we get

$$9y^2 - 5y - 4 = 0.$$

The solutions of this equation are $y_1 = 1$ and $y_2 = -\frac{4}{9}$. Solving the equation $x^2 = 1$ we obtain $x_{1,2} = \pm\sqrt{1} = \pm 1$. Solving the equation $x^2 = -\frac{4}{9}$, we obtain $x_{3,4} = \pm\sqrt{-\frac{4}{9}} = \pm\frac{2}{3}i$.

6.1.2   Equations Solved by Factoring

Example 6.1.3. Solve the equation

$$(x^2 - 9)^2 - (x + 5)(x - 3)^2 = 0.$$

Solution. Factoring the polynomial (see Example 4.5.39), we have

$$(x^2 - 9)^2 - (x + 5)(x - 3)^2 = (x - 3)^2(x + 4)(x + 1).$$

Hence, we solve

$$(x - 3)^2(x + 4)(x + 1) = 0.$$

This implies $(x - 3)^2 = 0$, $x + 4 = 0$, and $x + 1 = 0$. The solutions are $x = 3$ (double solution), $x = -4$, and $x = -1$.

Example 6.1.4. Solve the equation

$$x^3 + 2x^2 - 3 = 0.$$

Solution. Factoring the polynomial (see Example 4.5.42), we have

$$x^3 + 2x^2 - 3 = (x - 1)(x^2 + 3x + 3).$$

Hence, we solve the set of equations $x - 1 = 0$, and $x^2 + 3x + 3 = 0$. Thus, we obtain the solutions $x = 1$, $x = -\frac{3}{2} + \frac{\sqrt{3}}{2}i$ and $x = -\frac{3}{2} - \frac{\sqrt{3}}{2}i$.

Example 6.1.5. Solve the equation

$$x^4 + x^3 - 3x^2 - 5x - 2 = 0.$$

Solution. Factoring the polynomial (see Example 4.5.43), we have

$$x^4 + x^3 - 3x^2 - 5x - 2 = (x + 1)^3(x - 2).$$

Hence, we solve the set of equations $(x + 1)^3 = 0$, and $x - 2 = 0$. Thus, we obtain the solutions $x = -1$ (triple solution) and $x = 2$.

Example 6.1.6. Solve the equation

$$x^7 - x^5 - x^3 + x = 0.$$

Solution. Factoring the polynomial (see Example 4.5.37), we have

$$x^7 - x^5 - x^3 + x = x(x+1)^2(x-1)^2(x^2+1).$$

Hence, we solve the set of equations $x = 0$, $(x+1)^2 = 0$, $(x-1)^2 = 0$ and $x^2 + 1 = 0$. Thus, we obtain the solutions $x = 0$, $x = -1$ (double solution) $x = 1$ (double solution) and $x = \pm i$.

6.1.3   Symmetric Third-Degree Equations

A third-degree equation is called **symmetric** [2] if it is of the following form:

$$ax^3 + bx^2 + bx + a = 0 \quad (a \neq 0) \tag{6.1}$$

or

$$ax^3 + bx^2 - bx - a = 0 \quad (a \neq 0). \tag{6.2}$$

To solve (6.1) we factor the polynomial $ax^3 + bx^2 + bx + a$.
We have

$$ax^3 + bx^2 + bx + a = ax^3 + a + bx^2 + bx = a(x^3+1) + bx(x+1) =$$

$$a(x+1)(x^2-x+1) + bx(x+1) = (x+1)[a(x^2-x+1) + bx] =$$

$$(x+1)[ax^2 + (b-a)x + a].$$

Now (6.1) is equivalent to the equation

$$(x+1)[ax^2 + (b-a)x + a] = 0.$$

This equation is equivalent to the set of equations

---

[2]It can be proved that an equation is symmetric if and only if whenever the equation has root the number $r \neq \pm 1$ it also has root its reciprocal $\frac{1}{r}$ $(r \neq 0)$. For this reason symmetric equations are also called *reciprocal equations*.

$$x + 1 = 0 \quad \text{and} \quad ax^2 + (b - a)x + a = 0,$$

whose solutions can be easily found since they are a first-degree equation and a quadratic equation.

To solve (6.2), we work similarly, and we obtain

$$ax^3 + bx^2 - bx - a = 0 \iff (x - 1)[ax^2 + (b + a)x + a] = 0.$$

Hence

$$x - 1 = 0 \quad \text{and} \quad ax^2 + (b + a)x + a = 0,$$

Example 6.1.7. Solve the equation

$$x^3 + 5x^2 + 5x + 1 = 0.$$

Solution. We factor $x^3 + 5x^2 + 5x + 1$. We have

$$x^3 + 5x^2 + 5x + 1 = x^3 + 1 + 5x(x+1) = (x+1)(x^2 - x + 1) + 5x(x+1) = (x+1)(x^2 + 4x + 1).$$

So now, we must solve $(x + 1)(x^2 + 4x + 1) = 0$. This is equivalent to the equations

$$x + 1 = 0 \quad \text{and} \quad x^2 + 4x + 1 = 0,$$

The first equation has solution $x_1 = -1$. The second equation has two solutions $x_{2,3} = \frac{-4 \pm 2\sqrt{3}}{2} = -2 \pm \sqrt{3}$. Consequently, the original equation has three solutions

$$x_1 = -1, \quad x_2 = -2 + \sqrt{3} \quad \text{and} \quad x_3 = -2 - \sqrt{3}$$

Note that $x_3 = \frac{1}{x_2}$.

Example 6.1.8. Solve the equation

$$2x^3 + 3x^2 - 3x - 2 = 0.$$

Solution. We factor $2x^3 + 3x^2 - 3x - 2$. We have

$$2x^3 + 3x^2 - 3x - 2 = 2(x^3 - 1) + 3x(x - 1) = 2(x-1)(x^2 + x + 1) + 3x(x-1) =$$
$$(x - 1)[2(x^2 + x + 1) + 3x] = (x - 1)(2x^2 + 5x + 2).$$

So now, we solve $(x - 1)(2x^2 + 5x + 2) = 0$. This is equivalent to the equations

$$x - 1 = 0 \quad \text{and} \quad 2x^2 + 5x + 2 = 0,$$

The first equation has solution $x_1 = 1$. The second equation has two solutions $x_{2,3} = \frac{-5 \pm \sqrt{9}}{4} = \frac{-5 \pm 3}{4}$. Consequently, the original equation has three solutions

$$x_1 = 1, \quad x_2 = -2 \quad \text{and} \quad x_3 = -\tfrac{1}{2}$$

Note that $x_3 = \frac{1}{x_2}$.

6.1.4 Symmetric 4th and 5th Degree Equations

The 4th degree ***symmetric equation*** is an equation of the form

$$ax^4 + bx^3 - bx - a = 0 \quad (a \neq 0).$$

This equation is solved by factoring. In fact, we have

$$ax^4 + bx^3 - bx - a = 0 \; \Leftrightarrow \; a(x^4 - 1) + bx(x^2 - 1) = 0 \; \Leftrightarrow$$

$$a(x^2 - 1)(x^2 + 1) + bx(x^2 - 1) = 0 \; \Leftrightarrow \; (x^2 - 1)[ax^2 + bx + a] = 0.$$

The latter equation is equivalent to the equations

$$x^2 - 1 = 0 \quad \text{and} \quad ax^2 + bx + a = 0.$$

From which we obtain $x_{1,2} = \pm 1$ and two other solutions x_3, x_4 from the (symmetric) quadratic equation $ax^2 + bx + a = 0$.

The general 4th−**degree symmetric equation** is of the following form:

$$ax^4 + bx^3 + cx^2 + bx + a = 0 \quad (a \neq 0). \qquad (6.3)$$

To solve (6.3), first note that since $a \neq 0$ the equation has nonzero solutions, ie, we may assume that $x \neq 0$.

Dividing both sides of (6.3) by x^2, we obtain

$$ax^2 + bx + c + \frac{b}{x} + \frac{a}{x^2} = 0.$$

Regrouping the terms of this equation we have

$$a\left(x^2 + \frac{1}{x^2}\right) + b\left(x + \frac{1}{x}\right) + c = 0.$$

Now we set $w = x + \frac{1}{x}$. Then $x^2 + \frac{1}{x^2} = \left(x + \frac{1}{x}\right)^2 - 2x \cdot \frac{1}{x} = w^2 - 2$. Hence the equation becomes

$$a(w^2 - 2) + bw + c = 0 \Leftrightarrow aw^2 + bw + c - 2a = 0.$$

This latter quadratic equation has in general two solutions w_1, w_2.

Coming back to the substitution $w = x + \frac{1}{x} \Leftrightarrow x^2 - wx + 1 = 0$ with $w = w_1$ and $w = w_2$, we obtain two quadratic equations

$$x^2 - w_1 x + 1 = 0 \quad \text{and} \quad x^2 - w_2 x + 1 = 0$$

Finally, solving these two quadratic equations we obtain, in general, four solutions x_1, x_2, x_3, x_4 for the symmetric equation (6.3).

Example 6.1.9. Solve the equation

$$6x^4 - 35x^3 + 62x^2 - 35x + 6 = 0.$$

Solution. Dividing both sides of the equation by x^2 we get

$$6x^2 - 35x + 62 - \frac{35}{x} + \frac{6}{x^2} = 0 \Leftrightarrow$$

$$6\left(x^2 + \frac{1}{x^2}\right) - 35\left(x + \frac{1}{x}\right) + 62 = 0.$$

Using the substitution $w = x + \frac{1}{x}$, we have $x^2 + \frac{1}{x^2} = w^2 - 2$, and the equation becomes

$$6w^2 - 35w + 50 = 0.$$

Solving this quadratic equation we find $w_1 = \frac{10}{3}$ and $w_2 = \frac{5}{2}$. Hence we obtain the equations:

$$x^2 - w_1 x + 1 = 0 \Leftrightarrow 3x^2 - 10x + 3 = 0,$$

$$x^2 - w_2 x + 1 = 0 \Leftrightarrow 2x^2 - 5x + 2 = 0.$$

The first equation has two solutions $x_1 = 3$, $x_2 = \frac{1}{3}$. The second equation has also two solutions $x_3 = 2$, $x_4 = \frac{1}{2}$. Thus, the original equations has the four solutions $x_1 = 3$, $x_2 = \frac{1}{3}$, $x_3 = 2$ and $x_4 = \frac{1}{2}$.

The 5th—**degree symmetric equation** of the form

$$ax^5 + bx^4 + cx^3 + cx^2 + bx + a = 0 \quad (a \neq 0). \qquad (6.4)$$

is solved as follows:

$$ax^5 + bx^4 + cx^3 + cx^2 + bx + a = 0 \iff a(x^5 + 1) = bx(x^3 + 1) = cx^2(x+1) = 0 \iff$$

$$(x+1)\left[a(x^4 - x^3 + x^2 - x + 1) + bx(x^2 - x + 1) = cx^2\right] = 0.$$

Hence we obtain the equations

$$x + 1 = 0 \iff x = -1$$

and

$$ax^4 + (b - a)x^3 + (a - b + c)x^2 + (b - a)x + a = 0,$$

which is a 4th—degree symmetric equation and is solved as above.

Similarly, for the symmetric equation

$$ax^5 + bx^4 + cx^3 - cx^2 - bx - a = 0 \quad (a \neq 0) \qquad (6.5)$$

we obtain the equations,

$$x - 1 = 0 \iff x = 1$$

and

$$ax^4 + (a + b)x^3 + (a + b + c)x^2 + (a + b)x + a = 0,$$

which is again a 4th—degree symmetric equation.

Remark. In general, the symmetric equations of degree higher than the 5th, can not solved by reducing them to quadratic equations.

6.1.5   Exercises

A. Solve the equations.

1. $x^3 + 4x^2 - 21x = 0.$

2. $8x^3 - 10x^2 + 3x = 0.$

3. $x^3 - 2x^2 - x + 2 = 0.$

4. $x^3 - 3x + 2 = 0.$

5. $x^3 - 7x + 6 = 0$

6. $2x^3 - 3x^2 + 1 = 0.$

7. $x^4 - 13x^2 + 36 = 0.$

8. $4x^4 - 17x^2 + 18 = 0.$

9. $x^4 + 12x^2 - 64 = 0.$

10. $2x^4 - 9x^2 + 4 = 0.$

11. $(1+x)^3 = (1-x)^3.$

12. $(x^2 - 4)(x^2 - 9) = 7x^2.$

B. Solve the symmetric equations.

1. $3x^3 + 13x^2 + 13x + 3 = 0.$

2. $2x^3 - 3x^2 - 3x + 2 = 0.$

3. $x^4 - 6x^3 + 6x - 1 = 0.$

4. $x^4 + x^3 - 4x^2 + x + 1 = 0.$

5. $x^4 - 2x^3 + 2x^2 - 2x + 1 = 0.$

6. $2x^5 - 3x^4 - 5x^3 + 5x^2 + 3x - 2 = 0.$

C. Use the factor theorem to factor and solve the equations.

1. $x^3 + 2x^2 - 5x - 6 = 0.$

2. $2x^3 + 3x^2 - 8x + 3 = 0.$

3. $3x^3 + 4x^2 - 7x + 2 = 0.$

4. $x^4 - x^3 - 5x^2 - 7x + 12 = 0$.

5. $x^4 - 2x^3 + x^2 + 2x - 2 = 0$.

6. $4x^4 - 8x^3 - x^2 + 8x - 3 = 0$.

6.1.6   Binomial Equations

For $a > 0$, an equation of the form

$$x^n - a = 0 \quad \text{or} \quad x^n + a = 0$$

is called a ***binomial equation***.

We first solve the binomial equation in the special case when $a = 1$.

   1) The equation

$$x^n - 1 = 0. \tag{6.6}$$

For $n = 2$, this is the quadratic equation $x^2 - 1 = 0$, which has the solutions $x_1 = 1$ and $x_2 = -1$.

   When $n \geq 3$ the equation $x^n - 1 = 0$ has only one real solution $x = 1$ for *odd* n, and only two real solutions $x_1 = 1$ and $x_2 = -1$ for *even* n. The remaining solutions in both cases are complex numbers.

   Let us consider these two cases in more detail:

Case 1. Let n be even. Say $n = 2k$, where $k \in \mathbb{N}$. Then factoring we have

$$x^{2k} - 1 = (x^2)^k - 1 = (x^2 - 1)(x^{2(k-1)} + x^{2(k-2)} + ... + x^4 + x^2 + 1),$$

and the equation $x^{2k} - 1 = 0$ is equivalent to the set of equations

$$x^2 - 1 = 0, \quad x^{2(k-1)} + x^{2(k-2)} + ... + x^4 + x^2 + 1 = 0.$$

The first has solutions $x_1 = 1$ and $x_2 = -1$. The second equation has no real solutions, because clearly $x^{2(k-1)} + x^{2(k-2)} + ... + x^4 + x^2 + 1 > 0$ for all $x \in \mathbb{R}$.

   Case 2. Let $n = 2k + 1$ be odd. Factoring $x^{2k+1} - 1$, we have

$$x^{2k+1} - 1 = 0 \iff (x - 1)(x^{2k} + x^{2k-1} + ... + x^3 + x^2 + x + 1) = 0.$$

Hence we get the set of equations

$$x - 1 = 0, \quad x^{2k} + x^{2k-1} + \ldots + x^2 + x + 1 = 0.$$

The first equation gives $x = 1$. The second equation has no real solutions, because one can show[3] that $x^{2k} + x^{2k-1} + \ldots + x^2 + x + 1 > 0$ for all $x \in \mathbb{R}$.

Now, let $a > 0$ with $a \neq 1$, and consider the equation

$$x^n - a = 0.$$

We have

$$x^n - a = 0 \iff x^n - (\sqrt[n]{a})^n = 0 \iff \left(\frac{x}{\sqrt[n]{a}}\right)^n - 1 = 0.$$

Setting $y = \frac{x}{\sqrt[n]{a}}$, we obtain the equation $y^n - 1 = 0$. For even n, the equation $y^n - 1 = 0$ has only two real solutions $y_1 = 1$ and $y_2 = -1$, from which we get

$$x_1 = \sqrt[n]{a} \quad \text{and} \quad x_1 = -\sqrt[n]{a}.$$

For odd n, the equation $y^n - 1 = 0$ has only one real solution $y = 1$, from which we get $x = \sqrt[n]{a}$.

2) The equation

$$x^n + 1 = 0.$$

For any even $n = 2k$, the equation $x^{2k} + 1 = 0$ has no real solutions (since, $x^{2k} + 1 > 0$ for all $x \in \mathbb{R}$). For any odd $n = 2k + 1$, it can be shown as above that the equation $x^{2k+1} + 1 = 0$ has only one real solution $x = -1$.

Now, let $a > 0$ with $a \neq 1$, and consider the equation

$$x^n + a = 0.$$

For any even $n = 2k$, the equation $x^{2k} = -a$ has no real solutions (since, $x^{2k} \geq 0$ for all $x \in \mathbb{R}$). For any odd $n = 2k + 1$, the

---

[3]Consider the cases $x > 0$, $-1 < x < 0$ and $x < -1$ separetely.

equation $x^{2k+1} + a = 0 \Leftrightarrow x^{2k+1} = -a$ has only one real solution $x = \sqrt[2k+1]{-a} = -\sqrt[2k+1]{a}$.

Let us consider some examples.

Example 6.1.10. Solve the equation

$$x^3 + 8 = 0.$$

Solution. We have

$$x^3 + 8 = 0 \Leftrightarrow x^3 + 2^3 = 0 \Leftrightarrow (x+2)(x^2 - 2x + 4) = 0,$$

which is equivalent with the set of equations $x+2 = 0$ and $x^2 - 2x + 4 = 0$. Solving these equations we get $x_1 = -2$ and $x_2 = 1 + i\sqrt{3}$, $x_3 = 1 - i\sqrt{3}$.

Example 6.1.11. Solve the equation

$$x^5 - 81x = 0.$$

Solution. We have

$$x^5 - 81x = 0 \Leftrightarrow x(x^4 - 3^4) = 0 \Leftrightarrow x(x^2 + 9)(x^2 - 9) = 0,$$

which is equivalent with the equations $x = 0$, $x^2 + 9 = 0$, $x^2 - 9 = 0$. Solving these we get $x_1 = 0$, $x_2 = 3i$, $x_3 = -3i$, $x_4 = 3$, $x_5 = -3$.

Example 6.1.12. Solve the equation

$$x^5 - 32 = 0.$$

Solution. We have

$$x^5 - 32 = 0 \Leftrightarrow x^5 - 2^5 = 0 \Leftrightarrow (x-2)(x^4 + 2x^3 + 4x^2 + 8x + 16) = 0.$$

Hence we must solve $x - 2 = 0$ and $x^4 + 2x^3 + 4x^2 + 8x + 16 = 0$. The first one has solution $x_1 = 2$. To solve the second equation we divide by $x^2 \neq 0$, and we rewrite it as

$$\left(x^2 + \frac{16}{x^2}\right) + 2\left(x + \frac{4}{x}\right) + 4 = 0.$$

Setting $x + \frac{4}{x} = w$, we have $x^2 + \frac{16}{x^2} = w^2 - 8$, and substituting in the equation we get

$$w^2 + 2w - 4 = 0.$$

Solving this equation we find $w_1 = -1 + \sqrt{5}$, $w_2 = -1 - \sqrt{5}$. Now to find x we solve the equations

$$x + \frac{4}{x} = -1 + \sqrt{5} \iff x^2 - (\sqrt{5} - 1)x + 4 = 0,$$

and

$$x + \frac{4}{x} = -1 - \sqrt{5} \iff x^2 + (\sqrt{5} + 1)x + 4 = 0.$$

The first one has solutions

$$x_{2,3} = \frac{\sqrt{5} - 1 \pm i\sqrt{10 + 2\sqrt{5}}}{2},$$

and the second

$$x_{4,5} = \frac{-\sqrt{5} - 1 \pm i\sqrt{10 - 2\sqrt{5}}}{2}.$$

Example 6.1.13. Solve the equation

$$x^6 - 1 = 0.$$

Solution. The equation $x^6 - 1 = 0$ can be solved in several ways:

Way 1. $x^6 - 1 = 0 \iff (x^3)^2 - 1 = 0 \iff (x^3 + 1)(x^3 - 1) = 0.$

This gives the equations $x^3 + 1 = 0$ and $x^3 - 1 = 0$. Factoring, we get

$$x^3 + 1 = (x + 1)(x^2 - x + 1) = 0 \implies x_1 = -1, \; x_{2,3} = \frac{1 \pm i\sqrt{3}}{2},$$

and

$$x^3 - 1 = (x - 1)(x^2 + x + 1) = 0 \implies x_4 = 1, \; x_{5,6} = \frac{-1 \pm i\sqrt{3}}{2}.$$

Way 2. $x^6 - 1 = 0 \iff (x^2)^3 - 1 = 0 \iff (x^2 - 1)(x^4 + x^2 + 1) = 0.$

This gives the equations $x^2 - 1 = 0$ and $x^4 + x^2 + 1 = 0$, which is biquadratic (see, Section 6.1.1).

Way 3. $x^6 - 1 = 0 \Leftrightarrow (x - 1)(x^5 + x^4 + x^3 + x^2 + x + 1) = 0$.

This gives the equations $x - 1 = 0$ and $x^5 + x^4 + x^3 + x^2 + x + 1 = 0$, which is a $5^{th}-$ degree symmetric equation, (see, Section. 6.1.4).

Example 6.1.14. Solve the equation

$$x^8 + 1 = 0.$$

Solution. We have

$$x^8 + 1 = 0 \Leftrightarrow x^8 + 2x^4 + 1 - 2x^4 = 0 \Leftrightarrow (x^4 + 1)^2 - (\sqrt{2}x^2)^2 = 0 \Leftrightarrow$$

$$(x^4 + \sqrt{2}x^2 + 1)(x^4 - \sqrt{2}x^2 + 1) = 0.$$

Thus, we obtain two biquadratic equations

$$x^4 + \sqrt{2}x^2 + 1 = 0, \ x^4 - \sqrt{2}x^2 + 1 = 0.$$

Solving we get eight complex roots

$$x_{1,2} = \pm\sqrt{\frac{-\sqrt{2} + i\sqrt{2}}{2}}, \ x_{3,4} = \pm\sqrt{\frac{-\sqrt{2} - i\sqrt{2}}{2}},$$

$$x_{5,6} = \pm\sqrt{\frac{\sqrt{2} + i\sqrt{2}}{2}}, \ x_{7,8} = \pm\sqrt{\frac{\sqrt{2} - i\sqrt{2}}{2}}.$$

6.1.7   Trinomial Equations

An algebraic equation of the form

$$ax^{2n} + bx^n + c = 0 \tag{6.7}$$

is called a ***trinomial*** equation provided $n \geq 2$ and $a \neq 0, b \neq 0, c \neq 0$.

For $n = 2$ the trinomial equation reduces to the biquadratic equation $ax^4 + bx^2 + c = 0$, which we studied in Section 6.1.1.

By making the substitution $y = x^n$, equation (6.7) reduces to the quadratic equation

$$ay^2 + by + c = 0.$$

Solving this quadratic equation we find two solution y_1, y_2. Returning to the substitution $x^n = y$, we obtain the binomial equations

$$x^n = y_1 \text{ and } x^n = y_2,$$

which are binomial equations and can be solved according to Section 6.1.6

Example 6.1.15. Solve the equation

$$x^6 + 3x^3 + 2 = 0.$$

Solution. We set $x^3 = y$. Then the equation becomes

$$y^2 + 3y + 2 = 0 \iff (y+1)(y+2) = 0.$$

Hence $y_1 = -1$ and $y_2 = -2$. Hence we must solve the two binomial equations

$$x^3 = -1, \quad x^3 = -2.$$

Now solving

$$x^3 + 1 = 0 \iff (x+1)(x^2 - x + 1) = 0$$

we find $x_1 = -1$, $x_{2,3} = \frac{1}{2} \pm \frac{i\sqrt{3}}{2}$. Solving

$$x^3 + 2 = 0 \iff (x + \sqrt[3]{2})(x^2 - \sqrt[3]{2}x + \sqrt[3]{4}) = 0$$

we find $x_4 = -\sqrt[3]{2}$, $x_{5,6} = \frac{\sqrt[3]{2}}{2}(1 \pm i\sqrt{3})$.

Example 6.1.16. Solve the equation

$$x^{12} - 33x^7 + 32x^2 = 0.$$

Solution. We have

$$x^{12} - 33x^7 + 32x^2 = 0 \iff x^2(x^{10} - 33x^5 + 32) = 0.$$

This implies $x^2 = 0$ and $x^{10} - 33x^5 + 32 = 0$. The first equation gives $x_1 = x_2 = 0$. To solve the second we set $x^5 = y$. Then we solve $y^2 - 33y + 32 = 0$ to obtain $y_1 = 32$, $y_2 = 1$. Now we must solve the binomial equations $x^5 - 32 \iff x^5 - 32 = 0$ and $x^5 = 1 \iff x^5 - 1 = 0$. The equation $x^5 - 32 = 0$ was solved in Example 6.1.12. Since

$$x^5 - 1 = (x - 1)(x^4 + x^3 + x^2 + x + 1),$$

the equation $x^5 - 1 = 0$ gives $x - 1 = 0$ and $x^4 + x^3 + x^2 + x + 1 = 0$, which is a $4^{th}-$ degree symmetric equation, and we invite the reader to solve it as an exercise.

A more general form of a trinomial equation is

$$a(f(x))^{2p} + b(f(x))^p + c = 0,$$

where $f(x)$ is an expression in x, and $p \in \mathbb{Q}$.

Setting $y = f(x)^n$, this equation reduces to the quadratic

$$ay^2 + by + c = 0.$$

If y_1 and y_2 are the roots of the quadratic equation, then we solve $f(x)^n = y_1$ and $f(x)^n = y_2$.

Example 6.1.17. Solve the equation

$$2x^{\frac{2}{3}} + 3x^{\frac{1}{3}} - 2 = 0.$$

Solution. We set $x^{\frac{1}{3}} = y$. Then $x^{\frac{2}{3}} = y^2$, and the equation becomes

$$2y^2 + 3y - 2 = 0.$$

Solving this equation we get $y_1 = -2$, and $y_2 = \frac{1}{2}$. Now, we solve

$$x^{\frac{1}{3}} = -2 \iff x = (-2)^3 = -8$$

and

$$x^{\frac{1}{3}} = \frac{1}{2} \iff x = \left(\frac{1}{2}\right)^3 = \frac{1}{8}.$$

6.1.8   Exercises

Solve the equations.

1.

    (a) $x^3 - 8 = 0$ **(b)** $8x^3 + 27 = 0.$

2.

    (a) $64x^6 - x^3 = 0$ **(b)** $(2x - 1)^3 = 1.$

3.

    (a) $x^4 - 1 = 0$ **(b)** $x^4 + 1 = 0.$

4.

    (a) $x^5 - 27x^2 = 0$ **(b)** $x^9 - x^5 + x^4 - 1 = 0.$

5.

    (a) $x^6 - 5x^3 - 24 = 0$ **(b)** $x^8 - 80x^4 - 81 = 0.$

6.

    (a) $(x - 1)^6 - 9(x - 1)^3 + 8 = 0$ **(b)** $x + x^{\frac{1}{2}} - 20 = 0.$

7.

    (a) $6x + x^{\frac{1}{2}} - 1 = 0$ **(b)** $3x^{\frac{3}{2}} - 4x^{\frac{3}{4}} - 7 = 0.$

6.2   Rational Equations

A *rational equation* is an equation of the form

$$\frac{P(x)}{Q(x)} = 0,$$

where $P(x)$ and $Q(x)$ are polynomials. The fraction $\frac{P(x)}{Q(x)}$ is defined only for x for which $Q(x) \neq 0$.

To solve the equation $\frac{P(x)}{Q(x)} = 0$, we clear the denominator by multiplying both sides by $Q(x)$; then the equation will be equivalent to $P(x) = 0$. The derived equation $P(x) = 0$ will have all the roots of the given equation, and, if it has any roots besides these, which are also roots of the equation $Q(x) = 0$, these latter ones may readily be detected and are rejected.

Example 6.2.1. Solve the equation

$$\frac{x^2 - x - 6}{x + 1} = 0.$$

Solution. The fraction is defined for all $x \neq -1$. To clear the denominator we multiplying both sides by $x + 1$. So we have

$$\frac{x^2 - x - 6}{x + 1} = 0 \ \Leftrightarrow \ x^2 - x - 6 = 0.$$

Solving this quadratic equation by factoring we have

$$x^2 - x - 6 = 0 \ \Leftrightarrow \ (x - 3)(x + 2) = 0 \ \Leftrightarrow \ x_1 = 3, x_2 = -2.$$

Example 6.2.2. Solve the equation

$$\frac{3}{x} + \frac{6}{x - 1} - \frac{x + 5}{x^2 - x} = 0.$$

Solution. The fractions are defined for all $x \neq 0$ and $x \neq 1$. Factoring the denominator $x^2 - x = x(x - 1)$, we see that the least common multiple of the denominators is $x(x-1)$. To clear the denominators we multiply both sides by $x(x-1)$. So we have

$$x(x-1)\left[\frac{3}{x} + \frac{6}{x-1} - \frac{x+5}{x^2-x}\right] = x(x-1)[0] \iff 3(x-1)+6x-(x+5) = 0.$$

Now we solve $3(x-1) + 6x - (x+5) = 0$. We have

$$3(x-1)+6x-(x+5) = 0 \iff 3x-3+6x-x-5 = 0 \iff 8x-8 = 0 \iff x = 1.$$

Since $x = 1$ is not a permissible value for x, we reject it. We say that the given equation has no solution.

Example 6.2.3. Solve the equation

$$\frac{1}{x+1} + \frac{5}{2x-4} = 1.$$

Solution. The fractions are defined for all $x \neq -1$ and $x \neq 2$. Multiplying both sides by the least common denominator $(x+1)(2x-4)$, we have

$$(x+1)(2x-4)\left[\frac{1}{x+1} + \frac{5}{2x-4}\right] = (x+1)(2x-4)[1] \iff$$

$$2x-4+5(x+1) = (x+1)(2x-4) \iff 2x-4+5x+5 = 2x^2-2x-4 \iff$$

$$2x^2 - 9x - 5 = 0 \iff x_1 = -\frac{1}{2}, x_2 = 5.$$

Example 6.2.4. Solve the equation

$$\frac{3}{x+2} - \frac{2x-1}{x+1} = \frac{2x+1}{x^2+3x+2}.$$

Solution. The fractions are defined for all $x \neq -2$ and $x \neq -1$. Multiplying both sides by the least common denominator $(x+1)(x+2)$, we obtain

$$3(x+1) - (2x-1)(x+2) = 2x+1.$$

Simplifying we get

$$x^2 + x - 2 = 0 \iff (x-1)(x+2) = 0 \iff x = 1, \ x = -2.$$

Since $x = -2$ is not a permissible value of x, the equation has only one solution $x = 1$.

Example 6.2.5. Solve the equation

$$\frac{x+1}{x+2} + \frac{x+6}{x+7} = \frac{x+2}{x+3} + \frac{x+5}{x+6}.$$

Solution. The fractions are defined for all $x \neq -2, -3, -6, -7$. Before we clear the denominators, it is more efficient here to reduce each fraction to a mixed expression. We have

$$\frac{x+2-1}{x+2} + \frac{x+7-1}{x+7} = \frac{x+3-1}{x+3} + \frac{x+6-1}{x+6} \iff$$

$$1 + \frac{-1}{x+2} + 1 + \frac{-1}{x+7} = 1 + \frac{-1}{x+3} + 1 + \frac{-1}{x+6} \iff$$

$$\frac{1}{x+2} + \frac{1}{x+7} = \frac{1}{x+3} + \frac{1}{x+6} \iff$$

$$\frac{1}{x+2} - \frac{1}{x+3} = \frac{1}{x+6} - \frac{1}{x+7} \iff$$

$$\frac{1}{x^2+5x+6} = \frac{1}{x^2+13x+42} \iff$$

$$x^2 + 13x + 42 = x^2 + 5x + 6 \iff 8x = -36 \iff x = -\frac{9}{2}.$$

6.2.1 Exercises

Solve the equations.

1.

    (a) $\dfrac{2x^2 - 5x - 3}{x-2} = 0$ **(b)** $x + 1 + \dfrac{x+2}{x-1} = \dfrac{3}{x-1}.$

2.

$$\text{(a)} \quad \frac{x-1}{x+3} + \frac{x-2}{x-3} = \frac{1-2x}{3-x} \qquad \text{(b)} \quad \frac{1}{x} + \frac{3}{x+2} = 2.$$

3.

$$\text{(a)} \quad \frac{x}{x+2} - 1 = \frac{3x+2}{x^2+4x+4} \qquad \text{(b)} \quad \frac{1}{x+1} + \frac{1}{x-2} = 1.$$

4.

$$\text{(a)} \quad \frac{4}{x-2} - \frac{1}{x-4} = \frac{4}{x^2-6x+8} \qquad \text{(b)} \quad \frac{2x+1}{x} + \frac{4x}{2x+1} = 5.$$

5.

$$\text{(a)} \quad \frac{3}{2(x^2-1)} + \frac{x}{4x+4} = \frac{3}{8} \qquad \text{(b)} \quad \frac{x+1}{x} + 1 = \frac{x}{x-1}.$$

6.

$$\text{(a)} \quad \frac{1}{4-x} - \frac{1}{4} = \frac{1}{x+2} \qquad \text{(b)} \quad \frac{1}{x-1} - \frac{x-2}{x^2-1} + \frac{3x^2+x}{1-x^4} = 0.$$

7.

$$\frac{4}{x-1} - \frac{1}{4-x} = \frac{3}{x-2} - \frac{2}{3-x}.$$

8.

$$\frac{2x-2}{x^2-36} - \frac{x-2}{x^2-6x} = \frac{x-1}{x^2+6x}.$$

9.

$$\frac{x+7}{2x^2-7x+3} + \frac{x}{x^2-2x-3} + \frac{x+3}{2x^2+x-1} = 0.$$

10.

$$\frac{x+6}{x-1} - \frac{x^2+17}{x^2+x+1} = \frac{x+36}{x^3-1} - \frac{x+1}{x^2+x+1}.$$

6.3   Radical Equations

An equation is called a ***radical equation*** (or ***irrational***) if it contains a radical expression $\sqrt[n]{A(x)}$ of the unknown. The solutions of a radical equation must belong in the domain of definition of all the radical expressions of the equation.

In general, ***to solve a radical equation we seek to reduce it to a rational equation, that is, to an equation not containing radicals, by raising both sides of the equation to the proper power***.

If n is *even*, then this transformation may lead to the appearance of ***extraneous solutions***. Therefore, after solving the reduced rational equation, care must be taken to check its solutions in the original equation before accepting them as solutions of the given equation.

We illustrate these with a number of examples.

Example 6.3.1. Solve the equation

$$\sqrt{x} + 2 = x.$$

Solution. We isolate the radical expression in one side of the equation, so that, we have $\sqrt{x} = x - 2$. Now we square both sides of the equation:

$$\left(\sqrt{x}\right)^2 = (x - 2)^2 \ \Rightarrow \ x = x^2 - 4x + 4 \ \Leftrightarrow \ x^2 - 5x + 4 = 0,$$

whence $x_1 = 4$ and $x_2 = 1$. Substituting $x = 4$ in the given equation we obtain the true equality $\sqrt{4} + 2 = 4$, while the substitution $x = 1$ yields $3 = 1$ which is not true, That is, $x = 1$ is an extraneous solution which does not satisfy the given equation and therefore we reject it. Thus, the radical equation $\sqrt{x} + 2 = x$ has only one solution $x = 4$.

Example 6.3.2. Solve the equation

$$\sqrt{x^2 - 2x + 6} = 2x - 3.$$

Solution. We square both sides of the equation:

$$\left(\sqrt{x^2 - 2x + 6}\right)^2 = (2x-3)^2 \ \Rightarrow \ x^2-2x+6 = 4x^2-12x+9 \ \Leftrightarrow \ 3x^2-10x+3 = 0,$$

whence $x_1 = 3$ and $x_2 = \frac{1}{3}$. Substituting $x = 3$ in the given equation we obtain the true equality $\sqrt{9} = 3$, while the substitution $x = \frac{1}{3}$ yields $\frac{7}{3} = -\frac{7}{3}$ which is not true. Thus, the given equation has only one solution $x = 3$.

Example 6.3.3. Solve the equation

$$\sqrt{3x + 1} - 2 = \sqrt{x - 1}.$$

Solution. We square both sides of the equation:

$$\left(\sqrt{3x + 1} - 2\right)^2 = \left(\sqrt{x - 1}\right)^2 \Rightarrow \left(\sqrt{3x + 1}\right)^2 - 4\sqrt{3x + 1} + 4 = x - 1 \Leftrightarrow$$

$$3x + 1 - 4\sqrt{3x + 1} + 4 = x - 1 \Leftrightarrow 2x + 6 = 4\sqrt{3x + 1}.$$

We square again to eliminate the radical.

$$(2x + 6)^2 = \left(4\sqrt{3x + 1}\right)^2 \Rightarrow 4x^2 + 24x + 36 = 16(3x + 1) \Leftrightarrow$$

$$x^2 - 6x + 5 = 0 \Leftrightarrow (x - 1)(x - 5) = 0.$$

Hence $x_1 = 1$ and $x_2 = 5$. Checking, we see that both $x = 1$ and $x = 5$ are solutions of the original equation.

Example 6.3.4. Solve the equation

$$\sqrt{x + 1} + \sqrt{x - 4} = \sqrt{2x + 9}.$$

Solution. We square both sides of the equation:

$$\left(\sqrt{x + 1} + \sqrt{x - 4}\right)^2 = \left(\sqrt{2x + 9}\right)^2 \Rightarrow$$

$$x + 1 + x - 4 + 2\sqrt{(x + 1)(x - 4)} = 2x + 9 \Leftrightarrow \sqrt{(x + 1)(x - 4)} = 6 \Rightarrow$$

$$(x + 1)(x - 4) = 36 \Leftrightarrow x^2 - 3x - 40 = 0.$$

Solving $x^2 - 3x - 40 = 0$ we find $x_1 = 8$, $x_2 = -5$. Checking $x = 8$ and $x = -5$, we see that only $x = 8$ is a solution of the given equation. Note that the radical expressions of the equation are all defined for $x \geq 4$. This gives us another reason for rejecting the extraneous solution $x = -5$.

Example 6.3.5. Solve the equation

$$\frac{\sqrt{x} + \sqrt{x-3}}{\sqrt{x} - \sqrt{x-3}} = 2x - 5.$$

Solution. Rationalizing the denominator in the first side of the equation and simplifying, we have

$$\sqrt{x(x-3)} = 2x - 6.$$

Squaring both sides and simplifying, we have

$$x^2 - 7x + 12 = 0.$$

Solving, we find $x_1 = 3$, $x_2 = 4$. Testing, we see that both 3 and 4, are the roots of the original equation.

Radical equations of index higher than the 2nd

If the index n of the radical $\sqrt[n]{A(x)}$ is *odd*, $n = 2k + 1$, then there is no need to check for extraneous solution, since the original and the reduced equations are equivalent. This is so because for any $a, b \in \mathbb{R}$ the following holds true:

$$a = b \quad \Leftrightarrow \quad a^{2k+1} = b^{2k+1}.$$

In particular, (for $k = 1$)

$$a = b \quad \Leftrightarrow \quad a^3 = b^3.$$

Example 6.3.6. Solve the equation

$$\sqrt[3]{x^3 + 7} = x + 1.$$

Solution. Raising to the third power both sides of the equation, we have

$$\left(\sqrt[3]{x^3 + 7}\right)^3 = (x+1)^3 \Leftrightarrow x^3 + 7 = x^3 + 3x^2 + 3x + 1 \Leftrightarrow$$

$$x^2 + x - 2 = 0 \Leftrightarrow (x+2)(x-1) = 0 \Leftrightarrow x_1 = -2, \ x_2 = 1.$$

Example 6.3.7. Solve the equation

$$\sqrt[4]{8x^2 - 1} = 2x.$$

Solution. Raising to the fourth power both sides of the equation, we have

$$\left(\sqrt[4]{8x^2 - 1}\right)^4 = (2x)^4 \;\Rightarrow\; 8x^2 - 1 = 16x^4 \;\Leftrightarrow$$

$$16x^4 - 8x^2 + 1 = 0 \;\Leftrightarrow\; (4x^2 - 1)^2 = 0 \;\Leftrightarrow\; x_1 = x_3 = \frac{1}{2},\; x_2 = x_4 = -\frac{1}{2}.$$

Testing, we see that the only solution is $x = \frac{1}{2}$.

6.3.1 Exercises

Solve the equations.

1.
      (a) $\sqrt{2x - 1} = 1$ **(b)** $\sqrt{x - 3} - 5 = 0.$

2.
      (a) $\sqrt{x + 3} + 1 = 3x$ **(b)** $\sqrt{-16x - 3} = 2\sqrt{x^2 + 3}.$

3.
      (a) $\sqrt{x + 2} = 2 - \sqrt{x}$ **(b)** $\sqrt{x - 8} + \sqrt{x - 5} = 3.$

4.
      (a) $\sqrt{x - 4} - 2 = \sqrt{x + 8}$ **(b)** $\sqrt{x - 8} + \sqrt{x - 5} = \sqrt{3x - 21}.$

5.
      (a) $\sqrt{5(x + 2)} - \sqrt{x + 1} = \sqrt{x + 6}$ **(b)** $(x + 3)\sqrt{x + 2} = (x + 2)\sqrt{x + 5}.$

6.
      (a) $\sqrt[3]{7x + 1} = 4$ **(b)** $\sqrt[3]{x^3 + 9x^2} = 3 + x.$

7.

$$\textbf{(a)} \quad \sqrt[4]{3x+1} = 2 \qquad \textbf{(b)} \quad \sqrt[4]{5x^2 - 6} = x.$$

8.

$$\textbf{(a)} \quad \frac{\sqrt{x}-3}{\sqrt{x}+3} = \frac{\sqrt{x}+1}{\sqrt{x}-2} \qquad \textbf{(b)} \quad \frac{\sqrt{2x-1}+\sqrt{3x}}{\sqrt{2x-1}-\sqrt{3x}} + 3 = 0.$$

6.4   Fundamental Theorem of Algebra

Let

$$f(x) = a_n x^n + a_{n-1} x^{n-1} + \dots + a_1 x + a_0, \quad (a_n \neq 0),$$

be a polynomial with complex coefficients of degree n. We recall that a *root* (or *zero*) of $f(x)$ is a number r such that $f(r) = 0$. In other words, r is a solution of the polynomial equation $f(x) = 0$.

A question arises as to whether every polynomial has a root. The answer to this question is supplied by the following theorem. This theorem is called the *Fundamental Theorem of Algebra* or *D'Alebert's theorem*[4] because he was the first who tried to prove it in 1746. The first rigorous proof was given by Gauss[5] in 1799.

Theorem 6.4.1. *(Fundanental Theorem of Algebra)*. *Every polynomial of degree $n \geq 1$ with complex coefficients has at least one complex root.*

We accept this theorem without proof.[6] Let us prove some of its immediate consequences.

---

[4]Jean D' Alembert (1717-1783), a remarkable mathematician and philosopher.

[5]Karl Fr. Gauss (1777-1855), a great mathematician and astronomer. In all Gauss gave four different proofs of this theorem.

[6] There is a number of proofs given in higher mathematics. The algebraic proof is rather difficult. Simpler proofs are given by means of analysis.

Theorem 6.4.2. *Every polynomial $f(x)$ of degree $n \geq 1$ with complex coefficients can be factored into n linear factors (not necessarily distinct) of the form*

$$f(x) = a_n(x - r_1)(x - r_2) \cdots (x - r_n), \tag{6.8}$$

with $r_1, r_2, ..., r_n$ complex numbers.

Proof. According to the Fundamental theorem of algebra, the polynomial $f(x)$ has a root, say r_1. Then by the Factor theorem (Theorem 4.4.11) the binomial $x - r_1$ is a factor of $f(x)$ and so

$$f(x) = (x - r_1)Q_1(x),$$

where $Q_1(x)$ is a polynomial of degree $n - 1$ whose leading coefficient is also a_n. The Fundamental theorem of algebra is again applicable to the polynomial $Q_1(x)$, and therefore $Q_1(x)$ has a root r_2. Then by the Factor theorem again, $Q_1(x)$ has the factor $x - r_2$ and so

$$Q_1(x) = (x - r_2)Q_2(x),$$

where $Q_2(x)$ is a polynomial of degree $n - 2$ whose leading coefficient is a_n.

Repeating this procedure n times, we arrive at a final quotient $Q_n(x)$ which is a polynomial of degree $n - n = 0$ with leading coefficient a_n, ie, $Q_n(x) = a_n x^0 = a_n$. Thus,

$$f(x) = a_n(x - r_1)(x - r_2) \cdots (x - r_n).$$

$\square$

Note that in (6.8) some of the numbers $r_1, r_2, ..., r_n$ may be equal to one another. Combining similar linear factors, we can rewrite equality (6.8) in the form

$$f(x) = a_n(x - r_1)^{k_1}(x - r_2)^{k_2} \cdots (x - r_m)^{k_m},$$

where $k_1 + k_2 + ... + k_m = n$, assuming that the numbers $r_1, r_2, ..., r_m$ are distinct.

It can be shown that this factorization of $f(x)$ is *unique* (apart the order of the factors) and consequently the positive integers $k_1, k_2, ..., k_m$ are the so-called *multiplicities* of the roots $r_1, r_2, ..., r_m$, respectively.

Corollary 6.4.3. *Any polynomial with complex coefficients of degree n has exactly n roots if we count each root as many times as is its multiplicity.*

Example 6.4.4. The polynomial

$$f(x) = x^6 + 3x^5 - 4x^4 - 6x^3 + x^2 + 3x + 2$$

factors out in the form

$$f(x) = (x+2)(x+1)^3(x-1)^2.$$

That is, has roots the numbers -2, -1 and 1 with multiplicities 1, 3 and 2, respectively.

Theorem 6.4.5. *(**Conjugate Pairs Theorem**). If the polynomial of degree $n \geq 2$*

$$f(x) = a_n x^n + a_{n-1}x^{n-1} + \ldots + a_1 x + a_0, \quad (a_n \neq 0),$$

with real coefficients $a_0, \ldots, a_n \in \mathbb{R}$ has a complex root $r = a + ib$, then it also has the root $\overline{r} = a - ib$.

Proof. Since $f(r) = 0$, we have

$$f(\overline{r}) = a_n(\overline{r})^n + a_{n-1}(\overline{r})^{n-1} + \ldots + a_1(\overline{r}) + a_0$$

$$= \overline{a_n} \cdot \overline{r^n} + \overline{a_{n-1}} \cdot \overline{r^{n-1}} + \ldots + \overline{a_1} \cdot \overline{r} + \overline{a_0}$$

$$= \overline{a_n r^n} + \overline{a_{n-1}r^{n-1}} + \ldots + \overline{a_1 r} + \overline{a_0}$$

$$= \overline{a_n r^n + a_{n-1}r^{n-1} + \ldots + a_1 r + a_0}$$

$$= \overline{f(r)} = \overline{0} = 0.$$

$\square$

Corollary 6.4.6. *Any polynomial with real coefficients of odd degree has at least one real root.*[7]

---

[7]Irrational roots of polynomials are located on the number line by approximation methods and graphing techniques of calculus.

Theorem 6.4.7. (**Factorization Theorem**). *Every polynomial with real coefficients of degree n can be factored into a product of linear factors and/or quadratic factors* $x^2 + px + q$, *where the quadratic factors have no real roots.*

Proof. Suppose

$$f(x) = a_n(x - r_1)(x - r_2) \cdots (x - r_n).$$

If one of the roots $r_1, r_2, ..., r_n$ is not real, say $r_1 = a + ib$, then, by Theorem 6.4.5, $\overline{r_1} = a - ib$ is also a root. Consequently the product

$$(x - r_1)(x - \overline{r_1}) = (x - a - ib)(x - a + ib) = (x - a)^2 - (ib)^2$$

$$= x^2 - 2ax + a^2 + b^2 = x^2 + px + q,$$

where $p = -2a$ and $q = a^2 + b^2$, will be a quadratic factor of $f(x)$ with no real roots. $\qquad\square$

In conclusion

We have shown that a polynomial of degree $n \geq 2$

$$f(x) = a_n x^n + a_{n-1} x^{n-1} + ... + a_1 x + a_0, \quad (a_n \neq 0),$$

can be written in the form:

1) $\quad f(x) = a_n(x - r_1)(x - r_2) \cdots (x - r_n),$

where $a_n,\ r_1, r_2, ..., r_n$ are complex numbers

2) $\quad f(x) = a_n(x - r_1) \cdots (x - r_k)(x^2 + p_1 x + q_1) \cdots (x^2 + p_m x + q_m),$

where $k + 2m = n$, the numbers $a_n,\ r_1, ..., r_k,\ p_1, q_1, ..., p_m, q_m$ are real, and the quadratic factors have no real roots.

Example 6.4.8. The polynomial

$$f(x) = x^4 - x^3 - 6x^2 + 14x - 12$$

is known to have a root $r_1 = 1 + i$ (check!). Factor it into a product of linear and quadratic polynomials with real coefficients, and find all its roots.

Solution. Since $f(x)$ has the complex root $r_1 = 1 + i$, it also has the conjugate root $\overline{r_1} = 1 - i$. Hence it has the quadratic factor

$$(x - 1 - i)(x - 1 + i) = (x - 1)^2 - i^2 = x^2 - 2x + 2.$$

In other words, $f(x)$ is divisible by $x^2 - 2x + 2$. Dividing $f(x)$ by this trinomial, we get

$$x^4 - x^3 - 6x^2 + 14x - 12 = (x^2 - 2x + 2)(x^2 + x - 6).$$

Factoring now the trinomial $x^2 + x - 6 = (x - 2)(x + 3)$, we obtain the desired factorization

$$f(x) = (x^2 - 2x + 2)(x - 2)(x + 3).$$

The roots are readily seen to be

$$r_1 = 1 + i, \quad r_2 = 1 - i, \quad r_3 = 2, \quad r_4 = -3.$$

Example 6.4.9. Find the roots of the polynomial

$$f(x) = x^3 + \frac{1}{2}x^2 + \frac{1}{2}x - \frac{1}{2}.$$

Solution. We multiply $f(x)$ by 8 to get the polynomial

$$g(x) = (2x)^3 + (2x)^2 + 2(2x) - 4.$$

Setting $t = 2x$, we get the polynomial with integer coefficients

$$P(t) = t^3 + t^2 + 2t - 4.$$

The factors of the constant term are ± 1, ± 2, ± 4. If $P(t)$ has an integer root it will be found in this list. Testing these values, we see that $P(1) = 0$. Hence by the Factor theorem $t - 1$ divides $P(t)$. We divide and we find the quotient to be the quadratic trinomial $t^2 + 2t + 4$. Thus,

$$P(t) = (t - 1)(t^2 + 2t + 4).$$

Solving the quadratic equation $t^2 + 2t + 4 = 0$, we find two complex roots $t = -1 + \sqrt{3}i$ and $t = -1 - \sqrt{3}i$.

Returning to x, we obtain the roots,

$$r_1 = \frac{1}{2}, \quad r_2 = -\frac{1}{2} + \frac{\sqrt{3}}{2}i, \quad r_3 = -\frac{1}{2} - \frac{\sqrt{3}}{2}i.$$

Theorem 6.4.10. *(Conjugate Irrational Pairs Theorem). Let*

$$f(x) = a_n x^n + a_{n-1} x^{n-1} + ... + a_1 x + a_0, \quad (a_n \neq 0)$$

be a polynomial of degree $n \geq 2$ with rational coefficients $a_0, ..., a_n \in \mathbb{Q}$. If the irrational number $r_1 = a + \sqrt{b}$, with $a, b \in \mathbb{Q}$ and the positive b not a perfect square, is a root of $f(x)$, then its conjugate $r_2 = a - \sqrt{b}$ is also a root of $f(x)$.

Proof. Consider the trinomial $g(x) = (x - r_1)(x - r_2) =$

$$[x-(a+\sqrt{b})][x-(a-\sqrt{b})] = [(x-a)-\sqrt{b}][(x-a)+\sqrt{b}] = (x-a)^2-b = x^2-2ax+a^2-b.$$

Since the coefficients of $f(x)$ and $g(x)$ are rational numbers, dividing $f(x)$ by $g(x)$ the division algorithm gives

$$f(x) = g(x) \cdot Q(x) + cx + d,$$

where c, d are rational numbers.

Since $f(a + \sqrt{b}) = 0$ and $g(a + \sqrt{b}) = 0$, it follows

$$0 = c(a + \sqrt{b}) + d \quad \Rightarrow \quad c\sqrt{b} = -(ca + d).$$

If $c \neq 0$, then $b = \left(\frac{ca+d}{c}\right)^2$, which contadicts the hypothesis for b. Hence, $c = 0$, and therefore also $d = 0$. Thus,

$$f(x) = g(x) \cdot Q(x).$$

But $g(a - \sqrt{b}) = 0$. Therefore $f(a - \sqrt{b}) = 0$.

$\square$

6.4.1   Vieta's Relations[8] (*)

Vieta's relations give the relations between the roots and the coefficients of a polynomial.

Theorem 6.4.11. *(Vieta's). If*

$$f(x) = a_n x^n + a_{n-1} x^{n-1} + \ldots + a_1 x + a_0, \quad (a_n \neq 0),$$

is a polynomial with complex coefficients and $r_1, r_2, \ldots, r_n$ are its roots, then the following equalities hold true:

$$r_1 + r_2 + \ldots + r_n = -\frac{a_{n-1}}{a_n}.$$

$$r_1 r_2 + r_1 r_3 + \ldots + r_1 r_n + r_2 r_3 \ldots + r_2 r_n + \ldots + r_{n-1} r_n = \frac{a_{n-2}}{a_n}$$

$$r_1 r_2 r_3 + r_1 r_2 r_4 + \ldots + r_1 r_2 r_n + \ldots + r_{n-2} r_{n-1} r_n = -\frac{a_{n-3}}{a_n}$$

$$\ldots \ldots \ldots \ldots \ldots \ldots$$

$$\ldots \ldots \ldots \ldots \ldots \ldots$$

$$r_1 r_2 \cdots r_n = (-1)^n \frac{a_0}{a_n}.$$

Proof. Since

$$f(x) = a_n (x - r_1)(x - r_2) \cdots (x - r_n),$$

it follows that

$$\frac{1}{a_n} f(x) = (x - r_1)(x - r_2) \cdots (x - r_n).$$

Performing the multiplications in the second side of this equality and collecting like terms, we obtain

$$x^n - (r_1 + \ldots + r_n) x^{n-1} + (r_1 r_2 + r_1 r_3 + \ldots + r_{n-1} r_n) x^{n-2} - \ldots + (-1)^n r_1 r_2 \cdots r_n.$$

---

[8]Vieta Francois (1540-1603) a French mathematician. His work in algebra was an important step towards modern algebra.

On the other hand

$$\frac{1}{a_n} f(x) = x^n + \frac{a_{n-1}}{a_n} x^{n-1} + \frac{a_{n-2}}{a_n} x^{n-2} + \dots + \frac{a_1}{a_n} x + \frac{a_0}{a_n}.$$

By the equality of polynomials, equating coefficients of equal powers, we obtain the desired relations. □

Remark 6.4.12. Note that Theorem 5.4.11 is a reformulation of Vieta's theorem for a second-degree polynomial

$$f(x) = ax^2 + bx + c.$$

If r_1, r_2 are the roots of $f(x)$, then

$$r_1 + r_2 = -\frac{b}{a},$$

$$r_1 r_2 = \frac{c}{a}.$$

For a third-degree polynomial

$$f(x) = ax^3 + bx^2 + cx + d$$

Vieta's theorem tells us, that if r_1, r_2, r_3 are its roots. Then

$$r_1 + r_2 + r_3 = -\frac{b}{a},$$

$$r_1 r_2 + r_1 r_3 + r_2 r_3 = \frac{c}{a},$$

$$r_1 r_2 r_3 = -\frac{d}{a}.$$

6.5   Review Exercises

A. Solve the polynomial equations.

1. $x^3 - 5x^2 - 14x = 0$.

2. $x^3 - x^2 - x + 1 = 0$.

3. $x^3 + 3x^2 - 4x - 12 = 0$.

4. $3x^3 - x^2 - 15x + 5 = 0$.

5. $15x^3 - 61x^2 - 2x + 24 = 0$.

6. $x^4 - 50x^2 + 49 = 0$.

7. $x^4 - a(a-1)x^2 - a^3 = 0$.

8. $x^4 - 16 = 0$.

9. $2x^4 - 3x^3 + 4x^2 - 3x + 2 = 0$.

10. $2x^4 - 5x^3 + 7x^2 - 5x + 2 = 0$.

11. $x^4 + x^3 + x^2 + 3x - 6 = 0$.

12. $x^4 + 10x^3 + 31x^2 + 30x + 5 = 0$.

13. $4x^4 - 8x^3 - x^2 + 8x - 3 = 0$.

14. $x^5 - 27x^2 = 0$.

15. $x^5 - 4x^4 + 3x^3 + 3x^2 - 4x + 1 = 0$.

16. $x^5 - 11x^4 + 36x^3 - 36x^2 + 11x - 1 = 0$.

17. $x^6 + 2x^4 + 2x^2 + 1 = 0$.

18. $x^4 + 2x^3 + 3x^2 + 4x + 4 = 0$.

19. $(x^2 - 4)^2 - (x + 2)^2(5x - 4) = 0$.

20. $(x - 3)(2x + 1)^2 - (x^2 - 9)(x + 3) = 0$.

B. Solve the rational equations.

1.

(a) $\dfrac{2}{x-2}+\dfrac{1}{x+1}=\dfrac{1}{x^2-x-2}$

(b) $\dfrac{3x+2}{x^2+x}-\dfrac{x-5}{x^2-1}=\dfrac{x-2}{x^2-x}.$

2.

(a) $\dfrac{1}{x}+\dfrac{x}{x+2}=\dfrac{x}{2}$

(b) $\dfrac{2(x^2+2)}{5}=\dfrac{(x^2+1)(x^2-2)+6}{x^2+1}.$

3.

(a) $x^2+\dfrac{1}{x^2}=a^2+\dfrac{1}{a^2}$

(b) $\dfrac{1}{x-8}+\dfrac{1}{x-6}+\dfrac{1}{x+6}+\dfrac{1}{x+8}=0.$

C. Solve the radical equations.

1.

(a) $x^{\frac{3}{2}}-x^{\frac{1}{2}}=0$

(b) $\sqrt{x+2}+1=\sqrt{2x+5}.$

2.

(a) $\sqrt{x+1}+\sqrt{3x}=\sqrt{5x+1}$

(b) $\sqrt{5x^2-6x+1}-\sqrt{5x^2+9x-2}=5x-1.$

3.

(a) $\sqrt{2+\sqrt{x-5}}=\sqrt{13-x}$

(b) $\sqrt[3]{x^3-7}+1=x.$

4.

(a) $\sqrt[3]{8x^3+61}=2x+1$

(b) $\dfrac{\sqrt{x-1}-\sqrt{x+1}}{\sqrt{x-1}+\sqrt{x+1}}=x-3.$

5.

(a) $\sqrt[3]{x}+\sqrt[3]{2-x}=2$

(b) $\sqrt[4]{x^3}-5\sqrt{x}+6\sqrt[4]{x}=0.$

6.

(a) $\sqrt{\dfrac{2x-5}{x-2}} - 3\sqrt{\dfrac{x-2}{2x-5}} + 2 = 0$ (b) $\dfrac{1}{\sqrt{x+1}} - \dfrac{1}{\sqrt{x-1}} + \dfrac{1}{\sqrt{x^2+1}} = 0.$

D.

1. One root of the equation $x^3 - 2x + 4 = 0$ is $1+i$. Find the others.

2. One root of the equation $x^3 - 5x^2 + 8x - 6 = 0$ is $1+i$. Find the others.

3. One root of the equation $x^4 - 2x^3 - 2x^2 + 8x - 8 = 0$ is $1+i$. Solve the equation.

4. One root of the equation $2x^4 - x^3 + 5x^2 + 13x - 5 = 0$ is $1-2i$. Solve the equation.

5. One root of the equation $2x^4 - 11x^3 + 17x^2 - 10x + 2 = 0$ is $2+\sqrt{2}$. Solve the equation.

6. Find the equation of the lowest degree with rational coefficients two of whose roots are $-5 + 2i$ and $-1 + \sqrt{5}$.

E. Solve the equations.

1.
$$x^3 - \frac{5}{6}x^2 - \frac{22}{3}x + \frac{5}{2} = 0.$$

2.
$$\frac{1}{10}x^3 + \frac{1}{2}x^2 + \frac{1}{5}x - \frac{4}{5} = 0.$$

3.
$$\frac{2}{x^3} + \frac{1}{x^2} - \frac{1}{x} + 1 = 0.$$

4.
$$\sqrt[3]{x} - 2\sqrt[6]{x} + \sqrt{x} = 2.$$

5.
$$\sqrt{x} + \sqrt{x - \sqrt{1 - x}} = 1.$$

6.
$$\frac{x - 4}{\sqrt{x} + 2} = x - 8.$$

7.
$$\sqrt[3]{16 - x^3} = 4 - x.$$

8.
$$\frac{x\sqrt[3]{x} - 1}{\sqrt[3]{x^2} - 1} - \frac{x\sqrt[3]{x} - 1}{\sqrt[3]{x^2} + 1} = 4.$$

9.
$$\sqrt{x} - \frac{4}{\sqrt{x + 2}} + \sqrt{2 + x} = 0.$$

10.
$$\frac{4}{x + \sqrt{x^2 + x}} - \frac{1}{x - \sqrt{x^2 + x}} = \frac{3}{x}.$$

11.
$$\frac{2 - \sqrt{x}}{2 - x} = \sqrt{\frac{2}{2 - x}}.$$

12.
$$4\sqrt{x + 2} = |x + 1| + 4.$$

Historical Note[9]

_____

[9]The origins of algebra go back about 4000 years ago in Babylonia. The literal designations which we use today in algebra were unknown to the Babylonians who formulated their equations rhetorically. The first syncopated notations for unknown quantities are encountered in the writings of the ancient Greek mathematician Diophantus (2nd to 3rd century). Neither the Babylonians nor the Greeks considered negative numbers. Chinese scholars were solving first-degree equations and systems of them and also quadratic equations 2000 years prior to the Christian era. After the decline and occupation of the Greek

cities (states) by the Romans, many Greek scholars moved away from Greece, to Alexandria (Egypt) and other Arabic and Hindu regions in Asia. Books of Greek mathematicians (Diophantus' *Arithmetica* and Euclid's *Elements*) were transalated to Arabic and studied in Asia.

The founder of algebra as a special branch of mathematics was the Central Asian scholar Mohammad of Khorezmi, more generally know as al-Khowarizmi. His algebraic work, composed in the 9th century A.D. bears the name *the science of transposition and cancellation.* " Transposition" denoted the transfer of a subtrahend from one side of the equation to the other where it becomes an addend. The Arabic for *"transposition"* is " *al-jabr"*. Whence the name *"Algebra"*. Al-Khowarizmi and those that followed him made extensive use of algebra in commercial and monetary computations.

In the 12th century, the "Algebra" of al-Khowarizmi was translated into Latin and studied in Europe. It marked the begining of the development of algebra in European countries, at first under the strong influence of the science of the East. Syncopated notation appeared for the unknowns and new problems involved in trading were solved, but no essential advances were made until the first third of the 16th century when the Italians Scipione del Ferro from Bolonga and Tartaglia (real name Niccolo Fontana from Brescia) found rules for solving cubic equations of the type $x^3 = px + q, x^3 + px = q$, $x^3 + q = px$, and Giralamo Cardano $(1501 - 1576)$, in 1545 demonstrated that any cubic equation can be reduced to one of these three types. At the same time, Ferrari, a student of Cardano, solved a quadric (fourth-degree) equation.

After solutions had been found for equations of the third and fourth degree, mathematicians strenuously sought the formulas for solving the quintic (fifth-degree) equation. However, Ruffini (Italy) proved, at the turn of the 19th century, that the literal fifth-degree equation $x^5 + ax^4 + bx^3 + cx^2 + dx + e = 0$ cannot be solved algebraically; more precicely, it is impossible to express any root of it in terms of the coefficients a, b, c, d, e using the six algebraic operations of addition, subtraction, multiplication, division, involution (raising to power) and evolution (extracting root). (Ruffini's proof containd some errors, and in 1824 Abel of Norway gave a correct proof). In 1830 Everiste Galois (France) demonstrated that no general equation whose degree exceeds 4 can be solved algebraically. Nevertheless, as we know from the *Fundamental theorem of algebra*, every nth degree equation has (if we consider complex numbers as well) n roots, some of which may be equal.

Chapter 7

Inequalities

In this chapter, we study the basic algebraic inequalities. We solve linear, quadratic, polynomial, rational and radical inequalities.

7.1   Inequalities and their properties

Two algebraic expressions A, B related by the symbol $>$ (reads *greater than*) or $<$ (reads *less than*) form a **strict inequality**.

$$A > B, \quad A < B.$$

The inequality $A \geq B$ means $A > B$ or $A = B$, while the inequality $A \leq B$ means $A < B$ or $A = B$. The inequalities $A \geq B$ and $A \leq B$ are usually said **weak inequalities**.

In Chapter 1 we saw that for real numbers a and b the inequality

$$a > b \quad \text{means that the difference} \quad a - b > 0.$$

Note that $b < a \Leftrightarrow 0 < a - b \Leftrightarrow a - b > 0 \Leftrightarrow a > b$

The direction ($>$ or $<$) in which the inequality sign points is the *sense* of the inequality. Two or more inequalities are called *inequalities*

233

of the same sense if they contain one and the same sign $>$ or $<$. For instance, the inequalities $a > b$ and $c > d$ are said to have the same sense, while the inequalities $a > b$ and $c < d$ are said to have opposite senses.

Any real number $x \in \mathbb{R}$ is either positive or zero or negative. That is,

$$x > 0, \quad \text{or} \quad x = 0, \quad \text{or} \quad x < 0.$$

Hence given any two real numbers a and b one of the following is true:

$$a > b, \quad \text{or} \quad a = b, \quad \text{or} \quad a < b.$$

Let us consider the basic properties of inequalities:

Properties of Inequalities

Let $a, b, c, d \in \mathbb{R}$. Then

1. If $a > b$ and $b > c$, then $a > c$.

    Indeed, since $a - b > 0$ and $b - c > 0$ we have

    $$a - c = (a - b) + (b - c) > 0 \iff a > c.$$

2. If $a > b$, then for any c, we have $a + c > b + c$.

    Indeed, since $a - b > 0$, we have

    $$(a + c) - (b + c) = a - b + c - c = a - b > 0 \iff a + c > b + c.$$

    In particular note that

    $$a + c > b \iff a + c + (-c) > b + (-c) \iff a > b - c.$$

3.   (a) If $a > b$ and $c > 0$, then $ac > bc$.

      Indeed, since $a - b > 0$ and $c > 0$, we have

$$ac - bc = c(a - b) > 0 \implies ac > bc.$$

      That is, multiplying both sides of an inequality by a **positive** number the **inequality remains unchanged**.

  (b) If $a > b$ and $c < 0$, then $ac < bc$.

      Indeed, since $a - b > 0$ and $c < 0$, we have

$$ac - bc = c(a - b) < 0 \implies ac < bc.$$

      That is, multiplying both sides of an inequality by a **negative** number the **inequality reverses**.

4.   (a) If $a > b$ and $c > d$, then $a + c > b + d$, ie, two inequalities having the same sense may be added side by side.

      Since $a - b > 0$ and $c - d > 0$, we have

$$(a + c) - (b + d) = (a - b) + (c - d) > 0 \implies a + c > b + d.$$

  (b) If $a > b$ and $c < d$, then $a - c > b - d$.

      By property 3(b) $c > d \implies -c > -d$. Now since $a > b$ and $-c > -d$, adding these inequalities side by side, we get

$$a + (-c) > b + (-d) \quad \text{or} \quad a - c > b - d.$$

5. If a, b, c, d are *positive numbers* and $a > b$, $c > d$, then $ac > bd$. That is, two inequalities of the same sence in which both sides are positive may be multiplied side by side.

Indeed,

$$ac - bd = (ac - bc) + (bc - bd) = (a - b)c + (c - d)b > 0 \implies ac > bd.$$

7.2   Solving Inequalities

An *inequality* in one unknown x is a relation of the form

$$A(x) > B(x) \quad \text{or} \quad A(x) < B(x),$$

where $A(x)$ and $B(x)$ are algebraic expressions in x.

A *solution* of an inequality is a value of the unkwon x for which this inequality reduces to a true numerical inequality. To *solve* an inequality means to find all its solutions.

Two or more inequalities are said to be **equivalent** if any solution of one of them is a solution of the other, and vice versa. That is, if they have the same solutions.

In the process of solving an inequality we replace the given inequality by one which is simpler than, but equivalent to, the given inequality. In doing so, we use the basic properties of inequalities:

1. Any term of an inequality may be transposed from one side of the inequality to the other with an opposite sign, the sense of the inequality remaining unchanged.

2. Both sides of an inequality may be multiplied or divided by one and the same number not equal to zero. If the number is positive the sense of the inequality remains unchanged. However we must always remember to **reverse** the sense of the inequality when multiplying or dividing both sides of an equality by a *negative* number.

7.2.1   Solving Linear Inequalities

A *linear inequality* is an inequality of the form

$$ax + b > 0 \quad \text{or} \quad ax + b < 0 \quad (a \neq 0)$$

The signs $>$, $<$, may also be replaced by $\geq$, $\leq$. Let us see some examples which illustrate how we solve linear inequalities.

Example 7.2.1. Solve the inequality

$$3x + 6 > 0.$$

Solution. We procced as with equations, separating knowns from unknowns

$$3x + 6 > 0 \iff 3x > -6$$

Now we divide both sides by 3 to get $x > -\frac{6}{3}$ or $x > -2$.

Using interval notation the solution set is

$$\{x \in \mathbb{R} : -2 < x\} = (-2, +\infty).$$

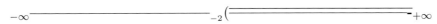

Example 7.2.2. Solve the inequality

$$3x + 6 \leq 0.$$

Solution. We have

$$3x + 6 \leq 0 \iff 3x \leq -6$$

Now we divide both sides by 3 to get $x \leq -\frac{6}{3}$ or $x \leq -2$.

Using interval notation the solution set is

$$\{x \in \mathbb{R} : x \leq -2\} = (-\infty, -2].$$

Example 7.2.3. Solve the inequality

$$-4x + 3 > 5.$$

Solution. We have

$$-4x + 3 > 5 \iff -4x > 5 - 3 \iff -4x > 2$$

Now we divide both sides by -4. Since $-4 < 0$, we reverse the inequality to get $x < \frac{2}{-4}$ or $x < -\frac{1}{2}$.

Using interval notation the solution set is

$$\left\{x \in \mathbb{R} : x < -\frac{1}{2}\right\} = (-\infty, -\frac{1}{2}).$$

Example 7.2.4. Solve the inequality

$$-2(x - 3) - 4 > 3(x - 2) - 4(x + 1).$$

Solution. Simplifying both sides of the inequality (by removing the parantheses and collecting like terms), we obtain

$$-2x + 6 - 4 > 3x - 6 - 4x - 4 \quad \Leftrightarrow \quad -2x + 2 > -x - 10 \quad \Leftrightarrow$$

$$-2x + x > -10 - 2 \quad \Leftrightarrow \quad -x > -12.$$

Multiplying both sides by -1, we get $x < 12$. Hence the solution set is the inetrval $(-\infty, 12)$.

Example 7.2.5. Solve the inequality

$$2(x - 1) + 1 > 3 - (1 - 2x).$$

Solution. Simplifying both sides of the inequality, we obtain

$$2x - 2 + 1 > 3 - 1 + 2x \quad \Leftrightarrow \quad 2x - 1 > 2 + 2x \quad \Leftrightarrow \quad 2x - 2x > 2 + 1 \quad \Leftrightarrow \quad 0x > 3.$$

That is, $0 > 3$ which is not true. Hence the inequality has no solution.

    Remark. Note that in case of the inequality $2(x-1)+1 < 3-(1-2x)$ we would have $0x < 3$ or $0 < 3$, which is always true, and hence any value of x is a solution of the inequality.

7.2.2 Solving Compound Inequalities

The statement that x is between 3 and 5, means that x is greater than 3 and x is less than 5, that is $x > 3$ and $x < 5$. These two inequalities may be written as the double inequality

$$3 < x < 5.$$

Such an inequality is said a *compound inequality*.

Example 7.2.6. Solve

$$3 \leq 2x - 5 < 9.$$

Solution. Since the compound inequality $3 \leq 2x - 5 < 9$ is a brief notation of two inequalities $3 \leq 2x - 5$ and $2x - 5 < 9$, we may apply the basic properties of inequalities. Adding the number 5 to each part of the inequality $3 \leq 2x - 5 < 9$, we obtain

$$3 + 5 \leq 2x - 5 + 5 < 9 + 5 \quad \Leftrightarrow \quad 8 \leq 2x < 14.$$

Dividing each part by 2, we find that

$$4 \leq x < 7.$$

Hence, the interval $[4, 7)$ is the set of solutions.

Example 7.2.7. Solve

$$-2 + x < 3x + 4 \leq 6x - 2.$$

Solution. The compound inequality $-2 + x < 3x + 4 \leq 6x - 2$ is a brief notation of two inequalities $-2 + x < 3x + 4$ and $3x + 4 \leq 6x - 2$. Because we can not isolate x between the inequality symbols, we solve each inequality separetely (as a *system of two inequalities*).

We solve the first inequality:

$$-2 + x < 3x + 4 \quad \Leftrightarrow \quad -3x + x < 4 + 2 \quad \Leftrightarrow \quad -2x < 6.$$

Dividing each side by -2, we find that

$$x > -3.$$

Then we solve the second inequality:

$$3x + 4 \leq 6x - 2 \quad \Leftrightarrow \quad 3x - 6x \leq -2 - 4 \quad \Leftrightarrow \quad -3x \leq -6.$$

Dividing each side by -3, we find that

$$x \geq 2.$$

Since we are looking for the x's which satisfy both inequalities, the solution set is the intersection (or overlap) of the intervals $(-3, \infty)$ and $[2, \infty)$, which is $[2, \infty)$

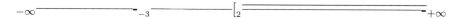

7.2.3   Inequalities Involving Absolute Value

1. The following inequalities are equivalent

$$|x| < k \quad \Leftrightarrow \quad -k < x < k,$$

where $k > 0$.

Indeed, if $x \geq 0$, then $|x| = x$, and the inequality $|x| < k$ becomes $x < k$. Consequently all numbers in the interval $[0, k)$ are solutions of the inequality $|x| < k$. And if $x < 0$, then $|x| = -x$, and the inequality $|x| < k$ becomes $-x < k \Leftrightarrow x > -k$. Consequently the numbers in the interval $(-k, 0)$ are also solutions of the inequality $|x| < k$. Combining the obtained results, we conclude that all numbers in the interval $(-k, k)$ satisfy inequality $|x| < k$. Conversely, suppose $-k < x < k$. There are two possibilities: $x \geq 0$ and $x < 0$. If $x \geq 0$, then $|x| = x < k$. If $x < 0$, then since $-k < x \Leftrightarrow k > -x$, we also have $|x| = -x < k$. Thus, in either case $|x| < k$.

Example 7.2.8. Solve the inequality

$$|2x - 3| < 7.$$

Solution.

The given inequality is equivalent to the compound inequality

$$-7 < 2x - 3 < 7 \quad \Leftrightarrow \quad -7 + 3 < 2x < 7 + 3 \quad \Leftrightarrow \quad -4 < 2x < 10.$$

Dividing each part by 2 we get $-2 < x < 5$ Hence the interval $(-2, 5)$ is the set of solutions.

2. The following inequalities are equivalent

$$|x| > k \quad \Leftrightarrow \quad x > k \ \text{ or } \ x < -k,$$

where $k > 0$.

Indeed, there are two possibilities: $x \geq 0$ or $x < 0$. If $x \geq 0$, then $|x| = x$, and the inequality $|x| > k$ becomes $x > k$. If $x < 0$, then $|x| = -x$, and $|x| > k$ becomes $-x > k \ \Leftrightarrow \ x < -k$. Consequently the numbers in the interval (k, ∞) or in the interval $(-\infty, -k)$ are solutions of the inequality $|x| > k$. The converse is now obvious.

Example 7.2.9. Solve the inequality

$$|2x - 7| > 5.$$

Solution. The given inequality is fulfilled if $2x - 7 > 5$ or $2x - 7 < -5$. Solving the first inequality we get

$$2x - 7 > 5 \ \Leftrightarrow \ 2x > 5 + 7 \ \Leftrightarrow \ 2x > 12.$$

Hence $x > 6$.

Solving the second inequality we get

$$2x - 7 < -5 \ \Leftrightarrow \ 2x < -5 + 7 \ \Leftrightarrow \ 2x < 2.$$

Hence $x < 1$. Thus, the given inequality is true if $x < 1$ or $x > 6$. The solution set is the union of the intervals $(-\infty, 1) \cup (6, \infty)$.

Example 7.2.10. Solve the inequality

$$|1 - 2x| > 3 - x.$$

Solution. There are two cases according to whether $1 - 2x \geq 0$ or $1 - 2x < 0$. If $1 - 2x \geq 0$, then $|1 - 2x| = 1 - 2x$. If $1 - 2x < 0$, then $|1 - 2x| = -(1 - 2x)$. Consequently the given inequality is equivalent to the set of two systems of inequalities:

$$\text{(I)} \quad 1 - 2x \geq 0 \ \text{ and } \ 1 - 2x > 3 - x$$

$$\text{(II)} \ 1 - 2x < 0 \ \text{ and } \ -(1 - 2x) > 3 - x.$$

To solve system (I) we need to solve each inequality and find the intersection of their solution sets. We solve the first inequality

$$1 - 2x \geq 0 \ \Leftrightarrow \ -2x \geq -1 \ \Leftrightarrow \ x \leq \frac{1}{2}.$$

We solve the second inequality

$$1 - 2x > 3 - x \ \Leftrightarrow \ -x > 2 \ \Leftrightarrow \ x < -2.$$

Hence the solution of system (I) is the intersection of the intervals $(-\infty, \frac{1}{2}]$ and $(-\infty, -2)$, which is $(-\infty, -2)$.

Similarly, we solve system (II): The first inequality gives

$$1 - 2x < 0 \ \Leftrightarrow \ -2x < -1 \ \Leftrightarrow \ x > \frac{1}{2}.$$

The second inequality gives

$$-(1 - 2x) > 3 - x \ \Leftrightarrow \ -1 + 2x > 3 - x \ \Leftrightarrow \ 3x > 4 \ \Leftrightarrow \ x > \frac{4}{3}.$$

Hence the solution of system (II) is the intersection of the intervals $(\frac{1}{2}, \infty)$ and $(\frac{4}{3}, \infty)$ which is $(\frac{4}{3}, \infty)$. Thus, the solution set of the original inequality is the interval $(-\infty, -2)$ or the interval $(\frac{4}{3}, \infty)$. That is, the union

$$(-\infty, -2) \cup \left(\frac{4}{3}, \infty\right).$$

Remark. To solve inequalities containing more than one absolute value, it is advisable to partition the number line into intervals so that in each interval we can write the inequality without using the sign of the absolute values. Next, we solve the inequality in each of these intevals. The solution of the original inequality is then the union of the solutions found in each interval. We illustrate this in the following example.

Example 7.2.11. Solve the inequality

$$|x + 1| - 3 < |x - 2|$$

Solution. To solve the inequality we must first drop the absolute value sign. To do so, we mark on the number line the points $x = -1$ and $x = 2$ at which the expressions standing inside the absolute values vanish. These points break the number line into three intervals

(I) $x \leq -1$, (II) $-1 \leq x \leq 2$, (III) $x \geq 2$.

Considering x consecutively on each of these intervals, we are able to drop the absolute values, and obtain that the original inequality is equivalent to the set of the following three systems:

(I). If $x \leq -1$, then $|x + 1| = -(x + 1$ and $|x - 2| = -(x - 2)$, and we must solve the system

$$x \leq -1 \quad \text{and} \quad -(x + 1) - 3 < -(x - 2)$$

We solve

$$-(x + 1) - 3 < -(x - 2) \iff -x - 1 - 3 < -x + 2 \iff -4 < 2,$$

which is always true. The solution set of this inequality is all real numbers. Hence the solution set of the system is $x \leq -1$.

(II). If $-1 \leq x \leq 2$, then $|x + 1| = x + 1$ and $|x - 2| = -(x - 2)$, and we must solve the system

$$-1 \leq x \leq 2 \quad \text{and} \quad x + 1 - 3 < -(x - 2)$$

We solve

$$x + 1 - 3 < -(x - 2) \iff x + 1 - 3 < -x + 2 \iff 2x < 4 \iff x < 2.$$

Hence the solution set of the system is $-1 \leq x < 2$.

(III). If $x \geq 2$, then $|x + 1| = x + 1$ and $|x - 2| = x - 2$, and we must solve the system

$$x \geq 2 \quad \text{and} \quad x + 1 - 3 < x - 2$$

We solve

$$x + 1 - 3 < x - 2 \iff x - 2 < x - 2 \iff -2 < -2,$$

which is not true. The inequality has no solution (the solution set is empty). Hence the system has no solution either.

Finally, the solution set of the original inequality is the union of the solution sets of the the three systems. That is, the union of the intervals $(-\infty, -1]$ and $[-1, 2)$, which is $(-\infty, 2)$, ie, $x < 2$.

7.2.4   Exercises

Solve the inequalities.

1.

    (a) $3x + 2 > 5$ **(b)** $7x - 12 < x - 18$.

2.

    (a) $4 - 2x > -5x - 8$ **(b)** $2(x + 3) \leq -2(x + 3)$.

3.

$$2(x - 1) + 3(2x + 4) - 7 < 5(2x - 1) - (x - 3).$$

4.

$$(x + 5)^2 - 2(3x - 6) > (x - 3)^2 - 3(2x + 5).$$

5.

    (a) $\dfrac{x - 1}{5} + \dfrac{2x + 3}{10} > \dfrac{3}{4}$ **(b)** $\dfrac{x - 3}{4} - \dfrac{x - 2}{3} > x - \dfrac{x - 1}{2}$.

6.

    (a) $\dfrac{x - 1}{2} + \dfrac{2x + 3}{4} < \dfrac{x}{6}$ **(b)** $\dfrac{x - 2}{2} + \dfrac{1 - 2x}{5} < \dfrac{x}{10} - \dfrac{2}{5}$.

7.

$\qquad$ (a) $4 < 2x - 6 < 10$ $\qquad$ (b) $3 \leq 2x + 2 < 6.$

8.

$\qquad$ (a) $2 < \dfrac{x-4}{3} \leq 9$ $\qquad$ (b) $-10 \leq \dfrac{5-x}{2} < 0.$

9.

$\qquad$ (a) $2 + x < 3x - 2 < 5x + 2$ $\qquad$ (b) $x > 2x + 3 > 4x - 7.$

10.

$\qquad$ (a) $|x - 5| < 1$ $\qquad$ (b) $|x + 2| \leq 4.$

11.

$\qquad$ (a) $|x - 3| > 6$ $\qquad$ (b) $|2x + 4| \geq 10.$

12.

$\qquad$ (a) $3 < |3x - 1| < 8$ $\qquad$ (b) $2 < \left| \dfrac{x+1}{3} \right| < 3.$

13.

$\qquad$ (a) $|x| + 8 < |3x| - 2$ $\qquad$ (b) $2|x| + x > 10.$

14.

$\qquad$ (a) $|x| < |x + 1|$ $\qquad$ (b) $|x + 2| < |x + 1|.$

15.

$\qquad$ (a) $3|x| + 4|x - 1| > 5$ $\qquad$ (b) $|x - 2| > |x + 1| - 3.$

16.

$\qquad$ (a) $|x - 1| + |x + 1| < 6$ $\qquad$ (b) $|x| + |x - 1| + |x - 2| > 9.$

7.3   Quadratic Inequalities

A *quadratic inequality* is an inequality of the form

$$ax^2 + bx + c > 0 \quad \text{or} \quad < 0,$$

where a, b, c are real numbers with $a \neq 0$. In other words, a quadratic inequality is an inequality whose left-hand side is a quadratic trinomial $f(x) = ax^2 + bx + c$ and the right-hand side member is zero.

From Theorem 5.4.16 we know

$$f(x) = ax^2 + bx + c = a(x - x_1)(x - x_2),$$

where x_1 and x_2 are the roots of the quadratic equation $ax^2 + bx + c = 0$.

Let $D = b^2 - 4ac$ be the discriminant of the quadratic trinomial.

The following cases are possible:

Case 1.

$$D < 0.$$

In this case the roots x_1, x_2 are complex numbers of the form

$$x_1 = \lambda + \mu i, \quad x_1 = \lambda - \mu i,$$

where $\lambda = -\frac{b}{2a}$, $\mu = \frac{\sqrt{|D|}}{2a}$.

Then, we have

$$f(x) = a(x - \lambda - \mu i)(x - \lambda + \mu i) = a[(x - \lambda)^2 + \mu^2].$$

Since $(x - \lambda)^2 + \mu^2 > 0$ for all $x \in \mathbb{R}$, we conclude

1. if $a > 0$, then $ax^2 + bx + c > 0$ for all $x \in \mathbb{R}$;

2. if $a < 0$, then $ax^2 + bx + c < 0$ for all $x \in \mathbb{R}$.

We may write this briefly as

$$D < 0 \quad \Leftrightarrow \quad af(x) > 0 \quad \text{for all } x \in \mathbb{R}.$$

Case 2.

$$D = 0.$$

In this case the roots are equal $x_1 = x_2 = -\frac{b}{2a}$.

Then, we have

$$f(x) = a(x - x_1)(x - x_2) = a\left(x + \frac{b}{2a}\right)^2.$$

Since $\left(x + \frac{b}{2a}\right)^2 > 0$ for all $x \neq -\frac{b}{2a}$, we conclude

1. if $a > 0$, then $ax^2 + bx + c > 0$ for all $x \in \mathbb{R}$ with $x \neq -\frac{b}{2a}$;

2. if $a < 0$, then $ax^2 + bx + c < 0$ for all $x \in \mathbb{R}$ with $x \neq -\frac{b}{2a}$.

We may write this briefly as

$$D = 0 \iff af(x) \geq 0 \text{ for all } x \in \mathbb{R} \text{ (with } af(x) = 0 \iff x = -\frac{b}{2a}\text{)}.$$

Case 3.

$$D > 0.$$

In this case the roots are real and distinct $x_1 \neq x_2$.

Without loss of generality we may assume $x_1 < x_2$. Then:

1.   (a) if $x < x_1 < x_2$, then $x - x_1 < 0$ and $x - x_2 < 0$, and so $(x - x_1)(x - x_2) > 0$.

    (b) if $x_1 < x_2 < x$, then $x - x_1 > 0$ and $x - x_2 > 0$, and so $(x - x_1)(x - x_2) > 0$.

    Hence in both these cases,

$$(x - x_1)(x - x_2) > 0.$$

    Therefore, if $a > 0$, then $f(x) > 0$ (if $a < 0$, then $f(x) < 0$) for all x *outside* the interval $[x_1, x_2]$.

2. if $x_1 < x < x_2$, then $x - x_1 > 0$ and $x - x_2 < 0$, and so

$$(x - x_1)(x - x_2) < 0.$$

    Therefore if $a > 0$, then $f(x) < 0$ (if $a < 0$, then $f(x) > 0$) for all x *inside* the interval (x_1, x_2).

Thus, when $D > 0$, we have

$$af(x) > 0 \text{ for } x \in (-\infty, x_1) \cup (x_2, \infty) \text{ and } af(x) < 0 \text{ for } x \in (x_1, x_2).$$

Example 7.3.1. Solve the inequality $x^2 - 2x + 3 > 0$.

Solution. The discriminant is $D = (-2)^2 - 4(1)(3) = -8 < 0$. Consequently, the equation $x^2 - 2x + 3 = 0$ has complex roots $x_1 = 1 + i\sqrt{2}$ and $x_2 = 1 - i\sqrt{2}$. Hence,

$$x^2 - 2x + 3 = (x - 1 - i\sqrt{2})(x - 1 + i\sqrt{2}) = (x - 1)^2 + 2.$$

Since $a = 1 > 0$, we see $x^2 - 2x + 3 > 0$ is true for all $x \in \mathbb{R}$.

Example 7.3.2. Solve the inequality $9x^2 - 6x + 1 > 0$.

Solution. We have $D = (-6)^2 - 4(9)(1) = 0$. Consequently the equation $9x^2 - 6x + 1 = 0$ has double root $x_1 = x_2 = \frac{1}{3}$. Hence,

$$9x^2 - 6x + 1 = 9\left(x - \frac{1}{3}\right)^2.$$

Since $a = 9 > 0$, we see $9x^2 - 6x + 1 > 0$ is true for all $x \in \mathbb{R}$ with $x \neq \frac{1}{3}$.

Example 7.3.3. Solve the inequality $2x^2 - 5x + 3 > 0$.

Solution. We have $D = (-5)^2 - 4(2)(3) = 1 > 0$. Consequently, the equation $2x^2 - 5x + 3 = 0$ has distinct real roots $x_1 = 1$ and $x_2 = \frac{3}{2}$. Hence

$$2x^2 - 5x + 3 = 2(x - 1)(x - \frac{3}{2}).$$

Since $a = 2 > 0$, we see $2x^2 - 5x + 3 > 0$ is true for $x < 1$ and $x > \frac{3}{2}$, ie, for all $x \in (-\infty, 1) \cup (\frac{3}{2}, \infty)$.

Example 7.3.4. Solve the inequality $x^2 - 2x - 3 < 0$.

Solution. We have $D = (-2)^2 - 4(1)(-3) = 16 > 0$. Consequently, the equation $x^2 - 2x - 3 = 0$ has distinct real roots $x_1 = -1$ and $x_2 = 3$. Hence,

$$x^2 - 2x - 3 = (x+1)(x-3).$$

Since $a = 1 > 0$, we see $x^2 - 2x - 3 < 0$ is true for $-1 < x$ and $x < 3$, ie, for all $x \in (-1, 3)$.

Example 7.3.5. Solve the inequality $x^2 - 4 > 0$.

Solution. We have $D = 0^2 - 4(1)(-4) = 16 > 0$. Consequently, the equation $x^2 - 4 = 0$ has roots $x_1 = -2$ and $x_2 = 2$. In fact,

$$x^2 - 4 = (x+2)(x-2).$$

Since $a = 1$, the solution of the inequality $x^2 - 4 > 0$ is

$$x \in (-\infty, -2) \cup (2, \infty).$$

Example 7.3.6. Solve the inequality $-3x^2 + 4x - 5 > 0$.

Solution. We have $D = (4)^2 - 4(-3)(-5) = -44 > 0$. Consequently, the equation $-3x^2 + 4x - 5 = 0$ has complex (conjugate) roots.
Since $a = -3 < 0$, we get $-3x^2 + 4x - 5 < 0$ for all $x \in \mathbb{R}$.
Thus, $-3x^2 + 4x - 5 > 0$ has no solution.

Example 7.3.7. Find the values of the parameter $k \in \mathbb{R}$ for which the quadratic equation

$$(k-1)x^2 - 2(k-3)x - k + 3 = 0$$

has:

1. Equal roots

2. Real distinct roots

3. Complex roots.

Solution. The nature of the roots of the equation $ax^2 + bx + c = 0$ is determined from the discriminant $D = b^2 - 4ac$. Here the coefficients are $a = k - 1$, $b = -2(k - 3)$ and $c = -k + 3$. Hence

$$D = [-2(k-3)]^2 - 4[(k-1)(-k+3)] = 4(k-3)^2 + 4(k-1)(k-3) = 8(k-3)(k-2).$$

The equation has:

1. Equal roots when $D = 0$. Therefore $(k - 3)(k - 2) = 0$, which implies $k = 2$ or $k = 3$.

2. Real distinct roots when $D > 0$. Therefore $(k - 3)(k - 2) > 0$, which implies $k < 2$ or $k > 3$.

3. Complex roots when $D < 0$. Therefore $(k - 3)(k - 2) < 0$, which implies $2 < k < 3$.

7.4　Polynomial Inequalities

A polynomial inequality $P(x) > 0$ (or < 0) of degree higher than two can be reduced to an equivalent inequality of the form

$$P(x) = f_1(x) \cdot f_2(x) \cdots f_n(x) > 0 \text{ (or } < 0),$$

where $f_1(x), f_2(x), ..., f_n(x)$ are first or second degree polynomials (see, Theorem 6.4.7).

　　The factors of the second degree (ie, the quadratic factors), have complex conjugate roots, and so they will have a fixed sign for all x. Consequently, they can be *dropped* from the inequality (*dropping* such a quadratic factor causes *no change in the sense of the inequality sign if this factor is positive*, and *reversing the inequality sign if it is negative*).

　　If a linear factor repeats, say k-times, then a factor of the form $(x - r)^k$ will be present. In this case, if k is odd the sign of $(x - r)^k$ is the same as the sign of $(x - r)$, and so we replace it (equivalently) by $(x - r)$. If k is *even*, then the sign of $(x - r)^k$ is *positive* for $x \neq r$, and so we may drop it from the product without affecting the sense of the inequality.

　　For intance, the inequality

$$(x^2 + x + 1)(x - 1)^2(x - 3)(x + 1)(x - 5)^3 > 0$$

is equivalent to

$$(x - 3)(x + 1)(x - 5) > 0,$$

because $x^2 + x + 1 > 0$ for all $x \in \mathbb{R}$, $(x - 1)^2 > 0$ for $x \neq 1$, $(x - 5)^3 = (x - 5)^2(x - 5)$ and $(x - 5)^2 > 0$ for $x \neq 5$.

　　Thus, a polynomial inequality $P(x) > 0$ (or < 0) can always take the form

$$(x - r_1)(x - r_2) \cdots (x - r_m) > 0 \ (< 0), \tag{7.1}$$

where $r_1, ..., r_m$ $(m \leq n)$ are the real distinct roots of the polynomial $P(x)$.

All these linear inequalities can be easily solved as in Section 7.2.1.

However, since we will have to encounter several cases regarding the sign of each linear factor, to solve inequality (7.1), it is more practical here to use the so-called **method of intervals**.

7.4.1   The Method of Intervals

The method of intervals is based on the following property of polynomials: if r_1 and r_2 (say, $r_1 < r_2$) are two neighbouring roots of a polynomial, that is, the polynomial has no other roots in the interval (r_1, r_2), then within this interval the sign of the polynomial remains unchanged, ie, for any x with $r_1 < x < r_2$ the polynomial takes on values having the same sign.

The method consists of the following steps:

(1) Arrange the real (distinct) roots $r_1, ..., r_m$ of the polynomial in increasing order: $r_1 < r_2 < ... < r_m$.

(2) On the number line mark the points $r_1, ..., r_m$ dividing the number line into intervals within each of which the polynomial does not change sign.

(3) Note that for any x in the rightmost interval (r_m, ∞), we have $r_1 < r_2 < ... < r_m < x$. Hence all factors of the product

$$(x - r_1)(x - r_2) \cdots (x - r_m)$$

are positive, and so the polynomial is positive for all $x > r_m$.

(4) Starting with plus $+$ sign in the rightmost interval (r_m, ∞), keep alternating the sign as you move from interval to interval (from right to left), and obtain all solutions of (7.1) depending on the inequality sign > 0 or < 0.

To clarify the method let us give some examples.

Example 7.4.1. Use the method of intervals to solve the inequality

$$x^2 + 2x - 3 > 0$$

$(x+3)(x-1)$	$+$	$-$	$+$
x	$x < -3$	$-3 < x < 1$	$1 < x$

Table 7.1:

Solution. We factor the quadratic polynomial:

$$x^2 + 2x - 3 = (x+3)(x-1).$$

Hence, we solve $f(x) = (x+3)(x-1) > 0$. The roots $r_1 = -3$ and $r_2 = 1$ break the number line into three intervals $(-\infty, -3)$, $(-3, 1)$ and $(1, \infty)$ Starting with a $+$ sing in the rightmost interval $(1, \infty)$, and alternating the sign as we move (from right to left) from interval to interval, we obtain the sign of $f(x)$ in the corresponding intervals. See, Table 7.1.

Therefore, $f(x) > 0$ for $x > 1$, $f(x) < 0$ for $-3 < x < 1$ and $f(x) > 0$ for $x < -3$. Thus, the solution of the inequality $x^2 - 3x + 2 < 0$ is $(-\infty, -3) \cup (1, \infty)$.

Example 7.4.2. Solve the inequality

$$x^3 - 3x + 2 \le 0.$$

Solution. We factor the polynomial:

$$x^3 - 3x + 2 = x^3 - x - 2x + 2 = x(x^2 - 1) - 2(x-1) = x(x-1)(x+1) - 2(x-1) =$$

$$(x-1)[x(x+1) - 2] = (x-1)(x^2 + x - 2) = (x-1)^2(x+2).$$

Since $(x-1)^2 \ge 0$, the inequality $(x-1)^2(x+2) \le 0$ is equivalent to $x + 2 \le 0$. Therefore, the solution is $x \le -2$.

Example 7.4.3. Solve the inequality

$$x^2 - 3x + 2 < 0$$

$f(x)$	$+$	$-$	$+$
x	$x < 1$	$1 < x < 2$	$2 < x$

Table 7.2:

by the method of intervals.

Solution. We factor the quadratic polynomial:

$$x^2 - 3x + 2 = (x - 1)(x - 2).$$

Hence, we solve $f(x) = (x - 1)(x - 2) < 0$. The roots $r_1 = 1$ and $r_2 = 2$ break the number line into three intervals $(-\infty, 1)$, $(1, 2)$ and $(2, \infty)$ Starting with a $+$ sing in the rightmost interval $(2, \infty)$, and alternating the sign as we move (from right to left) from interval to interval, we obtain the sign of $f(x)$ in the corresponding intervals. See Table 7.2.

Therefore, $f(x) > 0$ for $x > 2$, $f(x) < 0$, for $1 < x < 2$ and $f(x) > 0$, for $x < 1$. Thus, the solution of $x^2 - 3x + 2 < 0$ is $1 < x < 2$.

Example 7.4.4. Solve the inequality

$$x^3 - 3x^2 - 4x + 12 > 0.$$

Solution. First we factor the polynomial

$$x^3 - 3x^2 - 4x + 12 = x^2(x-3) - 4(x-3) = (x-3)(x^2-4) = (x-3)(x+2)(x-2).$$

Now, we solve

$$f(x) = (x - 3)(x + 2)(x - 2) > 0.$$

The roots of the first side are $r_1 = -2$, $r_2 = 2$, $r_3 = 3$, and they break the number line into four intervals $(-\infty, -2)$, $(-2, 2)$, $(2, 3)$ and $(3, \infty)$. Starting with a $+$ sing in the rightmost interval $(3, \infty)$, and alternating the sign as we move (from right to left) from interval to interval, we obtain the sign of $f(x)$ in the corresponding intervals. See, Table 7.3.

$f(x)$	$-$	$+$	$-$	$+$
x	$x < -2$	$-2 < x < 2$	$2 < x < 3$	$3 < x$

Table 7.3:

Selecting the intervals where the sign of $f(x)$ is $+$ (that is, $f(x) > 0$), we obtain the solution of the inequality to be the union of the intervals

$$(-2, 2) \cup (3, \infty).$$

Example 7.4.5. Solve the inequality

$$(x - 3)(x^2 - 2x + 1)(-x^2 + x - 2)(-2x^2 + 7x - 3)(4x - x^2) > 0.$$

Solution. First, we find the discriminant of each quadratic factor: We have

$$x^2 - 2x + 1 : \quad D = 0 \;\Rightarrow\; x^2 - 2x + 1 = (x - 1)^2 > 0 \text{ for } x \neq 1.$$

$$-x^2 + x - 2 : \quad D = -7 < 0 \;\Rightarrow\; -x^2 + x - 2 < 0 \text{ for all } x \in \mathbb{R}.$$

$$-2x^2 + 7x - 3 : \quad D = 25 > 0 \;\Rightarrow\; -2x^2 + 7x - 3 = -2(x - 3)\left(x - \tfrac{1}{2}\right).$$

$$-x^2 + 4x : \quad D = 16 \;\Rightarrow\; -x^2 + 4x = -x(x - 4).$$

Consequently, the inequality is

$$(x - 3)(x - 1)^2(-x^2 + x - 2)\left[-2(x - 3)\left(x - \frac{1}{2}\right)\right][-x(x - 4)] > 0$$

or

$$2(x - 3)^2(x - 1)^2(-x^2 + x - 2)\left(x - \frac{1}{2}\right)(x)(x - 4) > 0.$$

For all $x \neq 1$ and $x \neq 3$, the product $2(x - 3)^2(x - 1)^2(-x^2 + x - 2) < 0$, and so we may drop it reversing the sing of the inequality to get the equivalent inequality

$$f(x) = x\left(x - \frac{1}{2}\right)(x - 4) < 0.$$

$f(x)$	$-$	$+$	$-$	$+$
x	$x < 0$	$0 < x < \frac{1}{2}$	$\frac{1}{2} < x < 4$	$4 < x$

Table 7.4:

The roots of the first side are $r_1 = 0$, $r_2 = \frac{1}{2}$, $r_3 = 4$, and they break the number line into four intervals $(-\infty, 0)$, $\left(0, \frac{1}{2}\right)$, $\left(\frac{1}{2}, 4\right)$ and $(4, \infty)$ Staring with $+$ sign at the rightmost interval $(4, \infty)$, we get $f(x) > 0$ for $x > 4$. Now alternating the sign as we move from inteval to interval, we get $f(x) < 0$ for $\frac{1}{2} < x < 4$, $f(x) > 0$ for $0 < x < \frac{1}{2}$, and $f(x) < 0$ for $x < 0$. See, Table 7.4.

Thus, the solution of the original inequality is $x < 0$ or $\frac{1}{2} < x < 4$. In interval notation, $(-\infty, 0) \cup \left(\frac{1}{2}, 4\right)$.

7.4.2   Exercises

Solve the inequalities.

1.
   (a) $x^2 - x > 0$ **(b)** $5x^2 \leq 20x$.

2.
   (a) $x^2 + 2x - 3 > 0$ **(b)** $x^2 - 7x + 10 < 0$.

3.
   (a) $x^2 - 4x - 5 < 0$ **(b)** $2x^2 - x - 3 \geq 0$.

4.
   (a) $6x^2 - 11x + 4 > 0$ **(b)** $-16x^2 + 8x - 1 < 0$.

5.
   (a) $(x-4)(x^2-9x+14) > 0$ **(b)** $(3x^2-13x+4)(4x^2+12x+9) > 0$.

6.

(a) $(2x-5)(x^2-4)(x^2+8) > 0$ (b) $(2x-5)(x^2-4)(x^3+8) > 0$.

7.

(a) $(x+2)(x^2-2x+1)(3x^2+1) > 0$ (b) $(x^2-1)(5x^2+7)(2x^2-5x-7) < 0$.

8.

(a) $3x(x^2-1)(x-5)^2 > 0$ (b) $(x+1)(2-x)^2(x-3)^3 \geq 0$.

9.

(a) $x^3 + 2x^2 + 3x + 6 < 0$ (b) $x^4 - x^3 + x^2 - 3x - 6 \geq 0$.

10.

(a) $x^3 + 1 > x^2 + x$ (b) $x^4 - 1 > x^3 - x$.

11. Find the values of the parameter k for which the quadratice equation

$$kx^2 + 3kx + k + 5 = 0$$

has:

(a) equal roots.

(b) real distinct roots.

(c) complex roots.

7.5   Rational Inequalities

An inequality of the form

$$\frac{P(x)}{Q(x)} > 0 \text{ or } < 0,$$

where $P(x)$ and $Q(x)$ are polynomials is called a ***rational inequality***.
 Since the quotient of two numbers has the *same sign* as their product, it follows that for all x in the domain of definition of $\frac{P(x)}{Q(x)}$, ie for x such that $Q(x) \neq 0$, we have

$$\frac{P(x)}{Q(x)} > 0 \quad \Leftrightarrow \quad P(x) \cdot Q(x) > 0$$

and

$$\frac{P(x)}{Q(x)} < 0 \quad \Leftrightarrow \quad P(x) \cdot Q(x) < 0.$$

Thus, **to solve the inequality**

$$\frac{P(x)}{Q(x)} > 0 \text{ or } < 0,$$

we solve (equivalently) the inequality

$$P(x) \cdot Q(x) > 0 \text{ or } < 0.$$

For a complete understanding we give the some examples.

Example 7.5.1. Solve the inequality

$$\frac{x^2 - 4}{x - 3} < 0.$$

   Solution. The domain of definition is $x \neq 3$. The given inequality is equivalent to the inequality

$$(x^2 - 4)(x - 3) < 0 \quad \Leftrightarrow \quad (x + 2)(x - 2)(x - 3) < 0.$$

The roots of the product are -2, 2 and 3, and they break the number line into four intervals. Using the method of intervals we obtain the solution $x < -2$ or $2 < x < 3$, ie, $(-\infty, -2) \cup (2, 3)$.

Remark. Note that, if we were asked to solve

$$\frac{x^2 - 4}{x - 3} \leq 0,$$

then the solution would be $x \leq -2$ or $2 \leq x < 3$, ie $(-\infty, -2] \cup [2, 3)$.

Example 7.5.2. Solve the inequality

$$\frac{6}{x} \leq -2.$$

Solution. The domain of definition is $x \neq 0$. First we need to get 0 in the right-hand side. So we add 2 to both sides to get

$$\frac{6}{x} \leq -2 \quad \Leftrightarrow \quad \frac{6}{x} + 2 \leq 0 \quad \Leftrightarrow \quad \frac{6}{x} + \frac{2x}{x} \leq 0 \quad \Leftrightarrow \quad \frac{6 + 2x}{x} \leq 0.$$

The latter inequality is equivalent to the inequality

$$(6 + 2x)x \leq 0 \quad \Leftrightarrow \quad 2(x + 3)x \leq 0.$$

The roots of the product are $-3, 0$, and they break the number line into three intervals. Using the method of intervals we obtain the solution $-3 \leq x < 0$, ie, $[-3, 0)$.

Example 7.5.3. Solve the inequality

$$\frac{x^2 - 5}{x + 1} \geq -2.$$

Solution. The domain of definition is $x \neq -1$. First we need to get 0 in the right-hand side. Adding 2 to both sides to get

$$\frac{x^2 - 5}{x + 1} + 2 \geq 0 \quad \Leftrightarrow \quad \frac{x^2 - 5 + 2(x + 1)}{x + 1} \geq 0 \quad \Leftrightarrow \quad \frac{x^2 + 2x - 3}{x + 1} \geq 0 \quad \Leftrightarrow$$

$$\frac{(x + 3)(x - 1)}{x + 1} \geq 0.$$

This inequality is equivalent to the inequality

$$(x + 3)(x + 1)(x - 1) \geq 0.$$

The roots of the product are $-3, -1, 1$ and they break the number line into four intervals. Using the method of intervals we obtain the solution $-3 \leq x < -1$ or $x \geq 1$, ie, $[-3, -1) \cup [1, \infty)$.

Example 7.5.4. Solve the inequality

$$\frac{3}{x-1} + \frac{3}{x-2} < \frac{5}{x+3}$$

Solution. The domain of definition is $x \neq -3, 1, 2$. First we need to get 0 in the right-hand side. We have

$$\frac{3}{x-1} + \frac{3}{x-2} < \frac{5}{x+3} \quad \Leftrightarrow \quad \frac{3}{x-1} + \frac{3}{x-2} - \frac{5}{x+3} < 0 \quad \Leftrightarrow$$

$$\frac{x^2 + 24x - 37}{(x-1)(x-2)(x+3)} < 0.$$

This inequality is equivalent to the inequality

$$(x^2 + 24x - 37)(x-1)(x-2)(x+3) < 0.$$

The discriminant of the quadratic factor is $D = 724$, and its roots are $-12 \pm \sqrt{181}$. The roots of the product arranged in increasing order are $-12 - \sqrt{181}$, -3, 1, $-12 + \sqrt{181}$, 2 and they break the number line into six intervals. Using the method of intervals we obtain the solution $x < -12 - \sqrt{181}$ or $-3 < x < 1$ or $-12 + \sqrt{181} < x < 2$, ie,

$$(-\infty, -12 - \sqrt{181}) \cup (-3, 1) \cup (-12 + \sqrt{181}, 2).$$

7.5.1   Exercises

Solve the inequalities.

1.

      (a) $\dfrac{x-3}{x-5} > 0$ **(b)** $\dfrac{x-1}{x-2} \leq 0.$

2.

      (a) $\dfrac{3}{x} < 2$ **(b)** $\dfrac{8}{x} > 4.$

3.

$\qquad$ (a) $\dfrac{x}{x-3} > 1$ $\qquad$ (b) $\dfrac{x+1}{2x-3} < \dfrac{1}{2}$

4.

$\qquad$ (a) $\dfrac{x^2-4}{x+1} > 0$ $\qquad$ (b) $\dfrac{x^2}{x+1} < 2.$ Γ

5.

$\qquad$ (a) $\dfrac{x^2+5x+6}{x^2+x-6} \geq 0$ $\qquad$ (b) $\dfrac{6x^2-3x-3}{x^2-2x-8} < 0.$

6.

$\qquad$ (a) $x > \dfrac{15}{x+2}$ $\qquad$ (b) $x \leq \dfrac{6}{x-5}.$

7.

$\qquad$ (a) $2x^2 + \dfrac{1}{x} > 0$ $\qquad$ (b) $x - 17 \geq \dfrac{60}{x}.$

8.

$\qquad$ (a) $\dfrac{x^2(x+2)(x-3)^3}{(x+4)^2(x-5)^5} \leq 0$ $\qquad$ (b) $\dfrac{(x+3)^2(x^2+x+1)}{x(4-x)} \geq 0.$

9.

$\qquad$ (a) $\dfrac{3x^3-7x^2+4x}{x^3-6x^2+12x-8} \geq 0$ $\qquad$ (b) $\dfrac{x^3-3x^2-x+3}{x^2+3x+2} > 0.$

10.

$\qquad$ (a) $\dfrac{1}{x-2} + \dfrac{1}{x-1} > \dfrac{1}{x}$ $\qquad$ (b) $\dfrac{x-1}{x} - \dfrac{x+1}{x-1} < 2.$

11.

$\qquad$ (a) $\dfrac{2x+1}{x-2} \geq \dfrac{x+4}{2x+5}$ $\qquad$ (b) $\dfrac{x+1}{x} \leq \dfrac{x^2-5x+6}{x^2+5x+6}.$

12.

$\qquad$ (a) $\dfrac{2}{x^2-x+1} - \dfrac{1}{x+1} \geq \dfrac{2x-1}{x^3+1}$ $\qquad$ (b) $\dfrac{2}{2-x} + \dfrac{3}{2+x} \geq \dfrac{4x}{4-x^2}.$

7.6   Radical Inequalities (*)

Inequalities containing the unknown under the radical sign $\sqrt[n]{A(x)}$ are called **radical inequalities**.

As with the radical equations the idea is **to raise both sides of the inequality to the proper power to get rid of the radicals**.

However, to avoid *"common mistakes done often by students"* we must be careful when the index of the radical n is **even**. In this case we *can not freely raise both sides of the inequality to the even power, because this leads to errors*. For this reason, we shall study the two case according to whether the index n of the radical $\sqrt[n]{A(x)}$ is even or odd, separetly.

Case 1. Let $n = 2k+1$ be **odd**. Then for any real numbers $a, b \in \mathbb{R}$ the following holds true:

$$a < b \;\;\Leftrightarrow\;\; a^{2k+1} < b^{2k+1}.$$

This tells that the following inequalities are equivalent.

$$\sqrt[2k+1]{A(x)} < B(x) \;\;\Leftrightarrow\;\; A(x) < (B(x))^{2k+1}.$$

$$\sqrt[2k+1]{A(x)} > B(x) \;\;\Leftrightarrow\;\; A(x) > (B(x))^{2k+1}.$$

Let us consider an example.

Example 7.6.1. Solve the inequality

$$\sqrt[3]{1 - x + \sqrt[3]{x}} > 1, \;\; x \in \mathbb{R}.$$

Solution. Raising to the third power both sides of the inequality, we have

$$\left(\sqrt[3]{1 - x + \sqrt[3]{x}}\right)^3 > 1^3 \;\;\Leftrightarrow\;\; 1 - x + \sqrt[3]{x} > 1$$

$$\Leftrightarrow\;\; \sqrt[3]{x} > x \;\;\Leftrightarrow\;\; x > x^3 \;\;\Leftrightarrow\;\; x(1 - x^2) > 0.$$

Solving this inequality we obtain $x < -1, \quad 0 < x < 1$. In interval notation

$$(-\infty, -1) \cup (0, 1).$$

Case 2. Let $n = 2k$ be **even**. Here we must be careful when solving radical inequalities. First of all, we must *restrict our consideration* to only those values of x for which the radicals of the inequality are meaningful. Secondly, only for *nonegative* real numbers $a, b \geq 0$ there holds:

$$a < b \iff a^2 < b^2.$$

In general,

$$a < b \iff a^{2k} < b^{2k}.$$

2 a. Consider the inequality

$$\sqrt[2k]{A(x)} < B(x).$$

First we must have $A(x) \geq 0$. Since $\sqrt[2k]{A(x)} \geq 0$, we also must have $B(x) > 0$. Under these restictions then we have

$$\sqrt[2k]{A(x)} < B(x) \iff A(x) < (B(x))^{2k}.$$

Hence the given inequality is equivalent to the inequalities

$$A(x) \geq 0, \quad B(x) > 0, \quad A(x) < (B(x))^{2k}.$$

Thus, the solution of the original inequality is the intersection of the solution sets of the three above inequalities.

2 . Consider the inequality

$$\sqrt[2k]{A(x)} > B(x).$$

This inequality is equivalent to the following inequalities (systems of inequalities):

$$(1) \quad B(x) < 0 \quad \text{and} \quad A(x) \geq 0$$

or

$$(2) \quad B(x) \geq 0 \quad \text{and} \quad A(x) \geq (B(x))^{2k}.$$

Thus, the solution of the given inequality is the union of the solution sets of (1) and (2).

Let us consider some examples.

Example 7.6.2. Solve the inequality

$$\sqrt{6 - x} < x.$$

Solution. First of all we must have $6 - x \geq 0$ and $x > 0$. That is,

$$0 < x \leq 6.$$

Now squaring both sides of the original inequality, we have

$$\sqrt{6 - x} < x \quad \Leftrightarrow \quad 6 - x < x^2 \quad \Leftrightarrow \quad x^2 + x - 6 > 0.$$

Solving the quadratic inequality $x^2 + x - 6 > 0$ we find $x < -3$, $x > 2$.

Taking into account the restiction $0 < x \leq 6$, we see that the solution of the given inequality is $2 < x \leq 6$.

Example 7.6.3. Solve the inequality

$$\sqrt{x^2 - 4x} > x - 3.$$

Solution. First we solve the inequalities

$$x - 3 < 0 \quad \text{and} \quad x^2 - 4x \geq 0.$$

That is,

$$x < 3 \quad \text{and} \quad x(x - 4) \geq 0.$$

The solution of this system of inequalities is $x \leq 0$

Next we solve the inequalities

$$x - 3 \geq 0 \quad \text{and} \quad x^2 - 4x > (x - 3)^2.$$

That is,

$$x - 3 \geq 0 \quad \text{and} \quad 2x > 9.$$

Solving we obtain $x \geq 3$ and $x > \frac{9}{2}$. Hence $x > \frac{9}{2}$.

Thus, the solution of the original inequality is

$$(-\infty, 0] \cup (\frac{9}{2}, \infty).$$

Example 7.6.4. Solve the inequality

$$\sqrt{3x^2 + 19x + 20} > 4(x - 1).$$

Solution. First we solve the inequalities

$$4(x - 1) < 0 \quad \text{and} \quad 3x^2 + 19x + 20 \geq 0.$$

That is,

$$x \geq 1 \quad \text{and} \quad (x + 5)(3x + 4) \geq 0.$$

Solving we obtain

$$x < 1 \quad \text{and} \quad x \leq -5, \quad x \geq -\tfrac{4}{3}.$$

Hence

$$x \leq -5, \quad -\frac{4}{3} \leq x < 1.$$

Next we solve the inequalities

$$4(x - 1) \geq 0 \quad \text{and} \quad 3x^2 + 19x + 20 > 16(x - 1)^2.$$

That is,

$$x \geq 1 \quad \text{and} \quad (x-4)(13x+1) < 0.$$

Solving we obtain

$$x \geq 1 \quad \text{and} \quad -\tfrac{1}{13} < x < 4.$$

Hence

$$1 \leq x < 4.$$

Thus, the solution of the original inequality is

$$(-\infty, -5] \cup [-\tfrac{4}{3}, 1) \cup [1, 4) = (-\infty, -5] \cup [-\tfrac{4}{3}, 4).$$

Example 7.6.5. Solve the inequality

$$\sqrt{2x+1} - \sqrt{x-8} > 3.$$

Solution. For the radicals to be defined, we must have $2x+1 \geq 0$ and $x-8 \geq 0$. That is, $x \geq -\tfrac{1}{2}$ and $x \geq 8$. Both these inequalities are true if $x \geq 8$.

Now we solve

$$\sqrt{2x+1} > 3 + \sqrt{x-8} \quad \text{and} \quad x \geq 8.$$

Squaring both sides of the first inequality and simplifying, we obtain

$$x > 6\sqrt{x-8} \quad \text{and} \quad x \geq 8.$$

Squaring again and bringing all terms to the first side, we obtain

$$x^2 - 36x + 288 > 0 \quad \text{and} \quad x \geq 8.$$

Solving the quadratic inequality, we see that

$$x > 24, \ x < 12 \quad \text{and} \quad x \geq 8.$$

Therefore, the solution of the original inequality is

$$x > 24 \quad \text{and} \quad 8 \leq x < 12.$$

Example 7.6.6. Solve the inequality

$$\sqrt{\frac{x^3 + 8}{x}} > x - 2.$$

Solution.

First we solve the inequalities

$$x - 2 < 0 \quad \text{and} \quad \frac{x^3 + 8}{x} \geq 0.$$

That is,

$$x < 2 \quad \text{and} \quad x(x^3 + 8) \geq 0.$$

Factoring the second inequality

$$x < 2 \quad \text{and} \quad x(x + 2)(x^2 - 2x + 4) \geq 0.$$

Solving, we obtain

$$x < 2 \quad \text{and} \quad x \leq -2, \ 0 < x.$$

Hence

$$x \leq -2, \ 0 < x < 2, \text{ or } (-\infty, -2] \cup (0, 2).$$

Next we solve the inequalities

$$x - 2 \geq 0 \quad \text{and} \quad \frac{x^3 + 8}{x} > (x - 2)^2.$$

That is,

$$x \geq 2 \quad \text{and} \quad \frac{4x^2 - 4x + 8}{x} > 0$$

or

$$x \geq 2 \quad \text{and} \quad 4(x + 1)(x - 2)x > 0.$$

Solving we obtain

$$x \geq 2 \quad \text{and} \quad -1 < x < 0, \ x > 2.$$

Hence

$$x \geq 2, \text{ or } [2, \infty).$$

Thus, the solution of the original inequality is

$$x \leq -2, \ x > 0, \text{ or } (-\infty, -2] \cup (0, \infty).$$

7.6.1   Exercises

Solve the inequalities.

1.
   (a) $\sqrt{2x-1} < x - 2$ (b) $\sqrt{5-2x} < 6x - 1$.

2.
   (a) $\sqrt{x+61} < x + 5$ (b) $\sqrt{2x-x^2} < 5 - x$.

3.
   (a) $2\sqrt{4-x^2} < x + 4$ (b) $\sqrt{2x^2 - 3x - 5} < x - 1$.

4.
   (a) $x < \sqrt{2-x}$ (b) $x < \sqrt{9x - 20}$.

5.
   (a) $\sqrt{2x+5} > x + 1$ (b) $x + 2 < \sqrt{x + 14}$.

6.
   (a) $x < \sqrt{x^2 - 1}$ (b) $1 - x < \sqrt{x^2 - 2x}$.

7.
   (a) $x < \sqrt{x^2 + x - 2}$ (b) $\sqrt{x^2 + x} > 1 - 2x$.

8.
   (a) $\sqrt{x^2 - 3x - 10} < 8 - x$ (b) $\sqrt{x^2 - 3x - 4} > x - 2$.

9.
   (a) $x - 3\sqrt{x-3} > 1$ (b) $3\sqrt{x} - \sqrt{x+3} > 1$.

10.
   (a) $\sqrt{2x+4} < \sqrt{x+3} + \sqrt{x+2}$ (b) $\sqrt{x+3} - \sqrt{x-1} \geq \sqrt{2x-1}$.

11.
   (a) $\dfrac{1 - \sqrt{1 - 4x^2}}{x} > \dfrac{3}{2}$ (b) $\dfrac{2}{x} + 3 \leq \sqrt{41 - \dfrac{16}{x^2}}$.

7.7 Review Exercises

Solve the inequalities.

1.

    (a) $2(x+4) - (x+6) < 12 - x$ **(b)** $x + \dfrac{x}{6} + \dfrac{5}{3} \geq 2(1+x)$.

2.

    (a) $|x+8| < 3x - 1$ **(b)** $|3x - 2| < |3x + 1|$.

3.

    (a) $|2x + 1| + |6x| > 9$ **(b)** $|x - 1| + |x - 2| < 2x + 1$.

4.

    (a) $\dfrac{|x| + 1}{2} - \dfrac{2|x|}{3} > \dfrac{1 - |x|}{3}$ **(b)** $|2x+1| - 4|x-3| - |x-4| > 3$.

5.

    (a) $|x-1| + |x-2| + |x-3| < x+1$ **(b)** $x^2 + 5x - 14 > 0$.

6.

    (a) $2x^2 - 5x - 3 \leq 0$ **(b)** $5x - 20 \leq x^2 \leq 8x$.

7.

    (a) $x^3 - 7x - 6 > 0$ **(b)** $x^4 + 8x > 2x^3 + 3x^2 + 4$.

8.

    (a) $(x-1)(x^2+x-6)(2x^2+x+1) < 0$ **(b)** $(3-x)(2x^2+6x)(x^2+3) \leq 0$.

9.

    (a) $|x^2 - 5x| \geq 6$ **(b)** $x^2 - |5x + 6| > 0$.

10.

    (a) $x^8 - x^5 + x^2 - x + 1 > 0$ **(b)** $\dfrac{1}{x+2} < \dfrac{3}{x-3}$.

11.

    (a) $\left(x + \dfrac{1}{5}\right)^2 < \left(x - \dfrac{1}{3}\right)\left(x + \dfrac{1}{15}\right)$ **(b)** $\dfrac{2}{3x+1} > \dfrac{1}{x+1} - \dfrac{1}{x-1}$.

12.

    (a) $\dfrac{3}{x-2} - \dfrac{1}{2} < \dfrac{x+1}{x-2}$ **(b)** $\dfrac{5 - 4x}{3x^2 - x - 4} < 4$.

13.

    (a) $\dfrac{x^2 - 7x + 12}{x^2 - 2x - 3} \geq 0$ **(b)** $\dfrac{x^2 - 3x + 2}{x^2 + 3x + 2} \geq 1$.

14.

    (a) $\dfrac{2x - 5}{x^2 - 6x - 7} < \dfrac{1}{x-3}$ **(b)** $\dfrac{|x - 1|}{x + 2} < 1$.

15.

    (a) $x^2 + \dfrac{2}{2x - 1} \geq \dfrac{1}{x(2x - 1)}$ **(b)** $\dfrac{x^2 + 2}{\sqrt{x^2 + 1}} \geq 2$.

16.

    (a) $\left|\dfrac{x^2 - 1}{x + 2}\right| < 1$ **(b)** $\sqrt{\left(\dfrac{x + 1}{3 - 2x}\right)^2} > 1$.

17.

    (a) $\dfrac{|2x - 1|}{x^2 - x - 2} > \dfrac{1}{2}$ **(b)** $\sqrt{x^2 - 4x} + 3 > x$.

18.

    (a) $6 - 3x < \sqrt{8 + 2x - x^2}$ **(b)** $\dfrac{\sqrt{2x - 1}}{x - 2} < 1$.

19.

$$\text{(a)} \quad \sqrt{\frac{1}{x^2} - \frac{1}{4}} > \frac{1}{x} - \frac{1}{4} \qquad \text{(b)} \quad \sqrt{\frac{1}{x^2} - \frac{3}{x}} < \frac{1}{x} - \frac{1}{2}$$

20. Find the values of the parameter $k \in \mathbb{R}$ for which the quadratic equation

$$x^2 - (k+1)x + k + 4 = 0$$

has: (a) Equal roots. (b) Real distinct roots. (c) Complex roots.

21. Find the values of the parameter $\lambda \in \mathbb{R}$ for which the quadratic equation

$$(\lambda - 2)x^2 - 2(\lambda + 3)x + 2\lambda - 18 = 0$$

has (a) Positive roots. (b) Negative roots.

22. For what values of $\lambda \in \mathbb{R}$ is the inequality

$$x^2 - (\lambda - 3)x + \lambda + 6 > 0$$

is valid for all $x \in \mathbb{R}$?

23. For what values of $\lambda \in \mathbb{R}$ does the equation

$$(2 - x)(x + 1) = \lambda$$

has real and positive roots?

24. For what values of $\lambda \in \mathbb{R}$ do the roots of the equation

$$2x^2 + 6x + \lambda = 0$$

satisfy the condition

$$\frac{x_1}{x_2} + \frac{x_2}{x_1} < 2.$$

25. Solve the equation

$$|x^2 + 4x + 2| = \frac{5x + 16}{3}.$$

Hint: Consider the cases $x^2 + 4x + 2 \geq 0$ and $x^2 + 4x + 2 < 0$ separetely.

26. Solve the inequality

$$\sqrt{4 - \sqrt{x - 1}} - \sqrt{2 - x} > 0.$$

27. Solve the inequality

$$\sqrt{x + 2\sqrt{x - 1}} + \sqrt{x - 2\sqrt{x - 1}} > \frac{3}{2}.$$

28. Solve the inequality

$$\frac{2x^2}{1 - \sqrt{1 - x^2}} \leq 3.$$

29. Solve the inequality

$$\frac{1 - \sqrt{1 - 4x^2}}{x} < 3.$$

30. Solve the inequality

$$\sqrt{x + \frac{1}{x^2}} + \sqrt{x - \frac{1}{x^2}} > \frac{2}{x}.$$

31. Solve the inequality

$$|x^2 - 2x - 8| > 2x.$$

Chapter 8

Systems of Equations

In this chapter, we study systems of algebraic equations. We solve systems of linear equations and some simple systems of nonlinear algebraic equations.

A number of equations in two, three, four, etc, unknowns is called a ***system of equations***. A ***solution*** of a system is pair of numbers, a triple of numbers, a quatriple of numbers, etc, which *satisfy all* the given equations. For example,

$$\begin{aligned} x + y &= 2 \\ x - y &= 0 \end{aligned}$$

is a system of two equations in two unknowns x and y. The pair of numbers $(1, 2)$, that is, $x = 1$ and $y = 1$, satisfying both equations of the system is a solution of the system.

Definition 8.0.1. A system of equations is said:

1. ***Consistent*** if it has at least one solution;
   determined if it has a unique solution,
   indeterminate if it has more than one solution.

2. ***Inconsistent*** if it does not have any solution.

That is, systems of equations may have no solution, ie, be *inconsistent*, or may have solutions, ie, be *consistent*. Consistent systems break down into *determinate* systems and *inteterminate* systems.

Two systems are said to be *equivalent* if any solution of one system is a solution of the other, and vice versa. If both systems of equations have no solutions, then they are also regarded to be equivalent.

Here are some basic statements which make equivalent transition possible and are frequently used in the process of solving systems:

1. Solving one equation of the system for one unkown, say y in terms of x, and substituting the obtaind expression for $y = f(x)$ into the other equation. That is,

$$\begin{array}{ll} A(x,y) = B(x,y) \\ C(x,y) = D(x,y). \end{array} \quad \Leftrightarrow \quad \begin{array}{ll} y = f(x) \\ C(x,f(x)) = D(x,f(x)). \end{array}$$

2. Adding a (nonzero) multiple of one equation with a multiple of the other equation. That is, for $k, \lambda \neq 0$, $A = A(x,y)$, $B = B(x,y)$, $C = C(x,y)$ and $D = D(x,y)$, we have

$$\begin{array}{ll} A = B \\ C = D. \end{array} \quad \Leftrightarrow \quad \begin{array}{ll} A = B \\ kA + \lambda C = kB + \lambda D. \end{array}$$

8.1 Systems of Linear Equations

8.1.1 Systems of Two Linear Equations in Two Unknowns

A system of two linear equations in two unkwons x and y has the *general form*

$$\begin{array}{l} a_1 x + b_1 y = c_1 \\ a_2 x - b_2 y = c_2 \end{array},$$

where a_1, b_1, a_2, b_2, c_1 and c_2 are given numbers

Let us consider two basic methods for solving this system.

1. The Method of Substitution

This method consists of the following steps.

1. From one of the two equations express one unknown in terms of the other, for instance, express y in terms of x;

2. Substituting the found expression into the other equation of the system, obtain one equation in one unknown x;

3. Solving this equation, find the value of x;

4. Substituting the obtained value of x into the expression for y, find the value of y.

Example 8.1.1. Solve the system

$$
\begin{aligned}
2x + 3y &= 1 \\
x - y &= 3
\end{aligned}
$$

Solution. From the second equation we find $y = x - 3$. Substituting the expression for y into the first equation, we obtain

$$2x + 3(x - 3) = 1.$$

Solving this equation, we find

$$2x + 3x - 9 = 1 \quad \Leftrightarrow \quad 5x = 10 \quad \Leftrightarrow \quad x = 2.$$

Substituting $x = 2$ into the expression for y, we get

$$y = 2 - 1 = -1.$$

Thus, the solution is $x = 2$ and $y = -1$, or written in the form of an ordered pair of numbers $(2, -1)$

2. The Method of Addition

This method consists of the following steps.

1. Write the system in general form;

2. Multiply all the terms of one or both of the equations by constants chosen to make the coefficients of x (or y) differ only in sign;

3. Adding the obtained equations termiwise, find one of the unknowns.

4. Substituting the found value into one of the equations, find the second unknown.

This method is also known as the ***method of elimination***.

Let us solve the system of Example 8.1.1, using the addition method.

Example 8.1.2. Solve the system

$$
\begin{aligned}
3y &= 1 - 2x \\
x - y &= 3
\end{aligned}
$$

Solution. Writting the system in general form, we have

$$
\begin{aligned}
2x + 3y &= 1 \\
x - y &= 3
\end{aligned}
$$

Multiplying the second equation by 3, we obtain the equivalent system

$$
\begin{aligned}
2x + 3y &= 1 \\
3x - 3y &= 9
\end{aligned}
$$

Adding the equations to eliminate y, we find

$$5x = 10 \quad \Leftrightarrow \quad x = 2.$$

We now substitute $x = 2$ into either of the original equations and solve for y. If we use the first equation, we get

$$3y = 1 - 2(2) \quad \Leftrightarrow \quad 3y = -3 \quad \Leftrightarrow \quad y = -1.$$

The solution is $(2, -1)$.

Example 8.1.3. Solve the system

$$3x + 8y = 20$$
$$-2x + 3y = -55$$

Solution. We eliminate x. Multiplying both sides of the first equation second by 2 and both sides of the second equation by 3, we obtain the equivalent system

$$6x + 16y = 40$$
$$-6x + 9y = -165$$

Adding the equations to eliminate x, we find the equivalent system

$$3x + 8y = 20$$
$$25y = -125$$

Solving the second equation, we find

$$25y = -125 \iff y = -5.$$

Substituting now $y = -5$ into the first equation, we get

$$3x + 8(-5) = 20 \iff 3x = 60 \iff x = 20.$$

The solution is $(20, -5)$.

Example 8.1.4. Solve the system

$$2x - 3y = -4$$
$$5x + y = 7$$

Solution. Let us solve this system with both methods:

1. Method of substitution. From the second equation, we find $y = 7 - 5x$. Substituting this expression into the first equation of the system , we get

$$2x - 3(7 - 5x) = -4 \iff 17x = 17 \iff x = 1.$$

Therefore, $y = 7 - 5(1) = 2$. The solution is $x = 1$ and $y = 2$.

2. Method of addition. Multiplying both sides of the second equation by 3 (to eliminate y), we obtain

$$2x \;-3y = -4$$
$$15x \;+3y = 21$$

Adding these equations termwise, we get

$$17x = 17 \quad\Leftrightarrow\quad x = 1.$$

Then from the first equation of the original system, we find

$$2(1) - 3y = -4 \quad\Leftrightarrow\quad -3y = -6y \quad\Leftrightarrow\quad y = 2.$$

We found, of course, the same solution $x = 1$ and $y = 2$.

Example 8.1.5. Solve the system

$$3x \;-y = 1$$
$$12x \;-4y = 4$$

Solution. Note that the second equation is a multiple of the first. In fact,

$$12x - 4y = 4 \quad\Leftrightarrow\quad 4(3x - y) = 4(1) \quad\Leftrightarrow\quad 3x - y = 1.$$

That is, the given system is equivalent to one equation $3x - y = 1$. Consequently, it is satisfied by any pair of numbers x and $y = 3x - 1$. In other words, the system has (infinitely) many solutions

$$(x, 3x - 1), \text{ where } x \in \mathbb{R}.$$

Note. If we multiply both sides of the first equation by -4, and we add the resulting equation $-12x + 4y = -4$ to the second equation of the system, we obtain

$$0 = 0, \text{ which is always true.}$$

Consequently, we conclude as above, that the original system has an infinitude of solutions given by the formula $y = 3x - 1$, where x may be any number.

Example 8.1.6. Solve the system

$$\begin{aligned} x \ +y &= 5 \\ 2x \ +2y &= 8 \end{aligned}$$

Solution. This system of equations is not satisfied by any pair of numbers. Indeed, if we multiply the first equation by -2 and we add it to the second, we get $0 = 3$, which is impossible. Thus, the given system has no solution.

Now let us **investigate** the general system of two linear equations in two unknowns

$$\begin{aligned} a_1 x \ +b_1 y &= c_1 \\ a_2 x \ +b_2 y &= c_2 \end{aligned} \qquad (8.1)$$

Suppose that a_1, b_1, a_2, b_2, c_1 and c_2 are all non zero numbers. Multiplying the first equation by $-\frac{a_2}{a_1}$, we obtain the equivalent system

$$\begin{aligned} -a_2 x \ -\tfrac{a_2 b_1}{a_1} y &= -\tfrac{a_2 c_1}{a_1} \\ a_2 x \ +b_2 y \quad &= c_2 \end{aligned}$$

Adding the first equation to the second equation, we get

$$\left(b_2 - \frac{a_2 b_1}{a_1} \right) y = c_2 - \frac{c_1 a_2}{a_1} \quad \Leftrightarrow$$

$$(a_1 b_2 - a_2 b_1) y = a_1 c_2 - a_2 c_1. \qquad (8.2)$$

Consequently, if $a_1 b_2 - a_2 b_1 \neq 0$, then

$$y = \frac{a_1 c_2 - a_2 c_1}{a_1 b_2 - a_2 b_1}. \qquad (8.3)$$

Substituting this expression for y into the first equation of the given system, we find

$$x = \frac{b_2 c_1 - b_1 c_2}{a_1 b_2 - a_2 b_1}. \qquad (8.4)$$

Thus, if $a_1 b_2 - a_2 b_1 \neq 0$, then the system has a ***unique solution*** which can be found by the formulas (8.3) and (8.4).

Suppose now that $a_1b_2 - a_2b_1 = 0$. In this case, if $a_1c_2 - a_2c_1 \neq 0$, then equation (8.2), as well as the system (8.1), has **no solution**.

On the other hand, if $a_1c_2 - a_2c_1 = 0$, equation (8.2) becomes $0 = 0$. Consequently, the system (8.1) is equivalent to a single equation $a_1x + b_1y = c_1$. Hence y can be any number, and any pair of numbers (x, y), where

$$x = \frac{c_1}{a_1} - \frac{b_1}{a_1}y, \quad y \in \mathbb{R} \tag{8.5}$$

is a solution of the system. In other words, system (8.1) has **infinitely many solutions**.

Remark 8.1.7. Note that in this latter case, we have $a_1b_2 - a_2b_1 = 0 \Leftrightarrow \frac{a_1}{a_2} = \frac{b_1}{b_2}$ and $a_1c_2 - a_2c_1 = 0 \Leftrightarrow \frac{a_1}{a_2} = \frac{c_1}{c_2}$. Hence ,

$$\frac{a_1}{a_2} = \frac{b_1}{b_2} = \frac{c_1}{c_2}. \tag{8.6}$$

Conversely, if condition (8.6) holds, then system (8.1) has infinitely many solutions. Indeed, Setting $\frac{a_1}{a_2} = \frac{b_1}{b_2} = \frac{c_1}{c_2} = k \neq 0$, then $a_1 = ka_2$, $b_1 = kb_2$, and $c_1 = kc_2$. Therefore, the first equation of system (8.1) becomes $k(a_2x + b_2y) = kc_2$. That is, it is a multiple of the second equation of the system. Hence, the system reduces to one equation $a_2x + b_2y = c_2$, which has infinitely many solutions.

Definition 8.1.8. The expression

$$\begin{vmatrix} a_1 & b_1 \\ a_2 & b_2 \end{vmatrix} = a_1b_2 - a_2b_1$$

is called a **determinant of order** 2.

Using determinant notation, and denoting by

$$\Delta = \begin{vmatrix} a_1 & b_1 \\ a_2 & b_2 \end{vmatrix} = a_1b_2 - a_2b_1,$$

$$\Delta_x = \begin{vmatrix} c_1 & b_1 \\ c_2 & b_2 \end{vmatrix} = c_1b_2 - c_2b_1$$

and

$$\Delta_y = \begin{vmatrix} a_1 & c_1 \\ a_2 & c_2 \end{vmatrix} = a_1 c_2 - a_2 c_1$$

the solution (8.4) and (8.3) may be written in the form

$$x = \frac{\Delta_x}{\Delta}, \quad y = \frac{\Delta_y}{\Delta},$$

respectively. These formulas for the solution of the linear system (8.1) are known as **Cramer's Rule**[1]. This rule provides us with another method for solving the system (8.1).

Example 8.1.9. Solve the system using Cramer's rule.

$$\begin{aligned} 5x & - 3y = 1 \\ 3x & + 2y = 12 \end{aligned}$$

Solution. We have

$$x = \frac{\Delta_x}{\Delta} \iff x = \frac{\begin{vmatrix} 1 & -2 \\ 12 & 2 \end{vmatrix}}{\begin{vmatrix} 5 & -3 \\ 3 & 2 \end{vmatrix}} \iff \frac{2 + 36}{10 + 9} = \frac{38}{19} = 2.$$

and

$$y = \frac{\Delta_y}{\Delta} \iff y = \frac{\begin{vmatrix} 5 & 1 \\ 3 & 12 \end{vmatrix}}{\begin{vmatrix} 5 & -3 \\ 3 & 2 \end{vmatrix}} \iff \frac{60 - 3}{10 + 9} = \frac{57}{19} = 3.$$

Thus, the solution is $x = 2$ and $y = 3$.

---

[1]Gabriel Cramer (1704-1752) a Swiss mathematician, made contributions in geometry, analysis, and determinants.

Example 8.1.10. For what values of the parameter λ does the system

$$(\lambda + 1)x + 8y = 4\lambda$$
$$\lambda x + (\lambda + 3)y = 3\lambda - 1$$

have an infinite number of solutions?

Solution. For the system to have an infinite number of solutions we must have $\Delta = a_1 b_2 - a_2 b_1 = 0$, and $\Delta_y = a_1 c_2 - a_2 c_1 = 0$. We compute Δ and Δ_y.

$$\Delta = \begin{vmatrix} \lambda + 1 & 8 \\ \lambda & \lambda + 3 \end{vmatrix} = (\lambda + 1)(\lambda + 3) - 8\lambda = \lambda^2 - 4\lambda + 3.$$

$$\Delta_y = \begin{vmatrix} \lambda + 1 & 4\lambda \\ \lambda & 3\lambda - 1 \end{vmatrix} = (\lambda + 1)(3\lambda - 1) - 4\lambda^2 = (\lambda - 1)^2.$$

Threfore,

$$\Delta = 0 \iff \lambda^2 - 4\lambda + 3 = 0 \iff (\lambda - 3)(\lambda - 1) = 0 \iff \lambda = 3, \ \lambda = 1.$$

and

$$\Delta_y = 0 \iff (\lambda - 1)^2 = 0 \iff \lambda = 1.$$

Hence when $\lambda = 1$, then both $\Delta = 0$ and $\Delta_y = 0$. Thus, the given system has infinitely many solutions for $\lambda = 1$.

Example 8.1.11. Consider the system

$$x + \lambda y = 4$$
$$\lambda x + 9y = k$$

1. For which values of the parameter λ the system has a unique solution?

2. For which values of λ and k the system has infinitely many solutions?

Solution. First we compute the determinants of Δ, Δ_x and Δ_y. We have

$$\Delta = \begin{vmatrix} 1 & \lambda \\ \lambda & 9 \end{vmatrix} = 9 - \lambda^2,$$

$$\Delta_x = \begin{vmatrix} 4 & \lambda \\ k & 9 \end{vmatrix} = 36 - \lambda k,$$

$$\Delta_y = \begin{vmatrix} 1 & 4 \\ & k \end{vmatrix} = k - 4\lambda.$$

1. The system has a unique solution if and only if $\Delta \neq 0$. That is,

$$9 - \lambda^2 \neq 0 \iff \lambda \neq -3, \ \lambda \neq 3.$$

   The unique solution is $x = \frac{\Delta_x}{\Delta} = \frac{36 - \lambda k}{9 - \lambda^2}$ and $y = \frac{\Delta_y}{\Delta} = \frac{k - 4\lambda}{9 - \lambda^2}$.

2. The system has infinitely many solutions if and only if $\Delta = 0$ and $\Delta_y = 0$. Now $\Delta = 0 \iff \lambda = \pm 3$, and $\Delta_y = 0 \iff k = 4\lambda$.

   When $\lambda = 3$, then $k = 12$. When $\lambda = -3$, then $k = -12$.

The Homogeneous System

The system of linear equations

$$\begin{aligned} a_1 x &+ b_1 y = 0 \\ a_2 x &+ b_2 y = 0 \end{aligned} \tag{8.7}$$

is called the corresponding *homogeneous system* of system (8.1). Here $c_1 = c_2 = 0$. Therefore, $\Delta_x = 0$, $\Delta_y = 0$.

It is important to note that, if the determinant of the coefficients

$$\Delta = \begin{vmatrix} a_1 & b_1 \\ a_2 & b_2 \end{vmatrix} \neq 0,$$

then $x = \frac{0}{\Delta} = 0$ and $y = \frac{0}{\Delta} = 0$. That is, the homogeneous system (8.7) has *only the zero solution* $x = y = 0$. On the other hand, if

$$\Delta = \begin{vmatrix} a_1 & b_1 \\ a_2 & b_2 \end{vmatrix} = 0,$$

then the homogeneous system (8.7) has (infinitely) *many nonzero solutions*. In fact, from (8.5), we see that y may be any number and $x = -\frac{b_i}{a_1}y$. Thus, the nonzero solutions are

$$(-\frac{b_1}{a_1}y,\ y),\quad y \in \mathbb{R}.$$

Example 8.1.12. For what values of the parameter λ does the homogeneous system

$$(1 - \lambda)x + 3y = 0$$
$$2x + (\lambda + 4)y = 0$$

have nonzero solutions?

Solution. In order the system to have nonzero solutions, the determinant of the coefficients must be equal to zero, that is,

$$\Delta = \begin{vmatrix} 1 - \lambda & 3 \\ 2 & \lambda + 4 \end{vmatrix} = 0.$$

Expanding the determinant, we obtain

$$(1 - \lambda)(\lambda + 4) - (2)(3) = 0 \quad \Leftrightarrow \quad \lambda^2 + 3\lambda + 2 = 0.$$

Solving, this quadratic equation we find $\lambda = -1$ and $\lambda = -2$.

8.1.2 Systems of Three Linear Equations in Three Unknowns

The simplest way to solve these systems is by the elimination method. Here is an example

Example 8.1.13. Solve the system

$$\begin{aligned} x + y + z &= 6 \\ 2x + y + 3z &= 13 \\ 3x + y + z &= 8. \end{aligned}$$

Solution. Multiplying the first equation by -2 and adding it termwise to the second equation, we get

$$-y + z = 1 \quad \Leftrightarrow \quad y - z = -1.$$

Next we multiply the first equation by -3 and we add it to the third equation and obtain

$$-2y - 2z = -10.$$

Hence, we get the equivalent system

$$
\begin{array}{rrrl}
x & +y & +z & = 6 \\
& y & -z & = -1 \\
& -2y & -2z & = -10.
\end{array}
$$

Now, we multiply the second equation by 2 and we add it to the third equation and obtain

$$-4z = -12.$$

As a result, we get the equivalent system

$$
\begin{array}{rrrl}
x & +y & +z & = 6 \\
& y & -z & = -1 \\
& & -4z & = -12.
\end{array}
$$

Systems of this kind are called **triangular** (or **echelon**). They are easy to solve. Indeed, from the third, second and first equations we successively find $z = 3$, $y = z - 1 = 3 - 1 = 2$ and $x = 6 - y - z = 6 - 2 - 3 = 1$. Thus, the solution is $x = 1$, $y = 2$, $= 3$, or the triple $(1, 2, 3)$.

The method we used above is called the **Gaussian elimination method**. In Section 8.3, we elaborate this method using matrix notation. The Gaussian elimination method is so practical and effective that allows us to solve any general linear systems of m linear equations in n uknowns. Systems of n linear equations in n uknowns are called *square systems*.

8.1.3 Exercises

Solve the systems.

1.

(a) $\begin{array}{l} x - y = 1 \\ 2x - y = 0 \end{array}$ (b) $\begin{array}{l} 2x + y = 3 \\ 5x - y = 11 \end{array}$.

2.

(a) $\begin{array}{l} 5x - 3y = 12 \\ 2x - 3y = 3 \end{array}$ (b) $\begin{array}{l} 3x + 9y = 9 \\ -x - 5y = 3 \end{array}$.

3.

(a) $\begin{array}{l} 2x + 2y = 1 \\ 2x - 2y = 3 \end{array}$ (b) $\begin{array}{l} x + 2y = 1 \\ 3x + 6y = 3 \end{array}$.

4.

(a) $\begin{array}{l} x - 2y + 17 = 0 \\ 3x + y + 16 = 0 \end{array}$ (b) $\begin{array}{l} 2x - y - 6 = 0 \\ 4x - 2y - 12 = 0 \end{array}$.

5.

(a) $\begin{array}{l} x + y = 3 \\ 2x + 2y - 6 = 0 \end{array}$ (b) $\begin{array}{l} x + 3y = 2 \\ 3x - 5 = -9y \end{array}$.

6.

(a) $\begin{array}{l} 4x + y = 8 \\ 4x + 3y = 24 \end{array}$ (b) $\begin{array}{l} 2x - 5y = 10 \\ -x - \frac{5}{2}y = 5 \end{array}$.

7.

(a) $\begin{array}{l} \frac{x}{3} - \frac{y}{2} = 1 \\ 2x - 5y = -2 \end{array}$ (b) $\begin{array}{l} 2(x + y) - y = 1 \\ 3(x + 1) - y = -3 \end{array}$.

8.

(a) $\begin{array}{l} 3x = 4(2 - y) \\ 3(x - 2) = -4y \end{array}$ (b) $\begin{array}{l} \frac{2}{3}x + \frac{1}{9}y = 1 \\ \\ \frac{3}{2}x + \frac{1}{3}y = 2 \end{array}$.

9.

(a) $\begin{array}{l} \frac{3x - y + 2}{2} = \frac{x + 2y}{5} \\ \\ \frac{x - 2y - 3}{3} = \frac{2x - y}{2} \end{array}$ (b) $\begin{array}{l} \frac{x - 2y + 8}{3} + \frac{x + y - 6}{2} = \frac{x + 4}{3} \\ \\ x - 3y = \frac{3x}{4} - 5 \end{array}$

10.

$$\text{(a)} \quad \begin{aligned} x + y + z &= 3 \\ 2x + y + z &= 4 \\ 3x + y - z &= 5. \end{aligned} \qquad \text{(b)} \quad \begin{aligned} x + y + 2z &= 4 \\ 3x + 6y + 10z &= 17 \\ 2x + 3y + 6z &= 10. \end{aligned}$$

11.

$$\text{(a)} \quad \begin{aligned} x - 2y + 3z &= 2 \\ 2x - 3y + 8z &= 7 \\ 3x - 4y + 13z &= 8. \end{aligned} \qquad \text{(b)} \quad \begin{aligned} x + 2y + 3z &= 3 \\ 2x + 3y + 8z &= 4 \\ 5x + 8y + 19z &= 11. \end{aligned}$$

12.

$$\text{(a)} \quad \begin{aligned} x + y - z &= 6 \\ x + 3y - 2z &= 14 \\ 3x - 2y + z &= -5. \end{aligned} \qquad \text{(b)} \quad \begin{aligned} 2x - y - 4z &= 2 \\ 4x - 2y - 6z &= 5 \\ 6x - 3y - 8z &= 8. \end{aligned}$$

13. For what values of the parameter λ does the system

$$\begin{aligned} x + \lambda y &= 1 \\ \lambda x - 3y &= 2\lambda + 3 \end{aligned}$$

(a) have no solution?   (b) have many solutions?

14. Find the values of the parameters λ and k for which each system has infinitely many solutions

$$\text{(a)} \quad \begin{aligned} \lambda x + 3y &= 2 \\ 12x + \lambda y &= k \end{aligned} \qquad \text{(b)} \quad \begin{aligned} (2\lambda - 1)x + (4k+1)y &= 3 \\ (\lambda + 1)x + (k - 2)y &= 3. \end{aligned}$$

15. For what values of the parameter λ does the system

$$\begin{aligned} (5 - \lambda)x + 6y &= 0 \\ 3x + (2 + \lambda)y &= 0 \end{aligned}$$

have nonzero solutions?

8.2 Matrix Form of Systems of Linear Equations (*)

In discussing methods to solve systems of linear equations, it is convinient to introduce the concept of a matrix.[2] The word "matrix" was introduced by J. J. Sylvester in 1848, as a name of a rectangular array of numbers.

Definition 8.2.1. An $m \times n$ **matrix**, denoted by A, is a rectangular array of numbers arranged into m (horizontal) rows and n (vertical) columns of the form

$$A = \begin{pmatrix} a_{11} & a_{12} & ... & a_{1n} \\ a_{21} & a_{22} & ... & a_{2n} \\ . & . & ... & . \\ . & . & ... & . \\ . & . & ... & . \\ a_{m1} & a_{m2} & ... & a_{mn} \end{pmatrix}.$$

Briefly written as $A = (a_{ij})$ for $i = 1, 2, ..., m$ and $j = 1, 2, ..., n$.

The numbers a_{ij} are called the *entries* of A. If $m = n$, the matrix is called a *square matrix* of order n. Note that an 1×1 matrix is simply a number.

For instance, a 3×4 matrix is written in the general form as

$$A = \begin{pmatrix} a_{11} & a_{12} & a_{13} & a_{14} \\ a_{21} & a_{22} & a_{23} & a_{24} \\ a_{31} & a_{32} & a_{33} & a_{34} \end{pmatrix}.$$

---

[2]Matrices are studied in a course of Linear Algebra. Here we shall inrtoduce only those concepts which we shall need for solving systems of linear equations by the Gaussian elimination method. For solving *square systems* of linear equations, other methods are also used, such as the method of determinants (Cramer's Rule), or matrix methods, such as multiplication by the inverse matrix of the coefficient matrix. Nevertheless the student mastering Gauss's method of elimination will be in position of *solving* (and even *investigating*) a system of linear equations of any size!

Example 8.2.2. The matrix $A = \begin{pmatrix} 5 & -3 & 0 \\ 2 & 4 & 1 \\ 3 & 0 & -8 \end{pmatrix}$ is a 3×3 matrix. The

matrix $\begin{pmatrix} 1 & \frac{3}{4} \\ 2 & -3 \\ 0 & 9 \end{pmatrix}$ is a 3×2 matrix.

We denote by $R_i = (a_{i1} \ a_{i2} \ ... \ a_{in})$ the ith-*row* of A, and by

$$C_i = \begin{pmatrix} a_{1j} \\ a_{2j} \\ . \\ . \\ . \\ a_{mj} \end{pmatrix}$$ the jth- *column* of A for $i = 1, 2, ..., m$ and $j = 1, 2, ..., n$.

A $1 \times n$ matrix, ie, a matrix which consists of one row is said a *row matrix*. For example, $(-2 \ 5 \ 8)$ is an 1×3 row matrix.

An $m \times 1$ matrix, ie, a matrix which consists of one column is said

a *colun matrix*. For example, $\begin{pmatrix} 0 \\ 1 \\ -3 \end{pmatrix}$ is a 3×1 column matrix.

Definition 8.2.3. Let $A = (a_1 \ a_2 \ .... \ a_n)$ be an $1 \times n$ row matrix, and

$$X = \begin{pmatrix} x_1 \\ x_2 \\ . \\ . \\ . \\ x_n \end{pmatrix}$$ be an $n \times 1$ column matrix.

The **product** AX is defined by

$$AX = (a_1 \ a_2 \ .... \ a_n) \begin{pmatrix} x_1 \\ x_2 \\ . \\ . \\ . \\ x_n \end{pmatrix} = a_1 x_1 + a_2 x_2 + ... + a_n x_n.$$

Note that the result of this multiplication is a number.

Example 8.2.4. Multiply

$$
(2\ 3\ 0) \begin{pmatrix} 5 \\ -4 \\ 7 \end{pmatrix} = (2)(5) + (3)(-4) + (0)(7) = 10 - 12 + 0 = -2.
$$

Definition 8.2.5. The **product** AX of a $m \times n$ matrix $A = (a_{ij})$ and a $n \times 1$ column matrix X is defined by

$$
AX = \begin{pmatrix} a_{11} & a_{12} & ... & a_{1n} \\ a_{21} & a_{22} & ... & a_{2n} \\ . & . & ... & . \\ . & . & ... & . \\ . & . & ... & . \\ a_{m1} & a_{m2} & ... & a_{mn} \end{pmatrix} \begin{pmatrix} x_1 \\ x_2 \\ . \\ . \\ . \\ x_n \end{pmatrix} = \begin{pmatrix} a_{11}x_1 + & a_{12}x_2 + & ... & +a_{1n}x_n \\ a_{21}x_1 + & a_{22}x_2 + & ... & +a_{2n}x_n \\ . & . & ... & . \\ . & . & ... & . \\ . & . & ... & . \\ a_{m1}x_1 + & a_{m2}x_2 + & ... & +a_{mn}x_n \end{pmatrix}.
$$

That is, we multiply each row $R_i = (a_{i1}\ a_{i2}\ ...\ a_{in})$ of the matrix A by the column matrix X, resulting to an $m \times 1$ column matrix with entries $R_i X$, for $i = 1, 2, ..., m$.

Note that the result of this multiplication is an $m \times 1$ column matrix. We illustrate with the following example.

Example 8.2.6. Let $A = \begin{pmatrix} 2 & -3 & 5 \\ -1 & 4 & -2 \\ 6 & 0 & -2 \end{pmatrix}$ and $X = \begin{pmatrix} x \\ y \\ z \end{pmatrix}$. Then

$$
AX = \begin{pmatrix} 2 & -3 & 5 \\ -1 & 4 & -8 \\ 6 & 0 & -2 \end{pmatrix} \begin{pmatrix} x \\ y \\ z \end{pmatrix} = \begin{pmatrix} 2x - 3y + 5z \\ -x + 4y - 8z \\ 6x - 2z \end{pmatrix}.
$$

Definition 8.2.7. Two $m \times n$ matrices $A = (a_{ij})$ and $B = (b_{ij})$ are said to be **equal**, $A = B$, if and only if their corresponding entries are equal. That is, $a_{ij} = b_{ij}$ for all $i = 1, 2, ..., m$ and $j = 1, 2, ..., n$.

A system of m linear equations in n unknowns $x_1, x_2, ..., x_n$ is a set of equations of the general form:

$$
\begin{aligned}
a_{11}x_1 + a_{12}x_2 + ... + a_{1n}x_n &= b_1 \\
a_{21}x_1 + a_{22}x_2 + ... + a_{2n}x_n &= b_2
\end{aligned}
$$

$$. \quad . \quad . \quad . \quad . \quad . \quad . \quad . \tag{8.8}$$

$$a_{m1}x_1 + a_{m2}x_2 + ... + a_{mn}x_n = b_m,$$

where b_i and the coefficients a_{ij}, for $i = 1, 2, ..., m$ and $j = 1, 2, ..., n$, are given numbers. If all $b_i = 0$, then the system is said to be **homogeneous**, and **inhomogeneous**, otherwise. The values of the unknwons $x_1 = r_1, x_2 = r_2, ..., x_n = r_n$ which satisfy each of the equations of the system is called a **solution** of the system. A system of linear equations is said to be **consistent** if it has at least one solution, and **inconsistent** if it has no solution.

Matrix form of the linear system

Let $A = (a_{ij})$ be the matrix of coefficients of the system (8.8). Writting the unknowns $x_1, x_2, ..., x_n$ as $n \times 1$ column matrix X and the constants $b_1, b_2, ..., b_m$ as $m \times 1$ column matrix B, respectively, ie,

$$
X = \begin{pmatrix} x_1 \\ x_2 \\ . \\ . \\ . \\ x_n \end{pmatrix} \quad \text{and} \quad B = \begin{pmatrix} b_1 \\ b_2 \\ . \\ . \\ . \\ b_m \end{pmatrix},
$$

then, according to Definitions 8.2.5 and 8.2.7, we can write the linear system (8.8) in **matrix form** as follows:

$$AX = B.$$

If $B = 0 = \begin{pmatrix} 0 \\ 0 \\ . \\ . \\ . \\ 0 \end{pmatrix}$, then the corresponding homogeneous system is

$$AX = 0.$$

Definition 8.2.8. The matrix,

$$(A|B) = \begin{pmatrix} a_{11} & a_{12} & ... & a_{1n} & | & b_1 \\ a_{21} & a_{22} & ... & a_{2n} & | & b_2 \\ . & . & ... & . & . & . \\ . & . & ... & . & . & . \\ a_{m1} & a_{m2} & ... & a_{mn} & | & b_m \end{pmatrix}$$

is called the ***augmented matrix*** associated to the linear system

$$AX = B.$$

Example 8.2.9. The system

$$\begin{array}{rrrcr} 2x & -3y & +5z & = & 1 \\ x & +4y & -8z & = & 3 \\ 6x & & -2z & = & -4 \end{array}$$

written in matrix form is

$$AX = \begin{pmatrix} 2 & -3 & 5 \\ 1 & 4 & -8 \\ 6 & 0 & -2 \end{pmatrix} \begin{pmatrix} x \\ y \\ z \end{pmatrix} = \begin{pmatrix} 1 \\ 3 \\ -4 \end{pmatrix} = B.$$

Its associated augmented matrix is

$$(A|B) = \begin{pmatrix} 2 & -3 & 5 & | & 1 \\ 1 & 4 & -8 & | & 3 \\ 6 & 0 & -2 & | & -4 \end{pmatrix}.$$

8.3   Gaussian Elimination (*)

We shall now consider in some detail a systematic method of solving systems of linear equations. A basic idea for solving such systems is to transform the given system into a simpler one, without changing its solution set. This is achieved by the **_Gaussian method of elimination_**; that is, we eliminate some unknowns by applying the following three operations, known as **_elementary operations_**:

   1. Interchange two equations.

   2. Multiply both sides of the equation by a nonzero constant, c.

   3. Add a constant multiple of an equation to another equation.

Evidently, none of these operations alters a solution of the system. In fact, each of these three operations has its _inverse_ operation which is also an elementary operation. For example, the inverse of operation (2) is the operation $(2')$: multiply the equation by $\frac{1}{c}$, and the inverse of operation (3) is the operation $(3')$: add the negative of the same constant multiple of the equation to the other. Since each elementary operation has an inverse elementary operation, applying elementary operations we obtain equivalent systems. Note that in performing any of these three elementary operations, only the coefficients of the unkowns are involved in the operations, while the unknowns themselves $x_1, x_2, ..., x_n$ and the equal sign "=" are simply repeated. This suggests that in order to solve a system of linear equations, in a systematic way, one can perform these three elementary operations on the rows of the the augmented matrix associated to the linear system. In this case, the elementary operations are known as the **_elementary row operations_**:

   1. Interchange two rows;

We shall write $R_i \leftrightarrow R_j$.

   2. Multiply a row R_i by a nonzero constant c;

We shall write $R_i \to cR_i$ to indicate "Replace R_i by cR_i".

   3. Add a constant multiple of a row to another row;

We shall write $R_i \to cR_i + R_j$ to indicate "Replace R_i by $cR_i + R_j$."

Definition 8.3.1. Two matrices are said to be **_row-equivalent_** if one can be transformed to the other by a finite sequence of elementary row operations.

Definition 8.3.2. A matrix is said to be in ***echelon form*** if
(1) All zero rows, if any, are at the bottom of the matrix;
(2) Each nonzero entry in a row is to the right of the leading nonzero entry in the preceding row.
If in addition,
(3) The first nonzero entry in each nonzero row is equal 1;
(4) Each column of the matrix which contains the leading entry 1 of some row has all its other entries 0.
Then the matrix is said to be in ***row canonical form.***

The first nonzero entries in each row of an echelon matrix are called the *pivot elements* of the matrix. Thus, a matrix is in row-canonical form if the pivot elements are each equal to 1, and they are the only nonzero entries in their respective columns.

Example 8.3.3. The following matrix A is an echelon matrix, while the matrix B is a row canonical matrix;

$$A = \begin{pmatrix} 2 & 4 & -1 & 0 & -3 \\ 0 & 3 & 8 & 6 & 5 \\ 0 & 0 & 1 & 2 & -1 \\ 0 & 0 & 0 & 0 & 3 \\ 0 & 0 & 0 & 0 & 0 \end{pmatrix}, \quad B = \begin{pmatrix} 1 & 0 & 0 & 2 & 0 & 4 \\ 0 & 1 & 0 & 6 & 0 & -1 \\ 0 & 0 & 1 & 3 & 0 & 5 \\ 0 & 0 & 0 & 0 & 1 & -2 \\ 0 & 0 & 0 & 0 & 0 & 0 \end{pmatrix}$$

***Gaussian elimination* essentially consists of two parts:**
(a). We reduce the augmented matrix of the system to echelon form by applying elementary row operations, working our way down building zero enties column by column. First, we interchange rows (if needed) so that the 1st row R_1 starts with a nonzero entry. Then, we use R_1 to build zeros along the entries of the 1st column. After this is done, we use the 2nd row R_2 to build zeros along the 2nd column, and so on, untill we obtain an echelon matrix. This echelon matrix is the associated augmented matrix of an equivalent simpler system (in echelon form).

(b). Step-by-step back-substitution to find all the unknowns (the solution) of this simper system.

We illustrate the Gaussian elimination method by a number of examples.

Example 8.3.4. Solve the system of linear equations.

$$
\begin{aligned}
x_1 &+ x_2 &- 6x_3 &- 4x_4 &= 6 \\
3x_1 &- x_2 &- 6x_3 &- 4x_4 &= 2 \\
2x_1 &+ 3x_2 &+ 9x_3 &+ 2x_4 &= 6 \\
3x_1 &+ 2x_2 &+ 3x_3 &+ 8x_4 &= -7.
\end{aligned}
$$

Solution. The augmented matrix associated to the system is

$$
\left(\begin{array}{cccc|c}
1 & 1 & -6 & -4 & 6 \\
3 & -1 & -6 & -4 & 2 \\
2 & 3 & 9 & 2 & 6 \\
3 & 2 & 3 & 8 & -7
\end{array} \right).
$$

We will apply elementary row operations to transform this matrix in echelon form. To build zero entries along the 1st column, we apply the row operations $R_2 \to (-3)R_1 + R_2$, $R_3 \to (-2)R_1 + R_3$ and $R_4 \to (-3)R_1 + R_4$, that is, we *replace* R_2 by $(-3)R_1 + R_2$, R_3 by $(-2)R_1 + R_3$, and R_4 by $(-3)R_1 + R_4$ to obtain

$$
\left(\begin{array}{cccc|c}
1 & 1 & -6 & -4 & 6 \\
0 & -4 & 12 & 8 & -16 \\
0 & 1 & 21 & 10 & -6 \\
0 & -1 & 21 & 20 & -25
\end{array} \right).
$$

To build zero entries along the 2nd column, we apply $R_3 \to \frac{1}{4}R_2 + R_3$ and $R_4 \to (-\frac{1}{4})R_2 + R_4$ to obtain

$$
\left(\begin{array}{cccc|c}
1 & 1 & -6 & -4 & 6 \\
0 & -4 & 12 & 8 & -16 \\
0 & 0 & 24 & 12 & -10 \\
0 & 0 & 18 & 18 & -21
\end{array} \right).
$$

Next we apply $R_3 \to \frac{1}{24}R_3$ and then $R_4 \to (-18)R_3 + R_4$ to obtain the echelon matrix

$$\begin{pmatrix} 1 & 1 & -6 & -4 & | & 6 \\ 0 & -4 & 12 & 8 & | & -16 \\ 0 & 0 & 1 & \frac{1}{2} & | & -\frac{5}{12} \\ 0 & 0 & 0 & 9 & | & -\frac{27}{2} \end{pmatrix}.$$

The associated echelon system is

$$\begin{array}{rcl} x_1 + x_2 - 6x_3 - 4x_4 & = & 6 \\ -4x_2 + 12x_3 + 8x_4 & = & -16 \\ x_3 + \frac{1}{2}x_4 & = & -\frac{5}{12} \\ 9x_4 & = & -\frac{27}{2}. \end{array}$$

This completes part **(a)** of the Gaussian elimination. Next we aplly part **(b)** of the method; Solving the last equation for x_4, we find $x_4 = -\frac{3}{2}$. Now, by *back-substitution*, ie, substituting this value of x_4 into the third equation, we find $x_3 = \frac{1}{3}$. Substituting theese values of x_4 and x_3 into the second equation, we find $x_2 = 2$, and finally, substituting the values of x_4, x_3 and x_2 into the first equation, we find $x_1 = 0$. Thus, the solution of the system is

$$x_1 = 0, \ x_2 = 2, \ x_3 = \tfrac{1}{3}, \ x_4 = -\tfrac{3}{2}.$$

We can carry the reduction of the augmented matrix further to bring it into *row canonical matrix*. The whole process to obtain the row-reduced echelon form is called **Gauss-Jordan elimination**.

To do this apply elementary row operations to the echelon matrix working our way up. We first apply $R_4 \to \frac{1}{9}R_4$. Then the row operations $R_3 \to (-\frac{1}{2})R_4 + R_3$, $R_2 \to (-8)R_4 + R_2$, and $R_1 \to 4R_4 + R_1$, yield

$$\begin{pmatrix} 1 & 1 & -6 & 0 & | & 0 \\ 0 & -4 & 12 & 0 & | & -4 \\ 0 & 0 & 1 & 0 & | & \frac{1}{3} \\ 0 & 0 & 0 & 1 & | & -\frac{3}{2} \end{pmatrix}.$$

Applying $R_2 \to (-12)R_3 + R_2$ and $R_1 \to 6R_3 + R_1$ we obtain

$$\begin{pmatrix} 1 & 1 & 0 & 0 & | & 2 \\ 0 & -4 & 0 & 0 & | & -8 \\ 0 & 0 & 1 & 0 & | & \frac{1}{3} \\ 0 & 0 & 0 & 1 & | & -\frac{3}{2} \end{pmatrix}.$$

Finally, applying first $R_2 \rightarrow (-\frac{1}{4})R_2$ and then $R_1 \rightarrow (-1)R_2 + R_1$ we obtain the row canonical matrix

$$\begin{pmatrix} 1 & 0 & 0 & 0 & | & 0 \\ 0 & 1 & 0 & 0 & | & 2 \\ 0 & 0 & 1 & 0 & | & \frac{1}{3} \\ 0 & 0 & 0 & 1 & | & -\frac{3}{2} \end{pmatrix}.$$

The resulting system is very simple (in fact, we obtain the solution of the system):

$$\begin{aligned} x_1 &= 0 \\ x_2 &= 2 \\ x_3 &= \frac{1}{3} \\ x_4 &= -\frac{3}{2}, \end{aligned}$$

or written as a quatriple $(0, 2, \frac{1}{3}, -\frac{3}{2})$.

Remark 8.3.5. It is clear from this example that the Gauss-Jordan algorithm tranforming a matrix to row-reduced echelon form generalizes to any nonzero matrix. Thus, *every nonzero matrix is row equivalent to a row canonical matrix.*[3]

Example 8.3.6. Solve the system of linear equations.

$$\begin{aligned} x &-2y &+3z &= 2 \\ 2x &-3y &+8z &= 7 \,. \\ 3x &-4y &+13z &= 8 \end{aligned}$$

---

[3]However, what is not obvious, is that this row canonical matrix is, in fact, *unique*. This fact is proved in a course of Linear Algebra.

Solution. We reduce the augmented matrix to echelon form, by first applying the row operation $R_2 \to (-2)R_1 + R_2$ and $R_3 \to (-3)R_1 + R_3$

$$\begin{pmatrix} 1 & -2 & 3 & | & 2 \\ 2 & -3 & 8 & | & 7 \\ 3 & -4 & 13 & | & 8 \end{pmatrix} \to \begin{pmatrix} 1 & -2 & 3 & | & 2 \\ 0 & 1 & 2 & | & 3 \\ 0 & 2 & 4 & | & 2 \end{pmatrix}$$

Now, applying $R_3 \to (-2)R_2 + R_3$ we get

$$\begin{pmatrix} 1 & -2 & 3 & | & 2 \\ 0 & 1 & 2 & | & 3 \\ 0 & 0 & 0 & | & -4 \end{pmatrix}.$$

The corresponding echelon system is

$$\begin{aligned} x & -2y & +3z & = 2 \\ & y & +2z & = 3 \\ & & 0 & = -4. \end{aligned}$$

Since the last equation is impossible, the final system has no solution, and therefore neither has the original system.

Example 8.3.7. Solve the system of linear equations.

$$\begin{aligned} x & -2y & +3z & = 2 \\ 2x & -3y & +8z & = 7 \\ 3x & -4y & +13z & = 6. \end{aligned}$$

Solution. The coefficient matrix of the systm is the same one as in the above example. Applying the same row operations to the augmented matrix we get,

$$\begin{pmatrix} 1 & -2 & 3 & | & 2 \\ 2 & -3 & 8 & | & 7 \\ 3 & -4 & 13 & | & 12 \end{pmatrix} \to \begin{pmatrix} 1 & -2 & 3 & | & 2 \\ 0 & 1 & 2 & | & 3 \\ 0 & 2 & 4 & | & 6 \end{pmatrix} \to \begin{pmatrix} 1 & -2 & 3 & | & 2 \\ 0 & 1 & 2 & | & 3 \\ 0 & 0 & 0 & | & 0 \end{pmatrix}$$

The corresponding echelon system is

$$\begin{aligned} x & -2y & +3z & = 2 \\ & y & +2z & = 3. \end{aligned}$$

This system has two equations in three unknowns. It is now apparent that there are infinitely many solutions because if we let $t \in \mathbb{R}$ be any real number, and set $z = t$, then solving for y and x (in terms of t) by back substitution, we see that the solution set is

$$z = t, \; y = 3 - 2t \text{ and } x = 8 - 7t,$$

or written as a triple

$$(x, y, z) = (8 - 7t, 3 - 2t, t)), \;\; t \in \mathbb{R}.$$

This is the general solution of the system. To find some specific solutions, we can substitute numbers for t:

For example, if $t = 0$, then $z = 0$, $y = 3$, $x = 8$, ie,

$$(x, y, z) = (8, 3, 0).$$

If $t = 1$, then $(x, y, z) = (7, 1, 1)$ and if $t = 2$, then $(x, y, z) = (-6, -1, 2)$.

8.3.1   Homogeneous Systems

We now specialize the Gaussian elimination method to **homogeneous systems** of linear equations. Note that a homogeneous system $AX = 0$ *always has the zero solution*, $X = 0$, that is, the solution

$$x_1 = x_2 = \ldots = x_n = 0.$$

It turns out that, *if the number of equations is less than the numbers of the uknowns in a homogeneous system the system has non-zero solutions.*[4]

Example 8.3.8. Solve the homogeneous system of linear equations.

$$\begin{aligned} x \;\; -2y \;\; +3z \;\; &= 0 \\ 2x \;\; -3y \;\; +8z \;\; &= 0 \\ 3x \;\; -4y \;\; +13z &= 0. \end{aligned}$$

---

[4]This is proved in a course of Linear Algebra.

Solution. The coefficient matrix of the systm is the same one as in Example 8.3.7. The same reduction of the augmented matrix gives,

$$\begin{pmatrix} 1 & -2 & 3 & | & 0 \\ 2 & -3 & 8 & | & 0 \\ 3 & -4 & 13 & | & 0 \end{pmatrix} \to \begin{pmatrix} 1 & -2 & 3 & | & 0 \\ 0 & 1 & 2 & | & 0 \\ 0 & 2 & 4 & | & 0 \end{pmatrix} \to \begin{pmatrix} 1 & -2 & 3 & | & 0 \\ 0 & 1 & 2 & | & 0 \\ 0 & 0 & 0 & | & 0 \end{pmatrix}$$

The corresponding echelon system is

$$\begin{aligned} x - 2y + 3z &= 0 \\ y + 2z &= 0. \end{aligned}$$

Again as before, there are infinitely many solutions of this system (and so of the original). Indeed, let $t \in \mathbb{R}$ be any real number, and set $z = t$. Then $z = t$, $y = -2t$ and $x = -7t$, that is, the general solution of the system is

$$(x, y, z) = (-7t, -2t, t)), \quad t \in \mathbb{R}.$$

Note that for $t = 0$, we obtain the zero solution $(0, 0, 0)$. For $t = 1$, the solution $(-7, -2, 1)$.

Example 8.3.9. Solve the system of linera equations.

$$\begin{aligned} x_1 + x_2 - 6x_3 - 4x_4 &= 0 \\ 3x_1 - x_2 - 6x_3 - 4x_4 &= 0 \\ 2x_1 + 3x_2 + 9x_3 + 2x_4 &= 0 \\ 3x_1 + 2x_2 + 3x_3 + 8x_4 &= 0 \end{aligned}$$

Solution. This is the corresponding homogeneous system to the system in Example 8.3.4 The augmented matrix associated to the system is

$$\begin{pmatrix} 1 & 1 & -6 & -4 & | & 0 \\ 3 & -1 & -6 & -4 & | & 0 \\ 2 & 3 & 9 & 2 & | & 0 \\ 3 & 2 & 3 & 8 & | & 0 \end{pmatrix}.$$

Applying the same row operations as in Example 8.3.4, we obtain the row-reduced echelon matrix

$$\begin{pmatrix} 1\,0\,0\,0 \,|\, 0 \\ 0\,1\,0\,0 \,|\, 0 \\ 0\,0\,1\,0 \,|\, 0 \\ 0\,0\,0\,1 \,|\, 0 \end{pmatrix}.$$

This gives the trivial solution

$$\begin{aligned} x_1 &= 0 \\ x_2 &= 0 \\ x_3 &= 0 \\ x_4 &= 0 \end{aligned}.$$

8.3.2   Exercises

1. For each of the following system of linear equations, find all solutions or show that no solution exists:

(a)
$$\begin{aligned} x +2y &= 8 \\ 3x -4y &= 4 \end{aligned}$$

(b)
$$\begin{aligned} x +y +2z &= 1 \\ x +2y -z &= -2 \\ x +3y +z &= 5 \end{aligned}$$

(c)
$$\begin{aligned} 2x +3y -z &= 6 \\ 2x -y +2z &= -8 \\ 3x -y +z &= -7 \end{aligned}$$

(d)
$$\begin{aligned} 2x -2y +3z &= 6 \\ 4x -3y +2z &= 0 \\ -2x +3y -7z &= 1 \end{aligned}$$

(e)
$$\begin{aligned} x +2y +3z &= 3 \\ 2x +3y +8z &= 4 \\ 5x +8y +19z &= 11 \end{aligned}$$

(f)
$$\begin{aligned} x +2y -3z &= 4 \\ x +3y +z &= 11 \\ 2x +5y -4z &= 13 \\ 2x +6y +2z &= 22 \end{aligned}$$

(g)
$$\begin{aligned} 3x_1 -2x_2 -5x_3 +x_4 &= 3 \\ 2x_1 -3x_2 +x_3 +5x_4 &= -3 \\ x_1 -x_2 -4x_3 +9x_4 &= 22 \\ x_1 +2x_2 -4x_4 &= -3 \end{aligned}$$

(h)
$$\begin{aligned} x_1 +x_2 -2x_3 +3x_4 &= 2 \\ 2x_1 +4x_2 -3x_3 -4x_4 &= 5 \\ 5x_1 +10x_2 -8x_3 +11x_4 &= 12 \end{aligned}$$

2. For each of the following homogeneous system of linear equations, find all solutions:

(a)
$$\begin{aligned} x +y -z &= 0 \\ 2x -3y +z &= 0 \\ x -4y +2z &= 0 \end{aligned}$$
(b)
$$\begin{aligned} x +2y -3z &= 0 \\ 2x +5y +2z &= 0 \\ 3x - y -4z &= 0 \end{aligned}$$

(c)
$$\begin{aligned} x -y +2z &= 0 \\ 2x +y +z &= 0 \\ 5x +y +4z &= 0 \end{aligned}$$
(d)
$$\begin{aligned} x +2y +3z &= 0 \\ x +2y +z &= 0 \end{aligned}$$

(e)
$$\begin{aligned} 7x_1 +2x_2 +6x_3 -5x_4 &= 0 \\ 2x_1 +x_2 +4x_3 +x_4 &= 0 \\ 3x_1 +2x_2 -x_3 -6x_4 &= 0 \\ x_1 +8x_3 +7x_4 &= 0 \end{aligned}$$
(f)
$$\begin{aligned} x_1 +3x_2 +x_3 -2x_4 -9x_5 &= 0 \\ 2x_1 +9x_2 +5x_3 +2x_4 +x_5 &= 0 \\ x_1 +4x_2 +2x_3 -3x_5 &= 0 \end{aligned}$$

3. Investigate the following systems and find their general solution in relation to the value(s) of the parameter λ:

(a)
$$\begin{aligned} 3x +2y +z &= -1 \\ 7x +6y +5z &= \lambda \\ 5x +4y +3z &= 2 \end{aligned}$$
(b)
$$\begin{aligned} x +y +\lambda z &= 2 \\ x +\lambda y +z &= -1 \\ \lambda x +y +z &= -1 \end{aligned}$$

8.4 Simple Nonlinear Systems

A system of two or more equations is called a **system of higher degree than the first**, if at least one of its equations is of degree higher than the first. Such systems are also called **nonlinear systems**. There is a variety of such systems, and in general there is no unified way for solving them. Here we consider a number of simple such systems that appear frequently in mathematics and to which more involved system are often reduced. The general methods for solving these systems are: **substitution or addition methods, factorization methods** and intoduction of **new variables for the unkwons** seeking to reduce the system to a simpler system or simpler systems.

We shall consider a number of examples illustrating these methods:

8.4.1 Substitution and addition method

Example 8.4.1. Solve the system

$$x + y = p$$
$$xy = q$$

Solution. One way to solve this system is by the substitution method. Solving the first equation for $y = p - x$ and substituting in the second equation, we get

$$x(p - x) = q \iff x^2 - px + q = 0.$$

If x_1, x_2 are the roots of this quadratic equation, then the solution of the system is $x = x_1$, $y = p - x_1$, and $x = x_2$, $y = p - x_2$.

Another way is to use Vietta's relations for the sum and product of roots of a quadratic equation(see ?). Then x , y are the roots t_1, t_2 of the quadratic equation

$$t^2 - pt + q = 0.$$

That is, $x = t_1$, $y = t_2$, and $x = t_2$, $y = t_1$.

For instance, for the system

$$x + y = 5$$
$$xy = 6$$

the quadratic equation is

$$t^2 - 5t + 6 = 0 \iff (t - 2)(t - 3) = 0,$$

whose roots are $t_1 = 2$, $t_2 = 3$.

Thus, the system has solution $x = 2$, $y = 3$, and $x = 3, y = 2$.

Example 8.4.2. Solve the system

$$x^2 + y^2 = a^2$$
$$x + y = b$$

Solution. The system may be solved by the substitution method, and we leave this to the reader as an exarcise. We choose to use the identity

$$x^2 + y^2 = (x + y)^2 - 2xy.$$

Then the system becomes

$$(x + y)^2 - 2xy = a^2$$
$$x + y = b.$$

This system is equivalent to

$$\begin{array}{cc} b^2 - 2xy = a^2 \\ x + y = b \end{array} \iff \begin{array}{c} xy = \frac{b^2 - a^2}{2} \\ x + y = b, \end{array}$$

which is a system of the form of Example 8.4.1.

Example 8.4.3. Solve the system

$$x^2 - y^2 - 2x - 2y - 4 = 0$$
$$2x - y - 7 = 0$$

Solution. From the second equation, we have $y = 2x - 7$. Substituting this in the first equation, we obtain

$$3x^2 - 22x + 39 = 0 \quad \Leftrightarrow \quad (x-3)(3x-13) = 0 \quad \Leftrightarrow \quad x = 3, \; x = \frac{13}{3}.$$

Substituting these values for x into $y = 2x - 7$, we find $y = -1$, $y = \frac{5}{3}$.
Thus, the solutions of the system are the pairs $(3, -1)$ and $(\frac{13}{3}, \frac{5}{3})$.

Example 8.4.4. Solve the system

$$\begin{aligned} x^2 + y^2 - 4x - 3y + 5 &= 0 \\ 3x^2 + 3y^2y - 11x - 7y + 10 &= 0 \end{aligned}$$

Solution. Multiplying the first equation by -3 and adding to the second, we get
$$\begin{aligned} x^2 + y^2 - 4x - 3y + 5 &= 0 \\ x + 2y - 5 &= 0, \end{aligned}$$

which is a system of the form of Example 8.4.3. This system is equivalent to the system
$$\begin{aligned} 5y^2 - 15y + 10 &= 0 \\ x &= 5 - 2y. \end{aligned}$$

Solving this system we find the solutions $x = 1$, $y = 2$ and $x = 3$, $y = 1$.

Example 8.4.5. Solve the system

$$\begin{aligned} 4x^2 + y^2 - 3xy &= 1 \\ y^2 - 1 &= 4^2x + 4x \end{aligned}$$

Solution. From the second equation we have $y^2 = (2x+1)^2$. Hence

$$y = 2x + 1 \quad \text{or} \quad y = -2x - 1.$$

Consequently, the given system is equivalent to two systems

$$\begin{array}{cc} 4x^2 + y^2 - 3xy = 1 & \qquad 4x^2 + y^2 - 3xy = 1 \\ y = 2x + 1, & \qquad y = -2x - 1. \end{array}$$

Solving the first system by substitution, we find $x = 0$, $y = 1$ and $x = -\frac{1}{2}$, $y = 0$. Similarly, solving the second system, we find $x = 0$, $y = -1$ and $x = -\frac{1}{2}$, $y = 0$. Thus, the solutions of the system are

$$(0,1), \quad \left(-\frac{1}{2}, 0\right), \quad (0, -1).$$

Example 8.4.6. Solve the system

$$2x^2 + xy - y^2 = 20$$
$$x^2 - 3xy + 2y^2 = 2$$

Solution. The left hand sides of both equations of the system are ***homogeneous polynomials of degree*** 2. To solve them , we set $x = ty$. Then we have,

$$\begin{array}{c} 2t^2y^2 + ty^2 - y^2 = 20 \\ t^2y^2 - 3ty^2 + 2y^2 = 2 \end{array} \quad \Leftrightarrow \quad \begin{array}{c} y^2(2t^2 + t - 1) = 20 \\ y^2(t^2 - 3t + 2) = 2. \end{array}$$

Dividing the equations side by side, we obtain

$$\frac{2t^2 + t - 1}{t^2 - 3t + 2} = \frac{10}{1} \quad \Leftrightarrow \quad 8t^2 - 31t + 21 = 0.$$

Solving this quadratic equation, we find $t_1 = 3$ and $t_2 = \frac{7}{8}$.

Consequently, we must solve the systems:

$$\begin{array}{cc} x = 3y & x = \frac{7}{8}y \\ x^2 - 3xy + 2y^2 = 2, & x^2 - 3xy + 2y^2 = 2. \end{array}$$

Solving the first system, by substitution, we find $x = 3$, $y = 1$, and $x = -3$, $y = -1$. Solving simiraly, the second system, we find $x = \frac{7\sqrt{2}}{3}$, $y = \frac{8\sqrt{2}}{3}$, and $x = -\frac{7\sqrt{2}}{3}$, $y = -\frac{8\sqrt{2}}{3}$.

8.4.2   Factorization method

If one equation of the system can be factored, say

$$A = B \quad \Leftrightarrow \quad F \cdot G = 0,$$

then the given system is equivalent to two systems. That is,

$$\begin{array}{ccc} A = B & & F = 0 & G = 0 \\ C = D. & \Leftrightarrow & C = D, & C = D. \end{array}$$

Example 8.4.7. Solve the system

$$\begin{aligned} xy - y &= 0 \\ 3x - 8y + 5 &= 0 \end{aligned}$$

Solution. Factoring the first equation, we have

$$y(x - 1) = 0 \implies y = 0, \ x = 1.$$

Now the given system is equivalent to the following two systems:

$$\begin{array}{cc} y = 0 & x = 1 \\ 3x - 8y + 5 = 0, & 3x - 8y + 5 = 0. \end{array}$$

Solving the first system, we obtain $y = 0$ and $x = -\frac{5}{3}$. Solving the second system, we get $x = 1, \ y = 1$. Thus, the solutions of the original system are the pairs

$$\left(-\frac{5}{3}, 0\right), \ (1, 1).$$

Example 8.4.8. Solve the system

$$\begin{aligned} x^2 - 5xy + 6y^2 &= 0 \\ x^2 + y^2 + x - 11y - 2 &= 0 \end{aligned}$$

Solution. Factoring the first equation, we have

$$x^2 - 5xy + 6y^2 = 0 \iff (x - 2y)(x - 3y) = 0.$$

Hence, $x = 2y, \ x = 3y$.

Now the given system is equivalent to the following two systems:

$$\begin{array}{cc} x = 2y & x = 3y \\ x^2 + y^2 + x - 11y - 2 = 0, & x^2 + y^2 + x - 11y - 2 = 0. \end{array}$$

Solving these systems, we find all solutions of the given system, namely

$$(4, 2), \ \left(-\frac{2}{5}, -\frac{1}{5}\right), \ (3, 1), \ \left(-\frac{3}{5}, -\frac{1}{5}\right).$$

Example 8.4.9. Solve the system

$$x^2 - y^2 = 5$$
$$x^2 - xy + y^2 = 7$$

Solution. Multiplying the first equation bt 7 and the second by -5 and adding up the results, we obtain

$$2x^2 + 5xy - 12y^2 = 0 \iff (2x - 3y)(x + 4y) = 0.$$

Therefore, $y = \frac{2}{3}x$, $y = -\frac{1}{4}x$. Consequently, the given system is equivalent to the two systems

$$\begin{array}{ll} x^2 - y^2 = 5 & x^2 - y^2 = 5 \\ y = \frac{2}{3}x, & y = -\frac{1}{4}x. \end{array}$$

Solving each system by substitution, we find the solutions

$$(3, 2), \ (-3, -2), \ \left(\frac{4}{\sqrt{3}}, -\frac{1}{\sqrt{3}}\right), \ \left(-\frac{4}{\sqrt{3}}, -\frac{1}{\sqrt{3}}\right).$$

Example 8.4.10. Solve the system

$$10x^2 + 5y^2 - 27x - 4y + 5 = 0 = 0$$
$$x^2 + y^2 - 3x - y = 0$$

Solution. Multiplying the second equation by -5 and adding the resulting equation to the first equation, (to eliminate y^2), we get

$$5x^2 - 12x + y + 5 = 0.$$

Solving this equation for y, we get

$$y = -5x^2 + 12x - 5. \tag{8.9}$$

Substituting this expression for y into the second equation, we obtain

$$5x^4 - 24x^3 + 40x^2 - 27x = 6 = 0.$$

Factoring this fourth degree polynomial, using the factor theorem, we find

$$(x - 1)(x - 2)(5x^2 - 9x + 3) = 0.$$

Solving each factor we obtain, $x = 1$, $x = 2$, and $x = \frac{9 \pm \sqrt{21}}{10}$. Substituting these values of x in (8.9), we find, $y = 2$, $y = -1$, and $y = \frac{7 \pm 3\sqrt{21}}{10}$, respectively. Thus, the solutions of the given system are

$$(1, 2), \ (2, -1), \ (\frac{9 \pm \sqrt{21}}{10}, \frac{7 \pm 3\sqrt{21}}{10}).$$

8.4.3 Introducing new variables

By introducing new variables for the uknowns it is often possible to reduce a given system to a simpler system. Let us study some examples.

Example 8.4.11. Solve the system

$$3x + \frac{y}{x} = 6$$
$$7x - \frac{2y}{x} = 1$$

Solution. Introducing new variables, by setting $u = x$ and $v = \frac{y}{x}$, we obtain the system

$$3u + v = 6$$
$$7u - 2v = 1.$$

Solving this system of linear equations, (say by addition method), we find $u = 1$ and $v = 3$. Returning to x and y, we get $x = 1$ and $\frac{y}{x} = 3$. Therefore, $x = 1, y = 3$.

Example 8.4.12. Solve the system

$$x^2 + y^2 - xy = 3$$
$$x + y - xy = 1$$

Solution. Using $x^2 + y^2 = (x + y)^2 - 2xy$, the first equation becomes

$$(x + y)^2 - 3xy = 3.$$

Setting $u = x + y$ and $v = xy$, the system becomes

$$u^2 - 3v = 3$$
$$u - v = 1.$$

This system is equivalent to the system

$$\begin{array}{ccc} u^2 - 3u = 0 & & u(u - 3) = 0 \\ v = u - 1. & \Leftrightarrow & v = u - 1. \end{array}$$

Hence $u = 0$, $u = 3$, and so $v = -1$, $v = 2$ respectively.

Returning to x and y, we must solve the following systems:

$$\begin{array}{cc} x + y = 0 & x + y = 3 \\ xy = -1 & xy = 2. \end{array}$$

Both these systems are of the form of Example 8.4.1. The first system has two solutions: $(1, -1)$ and $(-1, 1)$. The second system has also two solutions: $(1, 2)$ and $(2, 1)$. Thus, the original system has solutions

$$(1, -1), \ (-1, 1), \ (1, 2), \ (2, 1).$$

Example 8.4.13. Solve the system

$$7x^2y^2 - x^4 - y^4 = 155$$
$$3xy - x^2 - y^2 = 5$$

Solution. Rewritting the system in the form

$$7x^2y^2 - (x^4 + y^4) = 155$$
$$3xy - (x^2 + y^2) = 5,$$

and using the identity $x^4 + y^4 = (x^2 + y^2)^2 - 2x^2y^2$, we see that it is convenient to introduce new variables $u = x^2 + y^2, \quad v = xy$. Now the system becomes

$$9v^2 - u^2 = 155$$
$$3v - u = 5.$$

Substituting $u = 3v - 5$ into the first equation, we get $30v = 180$. Hence, $v = 6$ and $u = 13$.

Returning to x and y, we solve the system

$$x^2 + y^2 = 13$$
$$xy = 6.$$

Using $x^2 + y^2 = (x+y)^2 - 2xy$, and $xy = 6$, the first equation becomes $(x+y)^2 = 25$. Hence $x + y = \pm 5$. Thus, we solve the two systems:

$$x + y = 5 \qquad x + y = -5$$
$$xy = 6, \qquad xy = 6.$$

The first system has two solutions: $(3, 2)$, $(2, 3)$. The second also has two solutions: $(-3, -2)$, $(-2, -3)$.

Example 8.4.14. Solve the system

$$\sqrt{\tfrac{x}{y}} - \sqrt{\tfrac{y}{x}} = \tfrac{3}{2}$$
$$x + xy + y = 9.$$

Solution. Set $t = \sqrt{\tfrac{x}{y}} \geq 0$. Then $\sqrt{\tfrac{y}{x}} = \tfrac{1}{t}$. Substituting in the first equation, we get the quadratic equation

$$2t^2 - 3t - 2 = 0.$$

The roots of this equation are $t_1 = 2$, $t_2 = -\tfrac{1}{2}$. Since $t \geq 0$, we reject the root $-\tfrac{1}{2}$.

Hence, we have

$$\sqrt{\frac{x}{y}} = 2 \;\Rightarrow\; \frac{x}{y} = 4 \;\Rightarrow\; x = 4y.$$

Now we obtain the system

$$x = 4y$$
$$x + xy + y = 9.$$

Solving this system, we find $x = 4$, $y = 1$, and $x = -9$, $y = -\tfrac{9}{4}$. Checking these solutions in the original system, we see that both satisfy the system.

Example 8.4.15. Solve the system

$$\frac{x^2+17}{12} = y + \frac{2}{3}\sqrt{x^2 - 12y + 1}$$
$$\sqrt{\frac{x}{3y} + \frac{1}{4}} - \frac{x}{8y} = \frac{2}{3} + \frac{y}{2x}.$$

Solution. Multipying both sides of the first equation by 12, we get

$$x^2 + 17 = 12y + 8\sqrt{x^2 - 12y + 1} \quad \Rightarrow \quad x^2 - 12y + 1 + 16 = 8\sqrt{x^2 - 12y + 1}.$$

Setting $t = \sqrt{x^2 - 12y + 1}$, we get the quadratic equation

$$t^2 - 8t + 16 = 0 \quad \Leftrightarrow \quad (t-4)^2 = 0 \quad \Rightarrow \quad t = 4.$$

Hence,

$$x^2 - 12y + 1 = 16 \quad \Rightarrow \quad x^2 - 12y = 15.$$

Since $y \neq 0$, we multiplying both sides of the second equation by $\frac{2x}{y}$, we have

$$\frac{2x}{y}\sqrt{\frac{x}{3y} + \frac{1}{4}} - \frac{x^2}{4y^2} = 1 + \frac{4x}{3y} \quad \Leftrightarrow$$

$$2\left(\frac{x}{2y}\right)\sqrt{\frac{4x}{3y} + 1} - \left(\frac{x}{2y}\right)^2 = \frac{4x}{3y} + 1 \quad \Leftrightarrow$$

$$\left(\frac{x}{2y}\right)^2 - 2\left(\frac{x}{2y}\right)\sqrt{\frac{4x}{3y} + 1} + \left(\frac{4x}{3y} + 1\right) = 0.$$

That is,

$$\left(\frac{x}{2y} - \sqrt{\frac{4x}{3y} + 1}\right)^2 = 0.$$

This implies

$$\frac{x}{2y} = \sqrt{\frac{4x}{3y} + 1}. \tag{8.10}$$

Squaring both sides of (8.10), and multiplying both sides of the resulting equation by $12y^2$ to cancel the denominators, we obtain

$$3x^2 - 16xy - 12y^2 = 0 \quad \Leftrightarrow \quad (x - 6y)(3x + 2y) = 0.$$

Hence, $x = 6y$ and $x = -\frac{2}{3}y$. The second one does not satisfy (8.10), and we reject it.

Thus now, we must solve the system

$$x^2 - 12y = 15$$
$$x = 6y.$$

This system has two solution $x = -3$, $y = -\frac{1}{2}$ and $x = 5$, $y = \frac{5}{6}$. Checking these solutions in the original system, we see that both satisfy the system.

8.4.4 Exercises

Solve the systems.

1.

$$\text{(a)} \quad \begin{array}{l} x + y = 2 \\ 4xy = 3 \end{array} \qquad \text{(b)} \quad \begin{array}{l} x + y = 5 \\ xy + 36 = 0 \end{array}.$$

2.

$$\text{(a)} \quad \begin{array}{l} x^2 + y^2 = 13 \\ x^2 - y = 7 \end{array} \qquad \text{(b)} \quad \begin{array}{l} x^2 + y^2 = 157 \\ xy = 66 \end{array}.$$

3.

$$\text{(a)} \quad \begin{array}{l} x^2 + 2xy - y^2 + 4x - 6y = -7 \\ 2x + y = 3 \end{array} \qquad \text{(b)} \quad \begin{array}{l} x^2 - y^2 = 1 \\ x^3 - y^2 = x \end{array}.$$

4.

$$\text{(a)} \quad \begin{array}{l} x^2 + y^2 + x - 3y + 2 = 0 \\ x^2 + y^2 + x - y = 0 \end{array} \qquad \text{(b)} \quad \begin{array}{l} x^2 - 3xy + 2y^2 = 0 \\ x^2 + xy = 6 \end{array}.$$

5.

$$\text{(a)} \quad \begin{array}{l} xy - x^2 + 3 = 0 \\ 3xy - 4y^2 = 2 \end{array} \qquad \text{(b)} \quad \begin{array}{l} x^2 - xy = 14 \\ xy - y^2 = 10 \end{array}.$$

6.

$$\text{(a)} \quad \begin{array}{l} 2x^2 - 3xy + y^2 = 3 \\ x^2 + 2xy - 2y^2 = 6 \end{array} \qquad \text{(b)} \quad \begin{array}{l} x^2 + xy = 28 \\ y^2 + xy = -12 \end{array}.$$

7.

(a) $\begin{aligned} 2x^2 + 3y^2 - 4xy &= 3 \\ 2x^2 - y^2 &= 7 \end{aligned}$ (b) $\begin{aligned} x - xy^3 &= 7 \\ xy^2 - xy &= 3 \end{aligned}$.

8.

(a) $\begin{aligned} x^3 - y^3 &= 26 \\ x - y &= 2 \end{aligned}$ (b) $\begin{aligned} x^3 + y^3 &= 19 \\ x + y &= 1 \end{aligned}$.

9.

(a) $\begin{aligned} x^3 - y^3 &= 37 \\ x^2 + xy + y^2 &= 37 \end{aligned}$ (b) $\begin{aligned} (x + y)(x^3 + y^3) &= 432 \\ x^2 + y^2 &= 20 \end{aligned}$.

10.

(a) $\begin{aligned} x^2 - xy + y^2 &= 7 \\ x^4 + x^2y^2 + y^4 &= 133 \end{aligned}$ (b) $\begin{aligned} \frac{1}{x} + \frac{1}{y} &= \frac{3}{2} \\ \frac{1}{x^2} + \frac{1}{y^2} &= \frac{5}{4} \end{aligned}$.

11.

(a) $\begin{aligned} \frac{x}{y} + \frac{y}{x} &= \frac{13}{6} \\ x + y &= 5 \end{aligned}$ (b) $\begin{aligned} \frac{x}{y} - \frac{y}{x} &= \frac{5}{6} \\ x^2 - y^2 &= 5 \end{aligned}$.

12.

(a) $\begin{aligned} \sqrt{x} + \sqrt{y} &= 11 \\ x + y &= 65 \end{aligned}$ (b) $\begin{aligned} \sqrt{x} - \sqrt{y} &= x + y - 2\sqrt{xy} \\ \sqrt{x} + \sqrt{y} &= 5 \end{aligned}$.

13.

(a) $\begin{aligned} 5\sqrt{x} - x\sqrt{x} &= y\sqrt{y} + \sqrt{y} \\ x - y &= 3 \end{aligned}$ (b) $\begin{aligned} (x - y)\sqrt{y} &= \frac{\sqrt{x}}{2} \\ (x + y)\sqrt{x} &= 3\sqrt{y}. \end{aligned}$

14.

(a) $\begin{aligned} \sqrt{\frac{x-y}{x+y}} + \sqrt{\frac{x+y}{x-y}} &= \frac{10}{3} \\ x - y &= -3 \end{aligned}$ (b) $\begin{aligned} \sqrt{\frac{2x-1}{y+2}} + \sqrt{\frac{y+2}{2x-1}} &= 2 \\ x + y &= 12. \end{aligned}$

15.

(a) $\begin{aligned} 2|x - 1| + 3|y + 2| &= 17 \\ 2x + y &= 7 \end{aligned}$ (b) $\begin{aligned} |x - 1| + |y - 5| &= 1 \\ y - 5 &= |x - 1| \end{aligned}$.

8.5 Applications. Word Problems

Problems involving more than one uknown can be solved by means of systems of equations. We illustrate this with a number of examples.

Example 8.5.1. The sum of two numbers is 81. The difference of twice one and three times the other is 62. Find the two numbers.

Solution. Let x and y be the two numbers. From the given information, we get the system

$$x + y = 81$$
$$2x - 3y = 62.$$

We use elimination and multiply the first equation by -2 and then add the equations.

$$x + y = 81$$
$$-5y = -100.$$

Solving, we find $y = 20$ and $x = 61$. Thus, the required numbers are 61 and 20.

Example 8.5.2. If we decrease the length of a rectangle by $5\,cm$ and increase its width by $2\,cm$, then its area decreases by $20\,cm^2$. If insdead, we increase its length by $8\,cm$ and decrease its width by $3\,cm$, then its area remains unchanged. Find the dimensions of the rectangle.

Solution. Let x represent the length and y represent the width of the rectangle. Then its area is xy. From the data of the problem, we have

$$(x - 5)(x + 2) = xy - 20,$$

and

$$(x + 8)(x - 3) = xy.$$

Simplifying both equations, we get the system

$$2x - 5y = -10$$
$$-3x + 8y = 24.$$

Solving the system, we find $x = 40$ and $y = 18$. Thus the dimensions of the rectangle are $40\,cm$ and $18\,cm$.

Example 8.5.3. An airplane flies 600 miles with the wind for 3 hours and returns against the wind in 4 hours. Find the speed of the wind and the air speed of the plane in miles per hour.

   Solution. Let v represent the air speed of the plane and w represent the wind speed.

   Flying with the wind the ground speed of the plane is $v + w$. On the return, against the wind, its ground speed is $v - w$. Since the distance travelled is equal to (speed)·(time), from the information of the problem we get the equations $3(v + w) = 600$ (outbound trip) and $4(v - w) = 600$ (return trip). That is, we obtain the system

$$\begin{array}{ll} 3(v + w) = 600 & \\ 4(v - w) = 600. \end{array} \quad \Leftrightarrow \quad \begin{array}{l} v + w = 200 \\ v - w = 150. \end{array}$$

Solving the system, we find $v = 175$ miles per hour and $w = 25$ miles per hour.

Example 8.5.4. A lab technician is required to prepare 200 liters of a solution containing 42% of alcohol. The technician is planning to mix a solution that contains 30% alcohol with a solution that contains 50% alcohol. How much of each solution should she use?

   Solution. Let x be the number of liters of the 30% solution, and y be the number of liters of the 50% solution. The first equation is

$$x + y = 200.$$

The second equation concerns the percent of alcohol in each solution. Since the amount of alcohol remains the same before and after the mixing, we get the equation

$$0.3x + 0.5y = (0.42)(200) = 84.$$

Hence, we obtain the system

$$x + y = 200$$
$$0.3x + 0.5y = 84.$$

Multilying the second equation by 10, to clear it from the decimals, we get

$$x + y = 200$$
$$3x + 5y = 840.$$

Solving this system, we find $x = 80$ liters and $y = 120$ liters.

Example 8.5.5. Walking without stopping a postman went from point a A through a point B to a point C. The distance from A to B was covered with a speed of $3.5\,km/hr$ and from B to C of $4\,km/hr$. To get back from C to A in the same time following the same route with a constant speed he was to to walk $3.75\,km/hr$. However, after walking at that speed and reaching B he stopped for 14 minutes and then, in order to reach A at the appointed time he had to move from B to A walking $4\,km/hr$. Find the distances from A and B and between B and C.

Solution. Let x be the distance between A and B, and y be the distance between B and C. Then since the time of motion is the same in all the cases stated in the problem, we obtain the system of linear equations

$$\frac{x}{3.5} + \frac{y}{4} = \frac{x+y}{3.75}$$

$$\frac{x+y}{3.75} = \frac{14}{60} + \frac{y}{3.75} + \frac{x}{4}.$$

Simplifying and solving this system, we find $x = 14$ and $y = 16$. Thus, the distance from A to B is $14\,km$ and the distance from B to C is $16\,km$.

Example 8.5.6. In a certain number of three digits, the second digit is equal to the sum of the first and the third, the sum of the second and third is 8, and if the first and third digits be interchanged, the number is increased by 99. Find the number.

Solution. Let $x =$ hundreds digit, $y =$ tens digit, and $z =$ ones digit (where x, y, z are positive integers with $0 < x, y, z < 10$). Then the number is

$$100x + 10y + z.$$

From the information given in the problem, we have

$$x + z = y$$
$$y + z = 8$$
$$100z + 10y + x = 100x + 10y + z + 99.$$

Simplifying the third equation, we obtain the system

$$x + z - y = 0$$
$$y + z = 8$$
$$z - x = 1.$$

Solving this system, we find $x = 2$, $y = 5$, and $z = 3$. Thus the number is 253.

Example 8.5.7. In a certain number of two digits, the product of the digits is equal to 35. If the digits be interchanged the resulting number exceeds the product of the digits by 40. Find the number.

Solution. Let $x =$ tens digit, and $y =$ ones digit (where x, y are positive integers with $0 < x, y < 10$). Then the number is

$$10x + y.$$

From the information given in the problem, we have

$$xy = 35$$
$$10y + x = xy + 40.$$

Thus, we must solve the system

$$\begin{array}{lll} \begin{array}{l} xy = 35 \\ 10y + x = 75. \end{array} & \Leftrightarrow & \begin{array}{l} (75 - 10y)y = 35 \\ x = 75 - 10y. \end{array} & \Leftrightarrow & \begin{array}{l} 2y^2 - 15y + 7 = 0 \\ x = 75 - 10y. \end{array} \end{array}$$

Solving the quadratic equation, we find $y = 7$, and $y = \frac{1}{2}$. Rejecting $y = \frac{1}{2}$ because is not an integer, we find the solution $y = 7$, $x = 5$. Thus the number is 57.

Example 8.5.8. Two men together can do a certain work in 6 hours. If the first man alone did 60% of the entire work and then the second man alone completed the remaining part, then they would have spend 12 hours. How much time will it take each man to do this work alone?

Solution. Let x be the number of hours in which the first man alone can do this work, and y be the number of hours in which the second man alone can do this work. Then in 1 hour the first man alone does $\frac{1}{x}$ of the work and the second man alone does $\frac{1}{y}$ of the work. Since both men working together do the entire work in 6 hours, in 1 hour they do $\frac{1}{6}$ of the work. Hence we obtain the equation

$$\frac{1}{x} + \frac{1}{y} = \frac{1}{6}.$$

The second equation concerns the percent of the work done by each man. By the data of the problem, we get

$$0.6x + 0.4y = 12.$$

Therefore, we obtain the system

$$
\begin{array}{ccc}
\frac{1}{x} + \frac{1}{y} = \frac{1}{6} & & \frac{1}{x} + \frac{1}{y} = \frac{1}{6} \\
& \Leftrightarrow & \\
0.6x + 0.4y = 12. & & 3x + 2y = 60.
\end{array}
$$

From the second equation

$$y = \frac{60 - 3x}{2}.$$

Substituting this expression into the first equation, we get

$$\frac{1}{x} + \frac{2}{3(20 - x)} = \frac{1}{6}.$$

Clearing the denominators and simplifying, we obtain

$$x^2 - 22x + 120 = 0 \iff (x-12)(x-10) = 0 \iff x_1 = 12, \ x_2 = 10.$$

Finally, for $x_1 = 12$, we find $y_1 = 12$. For $x_2 = 10$, we find $y_2 = 15$.

The problem permits two answers: $x = 12$ hours, $y = 12$ hours and $x = 10$ hours, $y = 15$ hours.

Example 8.5.9. The perimeter of a right-angled triangle is 60, *cm*. The height to the hypotenuse is $12\,cm$. Find the sides lenght of the sides of the triangle.

Solution. Let x, y be the lenght of the perpendicular sides and z be the lenght of the hypotenuse of the triangle. Then, since its perimeter is $60\,cm$, we have

$$x + y + z = 60.$$

From the Pythogorean theorem we also have

$$x^2 + y^2 = z^2.$$

The area A of the trianle can be computed in two ways: $A = \frac{1}{2}xy$ and $A = \frac{1}{2}z(12)$. Therefore,

$$xy = 12z.$$

Thus, we obtain the system

$$\begin{aligned} x + y + z &= 60 \\ x^2 + y^2 &= z^2 \\ xy &= 12z. \end{aligned}$$

The first equation gives $x + y = 60 - z$. Using the identity

$$x^2 + y^2 = (x+y)^2 - 2xy,$$

and taking into account the third equation, the second equation becomes

$$(60 - z)^2 - 24z = z^2 \implies 3600 = 144z \implies z = 25.$$

Substituting this value of z into the system , we obtain the system

$$x + y = 35$$
$$xy = 300.$$

Solving this system, we find $x = 20$, $y = 15$.

Thus, the sides of the triangle are $20\,cm$, $15\,cm$, and $25\,cm$.

8.5.1 Exercises

1. The perimeter of a rectangle is $90\,cm$. If the length is twice the width, what are the dimensions of the rectangle?

2. A local municipality plans to expand a rectangular playground. Presently it takes 1050 feet of fencing to enclose the playground. The extension plans call to triple the width of the playground and to double the length. Then the municipality will require 2550 feet of fencing. What is the length and width of the current playground?

3. Find a number of two digits from the followind data: (a) twice the first digit plus three times the second equals 37. (b) if the order of the digits be reversed the number diminished by 9.

4. Saint Francis High School put on their annual musical. The students sold 650 tickets for a value of $\$4,375$. If orchestra seats cost $\$7.50$ and balcony seats cost $\$3.50$, how many of each kind of seat were sold?

5. Tank A contains a mixure of 10 gallons water and 5 gallons of achohol. Tank B contains 12 gallons water and 3 gallons alcohol. How many gallons should be taken from each tank and combined in order to obtain an 8 gallon solution containing 25% alcohol?

6. Two pounds of coffee and 3 pounds of butter cost $\$4.20$. A year later the price of coffee increased 10% and that of butter 20%, making the total cost of a similar order $\$4.86$. Find the original cost per pound of each.

7. A and B are alloys of silver and cooper. An alloy which is 5 parts A and 3 parts B is 52% silver. One which is 5 parts A and 11 parts B is 42% silver. What are the percentages of silver in A and B respectively?

8. A train left point A at noon sharp. Two hours later another train started from point A in the same direction. it overtook the first train at 8 : 00PM. Find the speeds of the trains if the sum of their speeds is 70 mi/hr.

9. A motor boat whose speed in still water is 10 mi/hr went 90 miles downstream and then reurned. Calculate the speed of the river flow if the entire trip took 20 hours.

10. It takes a motor boat 3 hours to travel a distance of 45 miles upstream and 2 hours to travel a distance of 50 miles downstream. What is the speed of the boat in still water and the speed of the river current?

11. A sum of money at simple interest rate amounts to $2,556 in 2 years and to $2,767 in 4 years. What is the sum of money, and what the rate of interest?

12. Two men working together, can do a piece of work in 12 days. If the first man does half the work and then the second man does the other half, the work will be completed in 25 days. How many days would it take each man to do the work, if he worked alone?

13. The perimeter of a rectangle is $16\,cm$ and its area is $15\,cm^2$. Find its dimensions.

14. Find the dimensions of a rectangle whose diagonal is 20 inches and its area is 192 square inches.

15. The hypotenuse of a right-angled triangle is longer than the two perpendicular sides by 3 and 24 inches respectively. Find the sides of the triangle.

16. The sum of two numbers is the same as their product, and the difference of their reciprocals is 3. Find the numbers.

17. The difference of cubes of two numbers is 218 and the cube of their difference is 8. Find the numbers.

8.6 Review Exercises

1. Solve the systems:

$$(a) \quad \begin{array}{l} \frac{x}{2} + \frac{y}{3} = 3 \\ \frac{x}{4} - \frac{2y}{3} = -1 \end{array} \qquad (b) \quad \begin{array}{l} 2x + y = 5 \\ \frac{5}{2} - x = \frac{y}{2}. \end{array}$$

2. Solve the systems:

$$(a) \quad \begin{array}{l} x + y + z = 4 \\ 4x + 8y - 3z = 35 \\ 2x + 3z = 3. \end{array} \qquad (b) \quad \begin{array}{l} x - y - z = 1 \\ 2x + 3y + z = 2 \\ 3x + 2y = 0. \end{array}$$

3. Solve the systems.

$$(a) \quad \begin{array}{l} 2x + y + 4z + 8w = -1 \\ x + 3y - 6z + 2w = 3 \\ 3x - 2y + 2z - 2w = 8 \\ 2x - y + 2z = 4. \end{array} \qquad (b) \quad \begin{array}{l} x + 8y - 7z = 12 \\ 2x + 5y - 8z = 8 \\ 4x + 3y - 9z = 9 \\ 2x + 3y - 5z = 7. \end{array}$$

4. Solve the systems.

$$(a) \quad \begin{array}{l} x + 2y - 3z = 0 \\ 2x + 5y + 2z = 0 \\ 3x - y - 4z = 0. \end{array} \qquad (b) \quad \begin{array}{l} x + y = 2 \\ y + z = 2 \\ x - y = 0. \end{array}$$

5. Find the value of the parameter $\lambda \in \mathbb{R}$ so that the system

$$\begin{array}{l} -4x + \lambda y = 3 + \lambda \\ (6 + \lambda)x + 2y = 1 + \lambda \end{array}$$

has no solution.

6. For what values of the parameter λ does the homogeneous system

$$\begin{array}{l} (2 - \lambda)x + 2y = 0 \\ x + (3 - \lambda)y = 0 \end{array}$$

have nonzero solutions?

7. Determine the values of λ for which the system

$$x + y - z = 2$$
$$x + 2y + z = 3$$
$$x + y + (\lambda^2 - 5)z = \lambda$$

has (a) unique solution, (b) no solution (c) many solutions.

8. Investigate the system

$$(2 - k)x + (2 - k)y + z = 1$$
$$(3 - 2k)x + (2 - k)y + z = k$$
$$x + y + (2 - k)z = 1.$$

9. Solve the systems;

$$(a) \quad \begin{array}{l} x + y = 3 \\ xy + 4 = 0. \end{array} \qquad (b) \quad \begin{array}{l} x^2 - y^2 = 16 \\ x + y = 8. \end{array}$$

10. Solve the systems;

$$(a) \quad \begin{array}{l} x^2 + y^2 = 41 \\ x + y = 9. \end{array} \qquad (b) \quad \begin{array}{l} 7x^2 - 6xy = 8 \\ 2x - 3y = 6. \end{array}$$

11. Solve the systems;

$$(a) \quad \begin{array}{l} x^3 - y^3 = 7 \\ x - y = 1. \end{array} \qquad (b) \quad \begin{array}{l} x^4 + y^4 = 82 \\ x - y = 2. \end{array}$$

12. Find the value of λ for which one root of the equation

$$x^2 + (2\lambda - 1)x + \lambda^2 + 2 = 0$$

is twice as large as the other.

13. For what values of λ do the roots x_1 and x_2 of the equation

$$x^2 - (3\lambda + 2)x + \lambda^2 = 0$$

satisfy the condition $x_1 = 9x_2$? Find the roots.

14. Find the values of p for which the roots x_1 and x_2 of the equation

$$x^2 - \frac{15}{4}x + p = 0$$

satisfy $x_1 = x_2^2$.

15. A motor boat sails downstream from point A to point B, which 10 miles away from A, and returns. If the actual speed of the boat (in still water) is 3 mi/hr, the trip from A to B takes 2 hours 30 minutes less than the trip from B to A. What must the actual speed of the boat be for the trip from B to A to take 2 hours?

16. Two points move at a constant speed along the circumference of a circle whose length is 150 feet. When they move in opposite directions they meet every 5 seconds; when they move in the same direction they are together every 25 seconds. Find their speeds.

17. Two tanks A and B contain mixtures of alcohol and water. A mixture of 3 parts of A and 2 parts of B will contain 40% of alcohol; a mixture of 1 part from A and 2 parts from B will contain 32% of alcohol. What are the persentages of alcohol in A and B respectively?

18. George owes \$5,000 and Peter owes \$3,000. George could pay all his debts if besides his own money had $\frac{2}{3}$ of Peter's; and Peter could pay all but \$100 of his debts if besides his own money he had $\frac{1}{2}$ of George's. How much money has each?

19. A man invested a principal at a certain simple interest rate and received \$1,248 in one year. Were the principal \$100 greater and the rate $1\frac{1}{2}$ times as great the amount at the end of 2 years would be \$1,456. Find the principal and the interest rate.

20. After walking a certain distance a pedestrian rests for 30 minutes. He then continues his journey, but at $\frac{7}{8}$ of his original speed and on reaching his destination finds that he has accomplished the entire distance, 20 miles, in 6 hours. If he had walked 4 miles further at the original speed and then tested as before, the journey would have taken $5\frac{6}{7}$ hours. What was his original speed, and how far from the starting point did he rest?

21. Two men working together complete a work in 5 days. If the first man worked twice as fast and the other worked half as fast, they would have to work for 4 days. How much time would it take the first man to do the work alone?

22. Two students spent 7 hours to prepare for an exam, reckoning from the moment when the first student began working. The second student began $1\frac{1}{2}$ hours later than the first. If the job was assigned to each student separately, then it would take the first student 3 hours more than the second to complete the job. How much time would it take each student to complete the job if they worked separately?

23. Two motor boats A and B, ply between two islands K and L which are 200 miles apart. Boat A can start from island K 1 hour later than boat B, overtake B in 2 hours and having reached island L and made a 4 hours' wait there, on its return trip meets B 10 miles from L. What are the speeds of the boats A and B.

24. The sum of three numbers is 20. The first plus twice the second plus three times the third equals 44, and the twice the sum of the first and second minus four times the third equals -14. Find the numbers.

25. The sum of three numbers is 51. If the first number be divided by the second, the quotient is 2 and the remainder 5; but if the second number be divided by the third the quotient is 3 and the remainder 2. Find the numbers.

26. The difference of two numbers is the same as their product, and the sum of their reciprocals is 5. Find the numbers.

27. Three men A, B and C set out to walk a certain distance. A walks $4\frac{1}{2}$ miles an hour and finishes the journey 2 hours before B. B walks 1 mile an hour faster than C and finishes the journey in 3 hours less time. What is the distance?

28. The numerator of a certain fraction exceeds its denominator by 2. The fraction itself exceeds its reciprocal by $\frac{24}{35}$. Find the fraction.

29. The square of the sum of two numbers less their product is 63, and the difference of their cubes is 189. Find the numbers.

30. The diagonal of a rectangle is 13 feet long. If each side were 2 feet longer than it is, the area would be 38 square feet greater than it is, What are the dimensions of the rectangle?

31. The perimeter of a right-angled triangle is $36\,cm$ and the area of the triangle is $54\,cm^2$. Find the lengths of the sides.

Chapter 9

Exponentials and Logarithms

In this final chapter, we study exponentials and logarithms and their properties. We also study exponential and logarithmic equations, inequalities and systems.

9.1   Exponentials

In Chapter 3, we learned how to form powers a^p of a real number a to a rational exponent $p \in Q$, and we proved their basic properties.

Based on that discussion, for $a > 0$ and $p = \frac{m}{n}$, we gave meaning to expressions of the form

$$a^p = a^{\frac{m}{n}} = \sqrt[n]{a^m}.$$

But what is the meaning of a^x when x is an *irrational number*? A rigorous definition of this requires methods discussed in calculus[1]. However, the basic idea of the definition is easy to understand: Since an irrational

---

[1]In fact, the notion of *limit of a sequence*. The study of limits is part of calculus, and as such it is beyond the scope of this book.

x can be approximated by a teminating decimal, ie by a rational number, p, it is reasonable to expect that a^x is approximated by a^p. That is,

$$\text{if }\ x \approx p, \quad \text{then}\quad a^x \approx a^p.$$

For example, take the irrational number $\sqrt{2} \approx 1.411421356237....$ Then an approximation to $a^{\sqrt{2}}$ is

$$a^{\sqrt{2}} \approx a^{1.41}.$$

Better approximations to $a^{\sqrt{2}}$ would be

$$a^{1.411}, \ a^{1.4114}, \ a^{1.41142}, \ a^{1.411421}, \ a^{1.4114213}, \ a^{1.41142135}, \ ...$$

Continuing this way, we can obtain approximations to $a^{\sqrt{2}}$ to any desired degree of accurancy.

In general, if x is irrational and p_n is a decimal approximation to x with n decimal digits, then, it can be proved (using the notion of limit) that when the number n of decimal digits increases indefinetely, the numbers a^{p_n} *approximate* to a definite number denoted by a^x.

In symbols we write, and we define

$$a^x = \lim_{n \to \infty} a^{p_n}.$$

Furthermore, for any irrational $x > 0$, we define

$$a^{-x} = \frac{1}{a^x}.$$

Thus, the so-called ***exponetial expression*** a^x has now meaning for all real numbers $x \in \mathbb{R}$. Note that $a^x > 0$ for all $x \in \mathbb{R}$, and $a^0 = 1$.

For the bases $a = 2, 3, 4, 5, ...$, we obtain the exponentials

$$2^x, \quad 3^x, \quad 4^x, \quad 5^x, ...$$

While, for the bases $a = \frac{1}{2}, \frac{1}{3}, \frac{1}{4}, \frac{1}{5}, ...$, we obtain the exponential

$$\left(\tfrac{1}{2}\right)^x, \quad \left(\tfrac{1}{3}\right)^x, \quad \left(\tfrac{1}{4}\right)^x, \quad \left(\tfrac{1}{5}\right)^x, ...$$

9.2   Properties of Exponentials

The familiar properties of powers with rational exponents also hold for irrational exponents. Thus, the following properties hold. Let $a > 0$, $b > 0$ and $x, y \in \mathbb{R}$. Then

1. $a^x \cdot a^y = a^{x+y}$

2. $\frac{a^x}{a^y} = a^{x-y}$

3. $(a^x)^y = a^{xy}$

4. $(a \cdot b)^x = a^x \cdot b^x$

5. $\left(\frac{a}{b}\right)^x = \frac{a^x}{b^x}$.

Depending on the base a of the exponential, the following inequalities hold:

1. If $a > 1$, then $a^x > 1$ for $x > 0$ and $a^x < 1$ for $x < 0$.

2. If $a > 1$ and $x < y$, then $a^x < a^y$.

3. If $0 < a < 1$, then $a^x < 1$ for $x > 0$ and $a^x > 1$ for $x < 0$.

4. If $0 < a < 1$ and $x < y$, then $a^x > a^y$.

Two important bases: $a = 10$ **and** $a = e$

In several computations an important role is played by the exponential 10^x, the exponential to the base $a = 10$.

Another important base for the exponential is $a = e$, where

$$e \approx 2.718281828459...$$

is **Euler's number**[2]. The exponential e^x plays a very important role in higher mathematics.

---

[2] Euler Leonard (1707-1783) a pioneering mathematician who made important contributions in various fields of mathematics. In calculus, one learns that

$$e = \lim_{n \to \infty} \left(1 + \frac{1}{n}\right)^n.$$

9.3   Logarithms

Definition 9.3.1. Let $a > 0$ and $a \neq 1$. For all $x > 0$, the **logarithm to the base** a, denoted by $\log_a x$, is defined by

$$\log_a x = y \quad \Leftrightarrow \quad a^y = x.$$

In words, the ***logarithm to the base*** a ***of a positive numbers*** x ***is the exponent indicating the power to which the base*** a ***must be raised to obtain*** x. That is, *taking the logarithm of a positive number is an operation inverse of exponentiating.*

Note that $\log_a x$ is defined for all *positive numbers*. All negative numbers and zero are said to have no logarithms.[3]

Here are some immediate properties of the logarithm:

1. $\log_a 1 = 0$, because $a^0 = 1$.

2. $\log_a a = 1$, because $a^1 = a$.

3. $\log_a a^x = x$, because $a^x = a^x$.

4. Since $y = \log_a x \ \Leftrightarrow \ a^y = x$, we obtain the following ***fundamental logarithmic identity***
$$a^{\log_a x} = x.$$

Various bases

For the bases $a = 2, 3, 4, 5, ...$, we obtain the logarithms

$$\log_2 x, \ \log_3 x, \ \log_4 x, \ \log_5 x, ...$$

For the logarithm to the base $a = 10$, we write

$$\log x \quad \text{instead} \quad \log_{10} x,$$

---

[3]In higher mathematics, one learns that the logarithms of negative numbers are complex numbers.

and we call it the **common logarithm**. Thus,

$$\log x = y \iff 10^y = x.$$

For the logarithm to the base $a = e$, where $e \approx 2.718...$ is the Euler number, we write

$$\ln x \quad \text{instead} \quad \log_e x,$$

and we call it the **natural logarithm** or the *Napierian logarithm*[4]
Thus,
$$\ln x = y \iff e^y = x.$$
For the bases $a = \frac{1}{2}$, $a = \frac{1}{3}$, and so on, we obtain the logarithms

$$\log_{\frac{1}{2}} x, \ \log_{\frac{1}{3}} x, \ \text{and so on.}$$

Let us now compute some logarithms.

Example 9.3.2. Compute.

1. $\log_2 8 = 3$, because $2^3 = 8$.
2. $\log_2 32 = 5$, because $2^5 = 32$.
3. $\log_2 \frac{1}{4} = -2$, because $2^{-2} = \frac{1}{4}$.
4. $\log_3 81 = 4$, because $3^4 = 81$.
5. $\log 100 = 2$, because $10^2 = 100$.

Example 9.3.3. Find

1. $\log 0.001 = -3$, because $10^{-3} = \frac{1}{1,000} = 0.001$.

2. $\log_4 2 = \frac{1}{2}$, because $4^{\frac{1}{2}} = 2$.

---

[4]John Napier (1550-1617) was the first mathematician to intoduced the term *logarithm*, by combining the Greek words *logos = ratio* and *arithmos = number*. The symbol *ln* comes from the initials of the Latin words: *logarithmus* and *naturalis*.

3. $\log_{\frac{1}{2}} 0.25 = 2$, because $(\frac{1}{2})^2 = \frac{1}{4} = 0.25$.

4. $\log_{\frac{1}{3}} 9 = -2$, because $(\frac{1}{3})^{-2} = 9$.

5. $\ln(\sqrt{e}) = \frac{1}{2}$, because $e^{\frac{1}{2}} = \sqrt{e}$.

Remark. The reader should be aware that most values of logarithms are irrational numbers. For example, *the common logarithms of numbers which are not powers of* 10 *with rational exponent are irrational.* Indeed, let x be positive and let $x \neq 10^p$, where $p \in \mathbb{Q}$. Suppose that $\log x$ is rational. This means $\log x = \frac{m}{n}$ for some integers m, n. Then $x = 10^{\frac{m}{n}}$, ie, $x = 10^p$ with $p \in \mathbb{Q}$, which contradicts our initial hypothesis about x.

This is also true in general for any $\log_a x$. Thus, *for any positive number x which is not a rational power of the base a, (viz, $x \neq a^p$, with $p \in \mathbb{Q}$), the logarithm $\log_a x$ is irrational.* For instance, $\log_2 3$ is irrational (why?).

9.4   Properties of Logarithms

Let us use the fundamental logarithmic identity

$$a^{\log_a x} = x$$

to prove the basic properties of logarithms. These properties are consequences of the corresponding properties of the exponents (see, Section 9.2).

1. *The logarithm of a product is equal to the sum of the logarithms.* That is,

$$\log_a(MN) = \log_a M + \log_a N,$$

where $M, N > 0$.

Proof. We have $a^{\log_a M} = M$ and $a^{\log_a N} = N$. Multiplying these equalities and using property (1) of the exponents, we get

$$MN = a^{\log_a M} a^{\log_a N} = a^{\log_a M + \log_a N}.$$

On the other hand, by the fundamental logarithmic identity, we also have

$$a^{\log_a MN} = MN.$$

Therefore,

$$a^{\log_a(MN)} = a^{\log_a M + \log_a N} \quad \Rightarrow \quad \log_a(MN) = \log_a M + \log_a N.$$

2. The logarithm of a quotient is equal to the difference of the logarithms. That is,

$$\log_a\left(\frac{M}{N}\right) = \log_a M - \log_a N,$$

where $M, N > 0$.

Proof. We have $a^{\log_a M} = M$ and $a^{\log_a N} = N$. Dividing these equalities and using property (2) of the exponents, we get

$$\frac{M}{N} = \frac{a^{\log_a M}}{a^{\log_a N}} = a^{\log_a M - \log_a N}.$$

On the other hand,

$$\frac{M}{N} = a^{\log_a \frac{M}{N}}.$$

Therefore,

$$a^{\log_a\left(\frac{M}{N}\right)} = a^{\log_a M - \log_a N} \quad \Rightarrow \quad \log_a \frac{M}{N} = \log_a M - \log_a N.$$

3. The logarithm of a power is equal to the exponent times the logarithms. That is,

$$\log_a M^k = k \log_a M,$$

where $M > 0$ and k any real number.

Proof. We have $M = a^{\log_a M}$. Raising both sides to the k power and using property (3) of the exponents, we get

$$M^k = \left(a^{\log_a M}\right)^k = a^{k \log_a M}.$$

On the other hand,

$$M^k = a^{\log_a(M^k)}.$$

Therefore,

$$a^{\log_a(M^k)} = a^{k \log_a M} \quad \Rightarrow \quad \log_a M^k = a^k \log_a M.$$

Change of Base Formula

Most calculators have both log and ln keys to calculate the common logarithm and the natural logarithm of a number. However, sometimes we may need to calculate logarithms with base different than 10 or e. Let us look an example to see how we calculate such a logarithm.

Example 9.4.1. Calculate $\log_2 5$.

Solution. Set $y = \log_2 5$. Then $2^y = 5$. Taking natural logarithms to both sides, we have

$$\ln 2^y = \ln 5 \quad \Leftrightarrow \quad y \ln 2 = \ln 5 \quad \Leftrightarrow \quad y = \frac{\ln 5}{\ln 2} \approx 2.321928.$$

This process works in general and leads to the following formula.

Change of Base in Logarithms: If a, b and x are positive real numbers with $a \neq 1$, $b \neq 1$, then

$$\log_b x = \frac{\log_a x}{\log_a b}.$$

One way to prove this formula is to repeat the argument in the above example, and we leave this to the reader as an exercise. Alternatively, we may use $b^{\log_b x} = x$ and upon taking $\log_a$ to both sides, we get

$$\log_a(b^{\log_b x}) = \log_a x \quad \Rightarrow \quad \log_b x \cdot \log_a b = \log_a x \quad \Rightarrow \quad \log_b x = \frac{\log_a x}{\log_a b}.$$

In particular, if $a = 10$, or $a = e$, we get

$$\log_b x = \frac{\log x}{\log b}, \quad or \quad \log_b x = \frac{\ln x}{\ln b}.$$

Note also

$$\log_a b \cdot \log_b a = 1.$$

Example 9.4.2. Use the change of base formula to calculate $\log_3 8$.

Solution. We have

$$\log_3 8 = \frac{\log 8}{\log 3} = \frac{0.90308987}{0.47712125} \approx 1.892789034.$$

To the four decimal places $\log_3 8 = 1.8928$

Summary of the Properties of Logarithms

1. $\log_a 1 = 0, \quad \log_a a^x = x, \quad a^{\log_a x} = x.$
2. $\log_a(MN) = \log_a M + \log_a N.$
3. $\log_a\left(\frac{M}{N}\right) = \log_a M - \log_a N.$
4. $\log_a M^k = k \log_a M.$

Example 9.4.3. Write each expression in terms of the logarithms of x, y and z.

1.
$$\log_a\left(x^2 y^3 \sqrt{z}\right) = \log_a x^2 + \log_a y^3 + \log_a \sqrt{z} =$$
$$2\log_a x + 3\log_a y + \frac{1}{2}\log_a z.$$

2.
$$\log_a \sqrt{\frac{x}{y^4 z^3}} = \frac{1}{2}\log_a\left(\frac{x}{y^4 z^3}\right) =$$
$$\frac{1}{2}\left[\log_a x - \log_a(y^4 z^3)\right] =$$
$$\frac{1}{2}\left[\log_a x - \log_a y^4 - \log_a z^3\right] =$$
$$\frac{1}{2}\log_a x - 2\log_a y - \frac{3}{2}\log_a z.$$

Example 9.4.4. Write each expression as a single logarithm.

1.

$$\log 5 + \log(x+2) - 3\log x = \log[5(x+2)] - \log x^3 = \log\left(\frac{5(x+2)}{x^3}\right).$$

2.

$$\frac{1}{3}\log_a(x+5) - 2\log_a(x+1) + 4\log_a z = \log_a(x+5)^{\frac{1}{3}} - \log_a(x+1)^2 + \log_a z^4 =$$

$$\log_a \frac{(x+5)^{\frac{1}{3}}}{(x+1)^2} + \log_a z^4 = \log_a\left(\frac{z^4(x+5)^{\frac{1}{3}}}{(x+1)^2}\right).$$

Example 9.4.5. Chemists use a number denoted by pH to measure the concentration of acidity of a chemical solution. If $[H^+]$ is the hydrogen ion concentration in gram-ion per liter, then

$$pH = -\log[H^+].$$

1. Find the pH of seawater: $[H^+] \approx 3.2 \times 10^{-9}$

2. Beer has $pH \approx 4.2$. Find its hydrogen ion concentration.

   Solution.

1.
$$pH = -\log[H^+] = -\log(3.2 \times 10^{-9}) \approx 8.5.$$

2. Since $pH \approx 4.2$, we have

$$4.2 = -\log[H^+] \ \Rightarrow \ \log[H^+] = -4.2 \ \Rightarrow \ [H^+] = 10^{-4.2} = 6.3 \times 10^{-5}.$$

9.5   Exercises

1. Write each exponential expression in logarithmic form.

   (a) $3^4 = 81$

   (b) $2^5 = 32$

(c) $2^{-3} = \frac{1}{8}$

(d) $\left(\frac{1}{2}\right)^{-4} = 16$

(e) $m^p = k$

(f) $e^x = 5$

2. Write each logarithmic expression in exponential form.

(a) $\log_2 8 = 3$

(b) $\log_3\left(\frac{1}{9}\right) = -2$

(c) $\log_{\frac{1}{5}}(125) = -3$

(d) $\log_{\frac{1}{3}}\left(\frac{1}{81}\right)) = 4$

(e) $\log_4\left(\frac{1}{64}\right) = -3$

(f) $\ln x = 4$

3. Find each value of x.

(a) $\log_3 81 = x$

(b) $\log 1000 = x$

(c) $\log_{\frac{1}{2}} 8 = x$

(d) $\log_{\frac{1}{7}} 49 = x$

(e) $\log_{\frac{1}{3}} 27 = x$

(f) $\log_3 9 = x$

4. Find each value of x.

(a) $\log_9 3 = x$

(b) $\log_3 x = -2$

(c) $\log x = 2$

(d) $\log_{\frac{1}{2}} x = 2$

(e) $\log_{\frac{1}{3}} x = -3$

5. Find each value of x.

(a) $\log_x 125 = 3$

(b) $\log_x \frac{1}{9} = -2$

(c) $\log_x \frac{1}{16} = 2$

(d) $\log_x \frac{27}{8} = 3$

(e) $\log_x \frac{\sqrt{3}}{3} = -\frac{1}{2}$

6. Find the domain of definition each logarithm.

   (a) $\log_2(x - 5)$

   (b) $\log_3(x^2 - 4)$

   (c) $\log(x^2 - x - 6)$

   (d) $\ln(x^2 + 1)$

   (e) $\log_a\left(\frac{x}{x-1}\right)$

7. Write each expression in terms of the logarithms of x, y and z.

   (a) $\log_a(3xy^2)$

   (b) $\log_a(x^2 y^3 \sqrt{z})$

   (c) $\log_a \sqrt{\frac{x}{y}}$

   (d) $\log_a\left(\frac{x^5}{yz^2}\right)$

   (e) $\log_a \sqrt[4]{\frac{x^3 y^2}{z^4}}$

8. Write each expression as a sum and/or difference of logarithms.

   (a) $\log(x\sqrt{1 + x^2})$

   (b) $\log_a\left(\frac{x^3}{2x-1}\right)$

   (c) $\log_a\left(\frac{x^2\sqrt{x+1}}{(x-5)^3}\right)$

   (d) $\log_a \sqrt{\frac{x^2+1}{(x-3)^4}}$

9. Write each expression as a single logarithm.

   (a) $\log(x - 2) - \log(x + 1)$

   (b) $3\log_a x + 2\log_a(x - 1)$

   (c) $\log_a x + \log_a(x + 3) - \frac{1}{2}\log(x + 1)$

   (d) $\log\left(\frac{x}{x-1}\right) + \log\left(\frac{x+1}{x}\right) - \log(x^2 - 1)$

   (e) $8\log_2 \sqrt{5x + 2} - \log_2\left(\frac{6}{x}\right) + \log_2 6$

(f) $\log(xy + y^2) - \log(xz + yz) + \log z$

10. Show that the following equalities are true.

(a) $\log_a 50 = \log_a 2 + 2\log_a 5$

(b) $\log 20 - 1 = \log 2$

(c) $\log 3 + 2\log 4 - \log 12 = 2\log 2$

(d) $3\log 2 + \log 5 - \log 4 = 1$

(e) $\frac{1}{2}\log 25 + \frac{1}{3}\log 8 + \frac{1}{5}\log 32 = 1 + \log 2$

11. Show that.

(a) $\log_2 3 + 2\log_2 4 - \log_2 12 = 2$

(b) $2\log_2(2 + \sqrt{2}) + \log_2(6 - 4\sqrt{2}) = 2$

(c) $4^{1 - \frac{1}{2}\log_2 3} = \frac{4}{3}$

(d) $9^{\frac{1}{2}\log_3 18 - 1} = 2$

(e) $\log_2 3 \cdot \log_3 4 \cdot \log_4 5 \cdot \log_5 6 \cdot \log_7 8 = 3$

12. Prove the following logarithmic identities.

(a) $a^{\log b} = b^{\log a}$

(b) $\log_a b \cdot \log_b a = 1$

(c) $\log_a x + \log_{\frac{1}{a}} x = 0.$

13. Show that

$$\log_2 \sqrt{32\sqrt{16\sqrt[5]{2}}} = \frac{71}{20}.$$

14. The hydrogen ion concentration of a solution is 6.31×10^{-4}. Find its pH.

15. Find the pH of a solution with hydrogen ion concentration of 1.7×10^{-5}.

16. The pH of a solution is 2.9 liters. Find the hydrogen ion concentration of the solution.

17. Find the hydrogen ion concentration of a calcium solution whose pH is 13.2.

18. A solution is concidered acid if $[H^+] > 10^{-7}$ and basic if $[H^+] < 10^{-7}$. Find the corresponding inequalities for the pH.

19. According to the Richter scale the intensity R of an earthquake with amplitude A (measured in micrometers) and period P (the time of one oscillation of the Earth's surface, measured in seconds) is given by

$$R = \log \frac{A}{P}.$$

   Find the intensity of an earthquake with amplitute of $8,000$ micrometers and period 0.008 seconds.

20. An eartquake with period of $\frac{1}{5}$ second has an amplitude of $5,000$ micrometers. Find its intensity on the Richter scale.

9.6   Applications

9.6.1   Compound Interest

In Section 2.4.3 we studied simple interest problems. We recall that *interest* is money paid for the use of money. Interest is calculated as a percent of the *principal* (the total amount deposited in a bank account or borrowed from a bank). This percent is called the *interest rate*.
In working with problems involving interest we use the term *payment period* denoted by n as follows: Annually, $n = 1$. Semiannually, $n = 2$. Quarterly, $n = 4$. Monthly, $n = 12$. Daily, $n = 365$.

When interest due at the end of a payment period is added to the principal so that the interest computed at the end of the next payment period is based on this new principal amount (old principal + interest) the interest is said to have been **compounded**.

That is, **compound interest** is interest paid on previously earned interest.

Compound Interest Formula

If a principal P dollars is deposited in an account earning interest at an annual rate r, compounded n times each year, the amount A in the account after t years is given by

$$A = P(1 + \frac{r}{n})^{nt} \tag{9.1}$$

In calculus one learns that the larger n gets, the closer $(1 + \frac{r}{n})^n$ gets to e^r. We write that as $n \to \infty$

$$(1 + \frac{r}{n})^n \to e^r.$$

Thus, no matter how frequent the compounding, the amount after 1 year has the definite ceiling Pe^r. In this case formula (9.1) becomes

$$A = Pe^{rt} \tag{9.2}$$

and we say that interest is **compounded continuously**.

Example 9.6.1. An initial deposit of $\$1,000$ earns an anuual interest 8%. How much will be in the account after 5 years if interest is compounded:

1. Annually?

2. Quarterly?

3. Monthly?

4. Daily?

5. Continuously?

Solution. The principal is $P = 1,000$, the interest rate is $r = 8\% = 0.08$ and the time is $t = 5$.

We substitute these values in the formula for compound interest and according to the compounding period we have

1. For $n = 1$ we get

$$A = P(1 + \frac{r}{n})^{nt} = 1,000(1 + 0.08)^{(1)(5)} = 1,000(1.08)^5 = 1,469.33.$$

2. For $n = 4$ we get

$$A = P(1 + \frac{r}{n})^{nt} = 1,000(1 + \frac{0.08}{4})^{(4)(5)} = 1,000(1.02)^{20} = 1,485.95.$$

3. For $n = 12$ we get

$$A = P(1+\frac{r}{n})^{nt} = 1,000(1+\frac{0.08}{12})^{(12)(5)} = 1,000(1.006667)^{60} = 1,489.88.$$

4. For $n = 365$ we get

$$A = P(1+\frac{r}{n})^{nt} = 1,000(1+\frac{0.08}{365})^{(365)(5)} = 1,000(1.000219)^{1825} = 1,491.27.$$

5. Using the formula for continuously compounded interest, we have

$$A = Pe^{rt} = 1,000e^{(0.08)(5)} = 1,000e^{0.4} = (1,000)(1.491824698) = 1,491.82.$$

Example 9.6.2. How much must be invested now at a rate of 6% per annum compounded quarterly to grow to $40,000 after 20 years?

Solution. The unknown is the principal P. We substitute $A = 40,000$, $r = 0.06$, $n = 4$ and $t = 20$ in the formula for compound interest, we have

$$40,000 = P(1 + \frac{0.06}{4})^{(4)(20)} = P(1.015)^{80} = P(3.290662787).$$

Solving for P we get

$$P = \frac{40,000}{3.290662787} = \$12,155.61.$$

Example 9.6.3. How long does it take for an investment to double in value if it is invested at an annual rate 8% compounded:

1. Monthly?

2. Continuously?

Solution. If P is the principal, we want P to double, that is, the amount to be $A = 2P$.

1. We use the formula for compounded interest with $r = 0.08$ and $n = 12$ to find t. We have

$$2P = P(1 + \frac{0.08}{12})^{12t} \quad 2 = (1.00667)^{12t}.$$

Taking log both sides, we get

$$\log 2 = 12t \log(1.00667) \Rightarrow$$

$$t = \frac{\log 2}{12 \log(1.00667)} = \frac{0.301029996}{0.034645516} \approx 8.69.$$

That is, the doubling time is 8.69 years or 104.3 months.

2. We use the formula for continuously compounded interest with $r = 0.08$ to find t. We have

$$2P = Pe^{(0.08)t} \quad 2 = e^{(0.08)t}.$$

Taking ln both sides, we get

$$\ln 2 = 0.08t \ln e \Rightarrow$$

$$t = \frac{\ln 2}{0.08} = \frac{0.693147118}{0.08} \approx 8.66.$$

That is, the doubling time is 8.66 years or 103.9 months.

9.6.2 Exponential Growth and Decay

Many natural phenomena have been found to follow an exponential law of change. That is, a certain quantity Q varries with time t according to the formula

$$Q(t) = Q_0 e^{kt},$$

where $k \neq 0$ is a constant and $Q_0 = Q(0)$ is the original amount of Q at $t = 0$.

If $k > 0$, then Q increases over time and it is said to express the low of **exponential growth**. If $k < 0$, then Q decreases over time and it is said to express the low of **exponential decay**.

The law of exponential change finds many applications in Sciences. For instance, populations grow exponentially, while radioactive substances decay exponentially. All radioactive substances have a specific *half-life*, which is the time required for half of the radioactive substance to decay.

Example 9.6.4. (Radioactive Decay). If the half-life of a radioactive substance is h years, show that the function expressing its exponential decay is

$$Q(t) = Q_0 2^{-\frac{t}{h}} \tag{9.3}$$

Solution. Since the half-life is h years the formula $Q(t) = Q_0 e^{kt}$ for $t = h$ gives

$$\frac{Q_0}{2} = Q_0 e^{kh} \Rightarrow \frac{1}{2} = e^{kt} \Rightarrow e^k = \left(\frac{1}{2}\right)^{\frac{1}{h}} \Rightarrow e^k = 2^{-\frac{1}{h}}.$$

Thus,

$$Q(t) = Q_0 2^{-\frac{t}{h}}.$$

Example 9.6.5. The half-life of radium is $1,600$ years. How much of a 5-gram sample will remain

1. after 600 years

2. after $2,000$ years.

Solution.

1. Substituting $Q_0 = 5$, $h = 1,600$ and $t = 600$ in the formula of radioactive decay, we have

$$Q = Q(600) = 5 \cdot 2^{-\frac{600}{1,600}} = 5 \cdot 2^{-0.375} \approx 3.86.$$

   Thus, after 600 years will remain 3.86 grams.

2. Substituting $t = 2,000$, we get

$$Q = Q(2,000) = 5 \cdot 2^{-\frac{2,000}{1,600}} = 5 \cdot 2^{-1.25} \approx 2.1.$$

   Thus, after $2,000$ years will remain 2.1 grams.

Example 9.6.6. (Population Growth). The population of a town grows exponentially at an annual rate of 3%. If the population in 2014 is $173,000$, how large the population will be in 2034?

Solution. Using the formula of exponential growth $Q(t) = Q_0 e^{kt}$ with $Q_0 = 173,000$, $k = 3\% = 0.03$ and $t = 20$ (since $2034 - 2014 = 20$ years), we get

$$Q(20) = 173,000 e^{(0.03)(20)} = 173,000 e^{0.6} = (173,000)(1.8221188) \approx 315,227.$$

Example 9.6.7. A city's population grows at the annual rate of 4%. How long will it take the population to double?

Solution. Substituting $k = 0.12$ in the doubling time formula, we heve

$$t = \frac{\ln 2}{k} = \frac{\ln 2}{0.04} = \frac{0.6931447181}{0.04} \approx 17.3.$$

Thus, the population will double in 17.3 years.

Example 9.6.8. (Bacterial Growth). A colony of bacteria grows according to the law $Q(t) = Q_0 e^{kt}$, with $k > 0$. If $10,000$ bacteria are present initially and after 2 hours the number of bacteria increases to $30,000$, how many bacteria there will be after 5 hours?

Solution. We have $Q_0 = 10,000$ at $t = 0$, and $Q(2) = 30,000$ at $t = 2$. Substituting these values in the formula $Q(t) = Q_0 e^{kt}$, we get

$$30,000 = 10,000 e^{k(2)} \quad \Rightarrow \quad 3 = e^{2k}.$$

Taking logarithms yields

$$\ln 3 = 2k \quad \Rightarrow \quad k = \frac{\ln 3}{2} \approx 0.5493.$$

Thus,

$$Q(t) = 10,000 e^{0.5493t}$$

and at $t = 5$, this gives

$$Q(5) = 10,000 e^{0.5493(5)} = (10,0000)(15.8797837) \approx 158,797.84$$

Example 9.6.9. (Doubling Time). If a population grows exponentially at a certain rate k, the time required for the population to double is called the **doubling time**. Show that doubling time is given by

$$t = \frac{\ln 2}{k}.$$

Solution. For $Q = 2Q_0$ the exponential growth formula gives

$$2Q_0 = Q_0 e^{kt} \ \Rightarrow \ 2 = e^{kt} \ \Rightarrow \ \ln 2 = kt \ \Rightarrow \ t = \frac{\ln 2}{k}.$$

9.6.3   Exercises

1. An initial deposit of $\$7,500$ earns interest of 5% compounded annually. What will be the amount after thirty years?

2. An initial deposit of $\$10,000$ earns interest of 8% compounded quarterly. How much will be in the account in ten years?

3. Find the amount of $\$5,500$ in twenty years at 3% compounded interest, the interest being compounded semiannually.

4. Find the amount of $\$1,000$ in $4\frac{1}{2}$ years at 9% compounded interest, the interest being compounded monthly.

5. What sum will amount to $\$1,250$ if invested at 4% compounded annually for fifteen years?

6. If $\$100$ had been invested on January 1, 1900 at 5% interest compounded annually, what would it be worth on January 1, 2050?

7. How much must be invested now at 6% interest compounded quarterly to grow to $\$40,000$ in twenty years?

8. An initial investment of $\$5,000$ earns 8.2% annual interest compounded continuously. What will the investment be worth in twenty years?

9. An account now contains $\$11,180$ and has been accumulating interest at 7% annual rate compounded continuously for 7 years. Find the initial deposit.

10. An initial investment of $5,000 grows at an annual rate of 8.5% for five years. Compare the balances resulting from annual compounding and continous compounding.

11. How long does it take for an investment to double in value if it is invested at an annual rate of 6% compounded (a) monthly? (b) continuously?

12. How many years will it take for an initial investment of $10,000 at 6% compounded continuously, to grow to $25,000?

13. Which deal is better? Place $1,000 in a bank account that pays 5.6% compounded continuously or at another bank that pays 5.9% compounded monthly for one year.

14. (**Radioactive Decay**). A certain radioactive substance has s half-life of 5,080 years. What percentage will remain after an elapse of 10,000 years?

15. (**Radioactive Decay**). The half-life of radioactive carbon-14 is 5,700 years. How much of an initial sample will remain after 3,000 years?

16. (Radioactive Decay). A certain radioactive substance decreases at the rate of 0.0021% per year. What is its half-life?

17. (**Radioactive Decay**). In 2 years 20% of a radioactive substance decays. Find its half-life.

18. (**Radioactive Decay**). The half-life of tritium is 12.4 years. How long will it take for 25% of a sample of tritium to decompose?

19. (**Carbon -14 Dating**). Only 70% of the carbon-14 in a wooden bowl remain. How old is the bowl?

20. (**Radioactive Decay**). Strontinum-90 is a radioactive substance that decys according to the law

$$Q(t) = Q_0 e^{-0.0244t}.$$

(a) What is the half-life of Strontinum-90?
(b) How long it takes for 100 grams of Strontinum-90 to decay to 10 grams?

21. (**Population Growth**). The size $Q(t)$ of a certain incent population at time t (in days) obeys the equation

$$Q(t) = 500e^{0.02t}.$$

   (a) How large will the population be when $t = 15$ days?
   (b) After how many days will the population reach 1000?
   (c) How long will it take the population reach 2000?

22. (**Bacterial Growth**). The number N of bacteria in a culture at time t (in hours) grows according the law

$$N(t) = 1000e^{0.01t}.$$

   After how many hours will the population equal $1,500$? $2,000$?

23. (**Bacterial Growth**). A colony of bacteria grows according to the law $Q(t) = Q_0e^{kt}$ with $k > 0$. If 900 bacteria are present initially, and the number of bacteria doubles in 3 hours, how many bacteria will be present after 8 hours? How long will it take for the size of the colony to triple?

24. (**Bacterial Growth**). The population of a colony of bacteria obey the law of exponential growth. If there are $1,000$ bacteria initially, and there are $1,800$ after 1 hour, what is the size of the colony after 3 hours? How long is it until there are $10,000$ bacteria?

25. (**Population Growth**). The population of a town follows the law of exponential growth. if the population doubled in size over an 18 month period and the current population is $10,000$, what will the population be 2 years from now?

26. (**Newton's Law of Cooling**). Water whose temperature is at 100^oC is left to cool in a room where the temperature is 60^oC. After 3 minutes, the water temperature is 90^oC. If the water temperature T is a function of time t given by

$$T = 60 + 40e^{kt}.$$

   a) Find the constant k. b) Find the time for the water temperature to reach 70^oC.

9.7 Exponential Equations

An equation containing the unknown x in the exponent is called an ***exponential equation***. The simplest exponential equation is

$$a^x = b,$$

where $a > 0$, $a \neq 1$, $b > 0$. Its solution is $x = \log_a b$.

The idea for solving exponential equations is based on the fact

$$a^{x_1} = a^{x_2} \quad \Leftrightarrow \quad x_1 = x_2. \tag{9.4}$$

In words, ***equality of exponentials with equal base implies the equality of their exponents, and vice versa.***

For more general exponential equations of the form

$$a^{f(x)} = b^{g(x)},$$

we express the exponentials to a common basis (if possible) and then we use (9.4). If this is not possible, since both sides of the equation are positive, we may take common logarithms (or natural logarithms) to both sides of $a^{f(x)} = b^{g(x)}$, and thus obtain the equation $f(x) \log a = g(x) \log b$, which we solve accordingly.

We illustrate these with a number of examples.

Example 9.7.1. Solve the equation

$$2^x = \frac{1}{64}.$$

Solution. Since $\frac{1}{64} = 2^{-6}$, the equation becomes

$$2^x = 2^{-6} \quad \Rightarrow \quad x = -6.$$

Example 9.7.2. Solve the equation

$$5^x = 2^x.$$

Solution. Dividing both sides by $2^x > 0$, we obtain

$$\left(\frac{5}{2}\right)^x = 1 \quad \Rightarrow \quad \left(\frac{5}{2}\right)^x = \left(\frac{5}{2}\right)^0 \quad \Rightarrow \quad x = 0.$$

Alternatively, we can take natural logarithms to both sides of the equation, and we have

$$\ln\left(5^x\right) = \ln\left(2^x\right) \quad \Rightarrow \quad x\ln 5 = x\ln 2 \quad \Rightarrow$$

$$(\ln 5 - \ln 2)x = 0 \quad \Rightarrow \quad \left(\ln\frac{5}{2}\right)x = 0 \quad \Rightarrow \quad x = 0,$$

since $\ln\frac{5}{2} \neq 0$.

Example 9.7.3. Solve the equation

$$3^{x^2-1} = 81^{1-x}.$$

Solution. Since $81 = 3^4$, we have

$$3^{x^2-1} = \left(3^4\right)^{1-x} \quad \Rightarrow \quad 3^{x^2-1} = 3^{4-4x} \quad \Rightarrow \quad x^2 - 1 = 4 - 4x \quad \Rightarrow$$

$$x^2 + 4x - 5 = 0 \quad \Rightarrow \quad (x+5)(x-1) = 0 \quad \Rightarrow \quad x = -5, \; x = 1.$$

Example 9.7.4. Solve the equation

$$3^{3x+2} = 5^{x-2}.$$

Solution. We take common logarithms to both sides of the equation, and we have

$$\log\left(3^{3x+2}\right) = \log\left(5^{x-2}\right) \quad \Rightarrow \quad (3x+2)\log 3 = (x-2)\log 5 \quad \Rightarrow$$

$$(3\log 3)x + 2\log 3 = (\log 5)x - 2\log 5 \quad \Rightarrow \quad (3\log 3 - \log 5)x = -2(\log 3 + \log 5) \quad \Rightarrow$$

$$x = \frac{-2(\log 3 + \log 5)}{3\log 3 - \log 5} = -3.212.$$

Example 9.7.5. Solve the equation

$$4^x - 3^{x-0.5} = 3^{x+0.5} - 2^{2x-1}.$$

Solution. We have

$$4^x - 3^x \cdot 3^{-0.5} = 3^x \cdot 3^{0.5} - 2^{2x} \cdot 2^{-1} \;\Rightarrow$$

$$4^x - 3^x \cdot \frac{1}{\sqrt{3}} = 3^x \cdot \sqrt{3} - 4^x \cdot \frac{1}{2} \;\Rightarrow$$

$$(1 + \frac{1}{2})4^x = (\sqrt{3} + \frac{1}{\sqrt{3}})3^x \;\Rightarrow\; \frac{3}{2}4^x = \frac{4}{\sqrt{3}}3^x \;\Rightarrow$$

$$\left(\frac{4}{3}\right)^x = \frac{2}{3} \cdot \frac{4}{\sqrt{3}} = \frac{4}{3} \cdot \frac{\sqrt{4}}{\sqrt{3}} \;\Rightarrow$$

$$\left(\frac{4}{3}\right)^x = \left(\frac{4}{3}\right)^{\frac{3}{2}} \;\Rightarrow\; x = \frac{3}{2} = 1.5$$

Example 9.7.6. Solve the equation

$$4 \cdot 9^{x-1} = 3\sqrt{2^{2x+1}}.$$

Solution. To reduce the number of calculations, it is more efficient here to take logarithms to the base 3. Taking $\log_3$ to both sides and using the properties of logarithms, we obtain the equation

$$\log_3 4 + (x-1)\log_3 9 = \log_3 3 + \frac{1}{2}(2x+1)\log_3 2 \;\Rightarrow$$

$$2\log_3 2 + 2(x-1) = 1 + \frac{1}{2}(2x+1)\log_3 2 \;\Rightarrow$$

$$(2 - \log_3 2)x = 3 - 2\log_3 2 + \frac{1}{2}\log_3 2 \;\Rightarrow$$

$$(2 - \log_3 2)x = 3 - \frac{3}{2}\log_3 2 \;\Rightarrow$$

$$(2 - \log_3 2)x = \frac{3}{2}(2 - \log_3 2),$$

whence $x = \frac{3}{2}$ (since $2 - \log_3 2 \neq 0$). [5]

---

[5]One could have taken instead logarithms to the base 2 and work similarly to find the solution. The reader is invited to check this as an exercise.

By introducing a new variable, some exponential equations are reduced to algebraic equations, say, a quadratic equation, and so on.

We illustrate this with some examples.

Example 9.7.7. Solve the equation

$$9^x - 8 \cdot 3^x - 9 = 0.$$

Solution. Since $9^x = (3^2)^x = 3^{2x}$, we write the equation

$$3^{2x} - 8 \cdot 3^x - 9 = 0 \iff (3^x)^2 - 8 \cdot 3^x - 9 = 0.$$

Setting $3^x = y$, we get

$$y^2 - 8y - 9 = 0 \iff (y - 9)(y + 1) = 0 \iff y = 9, \quad y = -1.$$

If $y = 9$, then $3^x = 9$ or $3^x = 3^2$, and consequently $x = 2$.
If $y = -1$, then $3^x = -1$ and there is no solution, since $3^x > 0$ for any x.

Thus, $x = 2$ is the only solution.

Example 9.7.8. Solve the equation

$$e^x - 2 \cdot e^{-x} + 1 = 0.$$

Solution. First note

$$e^x - 2 \cdot e^{-x} + 1 = 0 \iff e^x - \frac{2}{e^x} + 1 = 0.$$

Setting $e^x = y$, we get

$$y - \frac{2}{y} + 1 = 0 \iff y^2 + y - 2 = 0 \iff (y - 1)(y + 2) = 0.$$

Hence $y = 1$ and $y = -2$.
If $y = 1$, then $e^x = 1$ or $e^x = e^0$, and consequently $x = 0$.
If $y = -2$, then $e^x = -2$ is impossible, because $e^x > 0$ for any x.

Thus, $x = 0$ is the only solution.

Example 9.7.9. Solve the equation

$$2^{2x+1} - 5 \cdot 6^x + 3 \cdot 9^x = 0.$$

Solution. The equation may be written as

$$2 \cdot 2^x - 5 \cdot 2^x \cdot 3^x + 3 \cdot 3^{2x} = 0.$$

Dividing both sides by 3^{2x}, we obtain

$$2 \cdot \left(\frac{2}{3}\right)^{2x} - 5 \left(\frac{2}{3}\right)^x + 3 = 0.$$

Setting $\left(\frac{2}{3}\right)^x = y$, we get

$$2y^2 - 5y + 3 = 0 \quad \Leftrightarrow \quad (y-1)(2y-3) = 0.$$

Hence, $y = 1$, and $y = \frac{2}{3}$.

If $y = 1$, then $\left(\frac{2}{3}\right)^x = 1 = \left(\frac{2}{3}\right)^0$, consequently, $x = 0$.

If $y = \frac{2}{3}$, then $\left(\frac{2}{3}\right)^x = \frac{3}{2} = \left(\frac{2}{3}\right)^{-1}$, consequently, $x = -1$.

Thus, $x = -1$ and $x = 0$ are the solutions.

9.7.1 Exercises

Solve each exponential equation.

1. $3^{4x} = 81$

2. $4^{x+3} = 8^{2x}$

3. $3^{2x} = 4^2$

4. $2^{x^2+2x} = \frac{1}{2}$

5. $5^{x-1} = 25^{2x}$

6. $4^{x^2-2x} = 64$

7. $7^{x^2+3x} = \frac{1}{49}$

8. $2^x \cdot 8^{-x} = 4^x$

9. $2^{x+1} = 3^x$

10. $5^{2x} = 2^{5x}$

11. $3^{2x} = 5^{x-3}$

12. $e^x = \pi^{1-x}$

13. $400e^{0.2x} = 600$

14. $3^{x+3} + 3^x = 84$

15. $9^x + 3^{x+1} - 4 = 0$

16. $4^x - 10 \cdot 2^x + 16 = 0$

17. $3^{2x+1} - 10 \cdot 3^x + 3 = 0$

18. $5^{2x-1} + 3 \cdot 5^{x+1} = 80$

19. $4^x - 7 \cdot 2^x = 8$

20. $3^{x+1} - 2^x = 3^{x-1} + 2^{x+3}$

21. $3 \cdot 2^{x-4} - 2^{x-1} = 5^{x-2} - 6 \cdot 2^{x-3}$

22. $2 \cdot 25^x - 7 \cdot 10^x + 5 \cdot 4^x = 0$.   (Hint: Set $y = \left(\frac{5}{2}\right)^x$)

23. $3 \cdot 9^x - 5 \cdot 6^x + \cdot 2^{2x+1} = 0$.   (Hint: Set $y = \left(\frac{2}{3}\right)^x$)

24. $5^{x-1} = 2 + \frac{3}{5^{x-2}}$.

25. $3^{x-1} - \frac{15}{3^{x+1}} + 3^x - \frac{21}{3^{x+1}} = 0$.

26. $3^{2x} + 9^x = 4^{x+1} + 11 \cdot 4^{x-1}$

27. $4^{x-1} - 5\sqrt{4^{x-2}} + 1 = 0$

28. $\sqrt{2^{6x-13}} - 3^{2(x-2)} = \sqrt{8^{2x-3}} - 3^{2x-3}$

9.8 Logarithmic Equations

An equation is called a ***logarithmic equation*** if it contains a logarithmic expression of the unknown x. The solutions of a logarithmic equation must belong in the domain of definition of all the logarithmic expressions of the equation.

Let $a > 0$ $a \neq 1$ and $f(x) > 0$. The logarithmic equation

$$\log_a (f(x)) = g(x) \tag{9.5}$$

is (by definition) equivalent to the equation

$$f(x) = a^{g(x)}.$$

On the other hand, the logarithmic equation

$$\log_a (f(x)) = \log_a (g(x)) \tag{9.6}$$

is equivalent to the equation

$$f(x) = g(x),$$

whenever $f(x) > 0$ (or, which is the same $g(x) > 0$).

In general, to solve logarithmic equations, we use the properties of logarithms, and we seek to reduce them to the above forms (9.5) or (9.6). When we use the properties of logarithms in solving equations, we must check the values of the obtained unknown by substituting them into the original equation. Or, else, we must find for which of those values the expressions appearing under the logarithm sign in the original equation are positive. Doing so, we identify and reject extraneous solutions. However, we must be even more careful here to avoid the loss of solutions. We must bear in mind that when we replace a logarithmic expression by another, the domain of definition of the new logarithmic expression may be "wider", which may lead to *extraneous solutions*, or the new domain may be "narrower", which may lead to the *loss of solutions* (See, Examples 9.8.3 and 9.8.6).

Example 9.8.1. Solve the equation

$$\log_3(2x - 5) = 2.$$

Solution. We change the equation to exponential form, and we have

$$\log_3(2x - 5) = 2 \iff 2x - 5 = 3^2 \iff$$

$$2x - 5 = 9 \iff 2x = 14 \iff x = 7.$$

Check: Substituting $x = 7$ in the original equation we get

$$\log_3(2 \cdot 7 - 5) = 2 \iff \log_3(9) = 2 \iff 9 = 3^2 \iff 9 = 9.$$

Since 7 does check, it is a solution.

Alternatevely, the domain of definition of $\log_3(2x - 5)$ is all x for which $2x - 5 > 0 \iff x > \frac{5}{2}$. Since $7 > \frac{5}{2}$, $x = 7$ is a solution.

Example 9.8.2. Solve the equation

$$2\log_5 x = 3\log_5 4.$$

Solution. By property (4) of the logarithms, we have

$$2\log_5 x = 3\log_5 4 \iff \log_5 x^2 = \log_5 4^3 \iff x^2 = 64 \iff x = 8, \ x = -8.$$

Since the expression $2\log_5 x$ is defined for $x > 0$, the solution $x = -8$ is an extraneous solution and we reject it.

Thus, the equation has only one solution $x = 8$.

Example 9.8.3. Solve the equation

$$\log_2 x + \log_2(x - 1) = 1.$$

Solution. By property (2) of the logarithms, we have

$$\log_2 x + \log_2(x - 1) = 1 \iff \log_2 x(x - 1) = 1 \iff x(x - 1) = 2^1 \iff$$

$$x^2 - x - 2 = 0 \iff (x - 2)(x + 1) = 0 \iff x = 2, \ x = -1.$$

The expressions $\log_2 x$ and $\log_2(x-1)$ are defined for $x > 0$ and $x > 1$, respectively. Hence the first side of the equation is defined for $x > 1$. Of the two solutions obtained, only $x = 2$ satisfies $2 > 1$. The solution $x = -1$ is an extraneous solution and we reject it.

Thus, the equation has only one solution $x = 2$.

Example 9.8.4. Solve the equation

$$\log_3(2 - x) - \log_3(2 + x) = \log_3 x - 1.$$

Solution. By property (3) of the logarithms, we have

$$\log_3\left(\frac{2-x}{2+x}\right) = \log_3\left(\frac{x}{3}\right) \quad \Leftrightarrow \quad \frac{2-x}{2+x} = \frac{x}{3} \quad \Leftrightarrow$$

$$x^2 + 5x - 6 = 0 \quad \Leftrightarrow \quad (x + 6)(x - 1) = 0 \quad \Leftrightarrow \quad x = -6, \ x = 1.$$

Substituting $x = -6$ in the original equation we see that $\log_3(-4)$ and $\log_3(-6)$ are not defined, therefore $x = -6$ is not a solution. Substituting $x = 1$, we get $\log_3(1) - \log_3(3) = \log_3 1 - 1 \quad \Leftrightarrow \quad -1 = -1$, and therefore $x = 1$ is a solution the equation.

Example 9.8.5. Solve the equation

$$\frac{1}{2}\log(x + 2) + \log\sqrt{x - 3} = 1 + \log\sqrt{3}.$$

Solution. By the properties of the logarithms, we have

$$\frac{1}{2}\left[\log(x + 2) + \log(x - 3)\right] = \log 10 + \log\sqrt{3} \quad \Leftrightarrow$$

$$\frac{1}{2}\log(x + 2)(x - 3) = \log(10\sqrt{3}) \quad \Leftrightarrow$$

$$\log(x + 2)(x - 3) = \log 300 \quad \Leftrightarrow \quad (x + 2)(x - 3) = 300 \quad \Leftrightarrow$$

$$x^2 - x - 306 = 0 \quad \Leftrightarrow \quad (x - 18)(x + 17) = 0 \quad \Leftrightarrow \quad x = 18, \ x = -17.$$

Checking shows that out of the two solutions obtained only $x = 18$ is a solution of the original equation.

The next example illustrates the case in which careless use of the properties of logarithms may lead to the loss of the solution.

Example 9.8.6. Solve the equation

$$\frac{1}{4}\log(x-6)^4 = \log(2x).$$

Solution. By property (4) of the logarithms, we have

$$\log(x-6) = \log(2x). \tag{9.7}$$

This impies

$$x - 6 = 2x \iff x = -6$$

Checking shows that $x = -6$ is neither a solution of equation (9.7) nor a solution of the original equation. However, it will be *wrong to say that the original equation has no solution*! The argument given above is valid for the case $x - 6 > 0$. Since $(x-6)^4 = (6-x)^4$, we must consider also the case $x - 6 < 0$, that is, $6 - x > 0$. In this case by property (4), the original equation becomes

$$\log(x-6) = \log(2x). \tag{9.8}$$

This impies

$$6 - x = 2x \iff 3x = 6 \iff x = 2.$$

Checking shows that $x = 2$ is a solution of the original equation.

By introducing a new variable, it is sometimes possible to reduce some logarithmic equations to algebraic equations.

We illustrate this with a number of examples.

Example 9.8.7. Solve the equation

$$\ln(e^x - 3) + x = 2\ln 2.$$

Solution. We have

$$x = \ln 4 - \ln(e^x - 3) \iff x = \ln\left(\frac{4}{e^x - 3}\right) \iff e^x = \frac{4}{e^x - 3}.$$

Setting $y = e^x$, we get

$$y = \frac{4}{y - 3} \iff y^2 - 3y - 4 = 0 \iff (y + 4)(y - 4) = 0.$$

Whence, $y = -1$, $y = 4$. Returning to x, we obtain the basic wxponential equations $e^x = -1$, which has no solution (since $e^x > 0$), and $e^x = 4$, which implies $x = \ln 4$.

Example 9.8.8. Solve the equation

$$\log(1 + 2^x) - x \log 5 = \log 6 - x.$$

Solution. Transfering the term $-x$ to the first side, and using the properties of logarithms, we have

$$\log(1+2^x)+x-x \log 5 = \log 6 \iff \log(1+2^x)+x(\log 10 - \log 5) = \log 6 \iff$$

$$\log(1 + 2^x) + x \log 2 = \log 6 \iff \log(1 + 2^x) + \log 2^x = \log 6 \iff$$

$$\log(2^x(1 + 2^x)) = \log 6 \iff 2^x(1 + 2^x) = 6.$$

That is,

$$(2^x)^2 + 2^x - 6 = 0.$$

Setting $y = 2^x$, we obtain the quadratic equation

$$y^2 + y - 6 = 0 \iff (y + 3)(y - 2) = 0 \iff y = -3, \ y = 2.$$

If $y = -3$, then $2^x = -3$, which has no solution (because $2^x > 0$).
If $y = 2$, then $2^x = 2$, consequently, $x = 1$.

Example 9.8.9. Solve the equation

$$3\sqrt{\ln x} + 2 \ln \frac{1}{\sqrt{x}} = 2$$

Solution. Since $2 \ln \frac{1}{\sqrt{x}} = 2 \ln x^{-\frac{1}{2}} = -\ln x$, the equation becomes

$$3\sqrt{\ln x} - \ln x = 2.$$

Setting $y = \ln x$, we get

$$3\sqrt{y} - y = 2.$$

Setting $t = \sqrt{y}$, this equation becomes

$$3t - t^2 = 2 \quad \Leftrightarrow \quad t^2 - 3t + 2 = 0 \quad \Leftrightarrow \quad (t-2)(t-1) = 0.$$

Whence, $t = 2$ and $t = 1$. Therefore $y = 4$ and $y = 1$. Returning to x, we get $\ln x = 4$, which implies $x = e^4$, and $\ln x = 1$, which implies $x = e$.

Example 9.8.10. Solve the equation

$$\log_3 x - \log_9(3x) = 1.$$

Solution. We have

$$\log_3 x - \log_9 3 - \log_9 x = 1 \quad \Leftrightarrow \quad \log_3 x - \frac{1}{2} - \log_9 x = 1.$$

By the change of base formula, we pass to logarithms to base 3, and we get

$$\log_9 x = \frac{\log_3 x}{\log_3 9} = \frac{\log_3 x}{2}.$$

Setting $y = \log_3 x$, we write the equation in the form

$$y - \frac{1}{2} - \frac{y}{2} = 1 \quad \Leftrightarrow \quad \frac{1}{2}y = \frac{3}{2} \quad \Leftrightarrow \quad y = 3.$$

Accordingly, we find

$$\log_3 x = 3 \quad \Leftrightarrow \quad x = 27.$$

Checking shows that $x = 27$ is the solution.

Example 9.8.11. Solve the equation

$$\log_2 x = 2 + \log_x 8.$$

Solution. Changing to logarithms of base 2, we get

$$\log_x 8 = \frac{\log_2 8}{\log_2 x} = \frac{3}{\log_2 x}.$$

Setting $y = \log_2 x$, the original equation becomes

$$y = 2 + \frac{3}{y} \quad \Leftrightarrow \quad y^2 - 2y - 3 = 0 \quad \Leftrightarrow \quad (y-3)(y+1) = 0.$$

Hence, we find $y = 3$, $y = -1$. If $y = 3$, then $\log_2 x = 3 \quad \Leftrightarrow \quad x = 8$. If $y = -1$, then $\log_2 x = -1 \quad \Leftrightarrow \quad x = \frac{1}{2}$. Checking shows that both $x = 8$ and $x = \frac{1}{2}$ are solutions.

Example 9.8.12. Solve the equation

$$x^{\log 5x} = 2.$$

Solution. Evidently, we should have $x > 0$. Taking common logarithms to both sides of the equation, we get

$$\log(5x) \cdot \log x = \log 2 \quad \Leftrightarrow \quad (\log 5 + \log x)\log x = \log 2.$$

This gives
$$\log^2 x + \log 5 \cdot \log x - \log 2 = 0.$$

Setting $y = \log x$, we obtain the quatratic equation

$$y^2 + (\log 5)y - \log 2 = 0.$$

The quadratic formula gives

$$y = \frac{1}{2}\left(-\log 5 \pm \sqrt{\log^2 5 + 4\log 2}\right).$$

Taking into accountthat $\log 2 = 1 - \log 5$, we see that $\log^2 5 - 4\log 5 + 4 = (\log 5 - 2)^2$. Consequently,

$$y = \frac{1}{2}\left(-\log 5 \pm (2 - \log 5)\right).$$

Whence, $y = 1 - \log 5 = \log 2$, and $y = -1$.

If $y = \log 2$, then $= \log 2$, and consequently, $x = 2$.
If $y = -1$, then $= -1$, and consequently, $x = 10^{-1}$.

Thus, $x = 2$ and $x = \frac{1}{10}$ are the solutions.

9.8.1   Exercises

Solve each logarithmic equation.

1. $\log_2(2x - 6) = 3$

2. $\log_3(4x + 1) = 2$

3. $\log_a(x^2 + 4) = \log_a(3x + 8)$

4. $\log x + \log(x + 9) = 1$

5. $\log(2x - 3) - \log(x - 1) = 0$

6. $\log(x + 1) + 2\log\sqrt{5x} = 2$

7. $\log(x - 6) - \log(x - 2) = \log\frac{5}{x}$

8. $\log(x - 1) - \log(x - 2) = \log 6 - \log x$

9. $2\log_3 x - \log_3(x - 4) = 2 + \log_3 2$

10. $\log(x - 1) + \log x = 1 - \log 5$

11. $\log(x^2 - x) = 1 + \log(x - 1)$

12. $\frac{\ln(8x - 15)}{\ln x} = 2$

13. $\log_a(x - 2) - \log_a(x + 3) = \log_a(x - 1) - \log_a(x + 6)$

14. $\log x^2 = (\log x)^2$

15. $\log\left[\log(2x^2 + x - 11)\right] = 0$

16. $(4x)^{\log 2 + \log\sqrt{x}} = 100$

17. $x^{\log x} = x^2\sqrt{x}$

18. $x^{\log\left(\frac{3x}{10}\right)} = 9 \cdot (3x)^{\log 9x^2}$

19. $3^{\log x} = 54 - x^{\log 3}$

20. $2^{\ln x} + 2^{2 - \ln x} = 5$

21. $\frac{3\ln x}{\ln x - 2} = \ln x + 2$

22. $\ln(e^x + 2) = x + \ln 3$

23. $\log_3 x \cdot \log_9 x = 2$

24. $2\log_4(4 - x) = 4 - \log(-2 - x)$

25. $(1 - \log 2) \log_5 x = \log 3 - \log(x - 2)$

26. $\log_2 x + \log_4 x + \log_{16} x = 7$

27. $\log_2(x - 1) + 1 = \log_{(x-1)} 4$

28. $\log_5 x \cdot \log_3 x = \log_5 x + \log_3 x$

29. $2 \log_3(x + 1) = 3 + 2 \log_{(x+1)} 3$

30. $\log_{0.5} x^2 - 14 \log_{16x} x^3 + 40 \log_{4x} \sqrt{x} = 0.$

9.9 Exponential and Logarithmic Inequalities (*)

Using the following **basic inequalities of exponentials and logarithms**, we can manage to solve some exponential and logarithmic inequalities directly.

1. Let $a > 1$. If $s < t$, then $a^s < a^t$ and $\log_a s < \log_a t$

2. Let $0 < a < 1$. If $s < t$, then $a^s > a^t$ and $\log_a s > \log_a t$.

We illustrate with a number of examples.

Example 9.9.1. Solve the inequality

$$3^{x^2 - x} < 9.$$

Solution. Since the base $a = 3 > 1$, taking logarithms to the base 3, we have

$$3^{x^2 - x} < 3^2 \quad \Leftrightarrow \quad x^2 - x < 2 \quad \Leftrightarrow \quad x^2 - x - 2 < 0.$$

Solving the quadratic inequality, we find the solution $-1 < x < 2$.

Example 9.9.2. Solve the inequality

$$\left(\frac{1}{2}\right)^{\frac{1}{x}} > 4.$$

Solution. Since the base $a = \frac{1}{2} < 1$, taking logarithms to the base $\frac{1}{2}$, we have

$$\left(\frac{1}{2}\right)^{\frac{1}{x}} > 4 \quad \Leftrightarrow \quad \left(\frac{1}{2}\right)^{\frac{1}{x}} > \left(\frac{1}{2}\right)^{-2} \quad \Leftrightarrow$$

$$\frac{1}{x} < -2 \quad \Leftrightarrow \quad \frac{2x + 1}{x} < 0.$$

Solving the equivalent quadratic inequality $x(2x + 1) < 0$, we find

$$-\frac{1}{2} < x < 0$$

.

Example 9.9.3. Solve the inequality

$$3^x > 2^{\frac{1}{x}}.$$

Solution. Taking logarithms to the base 3, we have

$$\log_3(3^x) > \log_3\left(2^{\frac{1}{x}}\right) \quad \Leftrightarrow \quad x > \frac{1}{x}\log_3 2 \quad \Leftrightarrow$$

$$\frac{x^2 - \log_3 2}{x} > 0 \quad \Leftrightarrow \quad x(x^2 - \log_3 2) > 0.$$

Solving the last inequality, say, by the method of intervals, we find that its solutions are:

$$-\sqrt{\log_3 2} < x < 0, \quad x < \sqrt{\log_3 2}.$$

Example 9.9.4. Solve the inequality

$$\log_{\frac{1}{3}}\left(x^2 - 3x + 5\right) < -1.$$

Solution. Here we should consider only the values of x for which

$$x^2 - 3x + 5 > 0. \tag{9.9}$$

Since the base of the logarithm is $\frac{1}{3} < 1$, exponentiating both sides to the base $\frac{1}{3}$, we have

$$\log_{\frac{1}{3}}(x^2 - 3x + 5) < -1 \quad \Leftrightarrow \quad x^2 - 3x + 5 > \left(\frac{1}{3}\right)^{-1}.$$

It is evident that any solution of the last inequality satisfies the inequality (9.9). Now, we have

$$x^2 - 3x + 5 > \left(\frac{1}{3}\right)^{-1} \quad \Leftrightarrow \quad x^2 - 3x + 2 > 0.$$

Solving this last quadratic inequality, we find $x < 1, \;\; 2 < x$, that is, $(-\infty, 1) \cup (2, \infty)$.

Example 9.9.5. Solve the inequality

$$\log(x^2 - 16) \leq \log(4x - 11).$$

Solution. Since the base of the logarithm is 10, this inequality is equivalent to the system of inequalities

$$0 < x^2 - 16 \leq 4x - 11.$$

The set of solutions of the first inequality $0 < x^2 - 16$, is the union of two intervals, $(-\infty, -4) \cup (4, \infty)$. The set of solutions of the second inequality $x^2 - 16 \leq 4x - 11 \quad \Leftrightarrow \quad x^2 - 4x - 5 \leq 0$ is the interval $[-1, 5]$. The intersection of these two sets, the interval $(4, 5]$, is the solution of the system in question and, hence also the solution of the original equation.

Example 9.9.6. Solve the inequality

$$\log_{\frac{1}{2}} \frac{x + 2}{x - 2} > 0.$$

Solution. Since the base of the logarithm is less than 1, this inequality is equivalent to the system of inequalities

$$0 < \frac{x + 2}{x - 2} < \left(\frac{1}{2}\right)^0 = 1.$$

Solving the inequality

$$0 < \frac{x+2}{x-2} \quad \Leftrightarrow \quad (x+2)(x-2) > 0,$$

we find the solution $(-\infty, -2) \cup (2, \infty)$. Solving the second inequality

$$\frac{x+2}{x-2} < 1 \quad \Leftrightarrow \quad \frac{4}{x-2} < 0,$$

we find the solution $(-\infty, 2)$. The intersection of these two sets, the interval $(-\infty, -2)$, is the solution of the system, and hence also of the original equation.

Example 9.9.7. Solve the inequality

$$2 \log_2(x-1) > 1 + \log_2(5 - x).$$

Solution. The logarithms $\log_2(x - 1)$ and $\log_2(5 - x)$ are defined for $x > 1$ and $x < 5$, respectively. Consequently, we seek solutions of the inequality satisfying the condition

$$1 < x < 5.$$

The given inequality becomes

$$\log_2(x-1)^2 > 1 + \log_2(5 - x). \tag{9.10}$$

Notice that the domain of definition of the logarithm $\log_2(x - 1)^2$ in (9.10) is 'wider' than that of $\log_2(x - 1)$. Evidently, any solution of the original inequality is also a solution of inequality (9.10). The converse is true if and only if $1 < x < 5$.

Form inequality (9.10), we get

$$(x - 1)^2 > 2(5 - x) \quad \Leftrightarrow \quad x^2 - 9 > 0.$$

Solving the last inequality, we find the solutions $x < -3$, $3 < x$ or $(-\infty, -3) \cup (3, \infty)$. From these solutions, only the values of x belonging to the interval $(3, 5)$ satisfy the condition $1 < x < 5$. Thus the solutions of the original inequality are $3 < x < 5$.

By introducing a new variable some exponential and logarithmic inequalities can be reduced to algebraic inequalities.

We illustrate this with some examples.

Example 9.9.8. Solve the inequality

$$\ln^2 x < \ln x.$$

Solution. Evidently, we seek solution satisfying $x > 0$. Setting $y = \ln x$, the inequality becomes

$$y^2 < y \quad \Leftrightarrow \quad y^2 - y < 0 \quad \Leftrightarrow \quad y(y - 1) < 0.$$

Solving this last inequality, we find $0 < y < 1$.

Now returning to x we find that the original inequality will have solutions for all x that satisfy the double inequality

$$0 < \ln x < 1.$$

Since the exponential function is increasing, this implies

$$e^0 < e^{\ln x} < e^1 \quad \Leftrightarrow \quad 1 < x < e.$$

Example 9.9.9. Solve the inequality

$$\log_2^2 x + 3 \log_2 x \geq \frac{5}{2} \log_{4\sqrt{2}} 16.$$

Solution. Evidently, we seek solution satisfying $x > 0$.

Since $\frac{5}{2} \log_{4\sqrt{2}} 16 = 4$ (why?), the inequality becomes

$$\log_2^2 x + 3 \log_2 x \geq 4.$$

Setting $y = \log_2 x$, we get the quadratice inequality

$$y^2 + 3y - 4 \geq 0 \quad \Leftrightarrow \quad (y + 4)(y - 1) \geq 0.$$

Solving, we find $y \leq -4$ and $y \geq 1$.

Returning to x, we must solve the basic inequalities $\log_2 x \leq -4$ and $\log_2 x \geq 1$. The first inequality implies $x \leq 2^{-4}$ and the second $x \geq 2$.

Thus, the solution is $(0, \frac{1}{16}) \cup (2, \infty)$.

Example 9.9.10. Solve the inequality

$$\log_4 x - \log_x 4 \leq \frac{3}{2}.$$

Solution. Evidently, we must have $x > 0$. Changing to logarithms to base 4, we can write the inequality as

$$\log_4 x - \frac{1}{\log_4 x} \leq \frac{3}{2}.$$

Setting $y = \log_4 x$, we obtain

$$y - \frac{1}{y} \leq \frac{3}{2},$$

whose solutions are: $y \leq -\frac{1}{2}$ and $0 < y \leq 2$.

Returning to x, we must solve $\log_4 x \leq -\frac{1}{2}$ and $0 < \log_4 x \leq 2$. These basic inequalities imply $x \leq \frac{1}{2}$ and $1 < x \leq 16$. Thus, the solution is $(0, \frac{1}{2}] \cup (1, 16]$.

Example 9.9.11. Solve the inequality

$$-3 \leq \log_x 2 \leq 1.$$

Solution. Evidently, we must have $x > 0$ and $x \neq 1$.

   Fisrt Solution: Changing to logarithms to base 2, we can write

$$-3 \leq \frac{1}{\log_2 x} \leq 1.$$

Setting $y = \log_2 x$, we get the algebraic double inequality

$$-3 \leq \frac{1}{y} \leq 1.$$

This ie equivalent to the system of inequalities

$$\frac{1+3y}{y} \geq 0 \quad \text{and} \quad \frac{1-y}{y} \leq 0.$$

Solving these inequalities, say, by the method of intervals, we find: the first inequality has solution $(-\infty, -\frac{1}{3}] \cup [0, \infty)$, and the second inequality $(-\infty, 0] \cup [1, \infty)$. The intersection of these sets, $(-\infty, -\frac{1}{3}] \cup [1, \infty)$, is the solution of the system of these inequalities.

Returnig to x, we have to solve the basic logarithmic inequalities $\log_2 x \le -\frac{1}{3}$ and $\log_2 x \ge 1$. The first inequality gives $x \le 2^{-\frac{1}{3}}$, and the second $x \ge 2$.

Thus, taking into account that $x > 0$, the solution of the original inequality is $(0, \frac{1}{\sqrt[3]{2}}] \cup [2, \infty)$.

Second Solution: Here, we will solve the inequality directly without introducing a new variable. However, since the basic inequalities of logarithms (see, p. 365 1, 2) differ for bases greater than or less than unity, we consider two cases: $x > 1$ and $0 < x < 1$.

Case 1. Let $x > 1$. Then $\log_x 2 > 0$, and all the more so $\log_x 2 \ge -3$. So, it remains to solve the inequality $\log_x 2 \le 1$. As, $1 = \log_x x$, and $\log_x$ is increasing, we get $2 \le x$. Hence, in this case the solution set of the original inequality is $[2, \infty)$.

Case 2. Let $0 < x < 1$. Then $\log_x 2 < 0$ and all the more so $\log_x 2 \le 1$. So, it remains to solve the inequality $-3 \le \log_x 2$. As $-3 = \log_x x^{-3}$, and $\log_x$ is decreasing, we get $x^{-3} \ge 2$ or $x \le 2^{-\frac{1}{3}}$. Hence, in this case the solution set of the original inequality is $(0, \frac{1}{\sqrt[3]{2}}]$ (taking into account that $x > 0$).

Thus, the solution of the original inequality is $(0, \frac{1}{\sqrt[3]{2}}] \cup [2, \infty)$.

Example 9.9.12. Solve the inequality

$$\log_3(16^x - 2 \cdot 12^x) \le 2x + 1.$$

Solution. The inequality is equivalent to the system of inequalities

$$0 < 16^x - 2 \cdot 12^x \le 3^{2x+1}.$$

We rewrite this in the form

$$0 < 4^{2x} - 2 \cdot 4^x \cdot 3^x \le 3 \cdot 3^{2x}.$$

Dividing all sides by $3^{2x} > 0$, we get

$$0 < \left(\frac{4}{3}\right)^{2x} - 2 \cdot \left(\frac{4}{3}\right)^{x} \leq 3.$$

Setting $y = \left(\frac{4}{3}\right)^{x}$, we get

$$0 < y^2 - 2y \leq 3.$$

Solving the inequality $0 < y^2 - 2y \quad \Leftrightarrow \quad y(y-2) > 0$, we find $y < 0$, $y > 2$. Since $y > 0$, we accept only $y > 2$. Solving the inequality $y^2 - 2y \leq 3 \quad \Leftrightarrow \quad (y-3)(y+1) \leq 0$, we find $-1 \leq y \leq 3$. The intersection of these intervals, that is, the solution of the system of these inequalities, is $2 < y \leq 3$.

Returning to x, we obtain the basic exponential inequalities

$$2 < \left(\frac{4}{3}\right)^{x} \leq 3 \quad \Leftrightarrow \quad \log_{\frac{4}{3}} 2 < x \leq \log_{\frac{4}{3}} 3.$$

9.9.1   Exercises

Solve each exponential inequality.

1. $3^{2x-4} > 3^{x+1}$

2. $5^{x^2-5x+6} < 1$

3. $\left(\frac{1}{2}\right)^{3x} < \frac{1}{8}$

4. $5^x + 5^{1-x} < 6$

5. $\frac{3^x - 9}{5^x + 2} > 0$

6. $\frac{7}{9^x - 2} \geq \frac{2}{3^x - 1}$

7. $\frac{4^x}{4^x - 3^x} < 4$

8. $3^{x-1} \leq 6^{2x-1}$

9. $3^{\frac{6x-3}{x}} < \sqrt[4]{27^{2x-1}}$

Solve each logarithmic inequality.

1. $\log(x - 5) < \log(3x - 7)$

2. $\log(x^2 - 4) < \log 3x$

3. $\log_a(8 - x) \geq \log_a(x^2 + 2)$

4. $\log_2 x + \log_2(x + 1) \leq \log_2(2x + 6)$

5. $\log_9 \left(\frac{2x}{x+1} \right) > \frac{1}{2}$

6. $\log_{\frac{1}{2}}(2 - 3x) > -2$

7. $\log x^2 > (\log x)^2$

8. $x^{\log x} > 10$

9. $\log_3(x + 2) > \log_{x+2} 81$ (Hint: Change to base 3).

10. $\log_x(6x - 1) > \log_x(2x)$.

9.10 Exponential and Logarithmic Systems (*)

The methods for solving exponential and logarithmic systems are based on the properties of exponents and logarithms, and the techniques of solving exponential and logarithmic equations developed in Sections 9.7 and 9.8.

9.10.1 Exponential Systems

Example 9.10.1. Solve the system

$$2^x = 3y$$
$$3^x = 2y$$

Solution. Dividing the equations side by side, we obtain

$$\left(\frac{2}{3}\right)^x = \frac{3}{2} = \left(\frac{2}{3}\right)^{-1}.$$

Therefore, $x = -1$. Substituting this value of x in the first equation, we get $2^{-1} = 3y$, that is, $y = \frac{1}{6}$. Thus, the solution is $(-1, \frac{1}{6})$.

Example 9.10.2. Solve the system

$$2^{3x+y} = 32$$
$$3^{2x-y} = 1$$

Solution. Writting $32 = 2^5$ and $1 = 3^0$, the system becomes

$$\begin{matrix} 2^{3x+y} = 2^5 \\ 3^{2x-y} = 3^0. \end{matrix} \quad \Leftrightarrow \quad \begin{matrix} 3x + y = 5 \\ 2x - y = 0. \end{matrix}$$

Solving this system, we find $x = 1$, $y = 2$.

Example 9.10.3. Solve the system

$$3^{x+2} \cdot 3^{y-4} = 81$$
$$4^x \cdot 2^{y-2} = 64$$

Solution. Since $3^{x+2} \cdot 3^{y-4} = 3^{x+y-2}$ and $4^x \cdot 2^{y-2} = 2^{2x+y-2}$, the system becomes

$$\begin{matrix} 3^{x+y-2} = 3^4 \\ 2^{2x+y-2} = 2^6 \end{matrix} \quad \Leftrightarrow \quad \begin{matrix} x + y - 2 = 4 \\ 2x + y - 2 = 6. \end{matrix}$$

Solving this system, we find $x = 2$, $y = 4$.

Example 9.10.4. Solve the system

$$3^{x^y - y^x} = 1$$
$$x - y^2 = 0$$

Solution. An obvious solution of the system is $x = 1$, $y = 1$. Suppose $x > 0$, $y > 0$ and $x \neq 1$, $y \neq 1$. We can write the first equation as

$3^{x^y - y^x} = 3^0$. Therefore, $x^y - y^x = 0$, or $x^y = y^x$. The system now becomes

$$x^y = y^x \qquad (y^2)^y = y^{y^2} \qquad y^{2y} = y^{y^2}$$
$$x = y^2 \quad \Leftrightarrow \quad x = y^2 \quad \Leftrightarrow \quad x = y^2.$$

The first equation implies

$$2y = y^2 \iff y(y - 2) = 0 \iff y = 0, \ y = 2.$$

Since the base of any exponential is positive, the value $y = 0$ is rejected. For $y = 2$, the value of $x = 4$. Thus, the solutions of the system are $(1, 1)$ and $(4, 2)$.

Example 9.10.5. Solve the system

$$x^y = y^x$$
$$x^3 = y^2$$

Solution. An obvious solution of the system is $x = 1$, $y = 1$. For $x > 0$, $y > 0$ with $x \neq 1$ and $y \neq 1$, taking common logarithms to both sides of each equation, we get the equivalent system

$$y \log x = x \log y$$
$$3 \log x = 2 \log y.$$

Dividing these equation side by side, we obtain

$$\frac{y}{3} = \frac{x}{2} \iff y = \frac{3}{2}x.$$

Substituting this expression of y into the second equation of the original system, we get

$$x^3 = \left(\frac{3}{2}x\right)^3 \iff x^3 = \frac{9}{4}x^2 \iff x^2\left(x - \frac{9}{4}\right) = 0.$$

Since $x > 0$, we get $x = \frac{9}{4}$. Therefore, $y = \frac{3}{2} \cdot \frac{9}{4} = \frac{27}{8}$.

Thus, the solutions of the given system are

$$(1, 1), \quad \left(\frac{9}{4}, \frac{27}{8}\right).$$

Example 9.10.6. Solve the system

$$x^{x+y} = y^3$$
$$y^{x+y} = x^3$$

Solution. Evidently, $x = 1, y = 1$ is a solution of the system.

First Solution: Suppose $x, y > 0$ and $x \neq 1$, $y \neq 1$. From the first equation we get

$$x = y^{\frac{3}{x+y}}.$$

Substituting this expresion of x into the second equation, we habe

$$y^{x+y} = y^{\frac{9}{x+y}} \quad \Leftrightarrow \quad x + y = \frac{9}{x+y} \quad \Leftrightarrow \quad (x+y)^2 = 9.$$

Since x, y are positive this implies $x + y = 3$. Substituting this into the first equation, we get

$$x^3 = y^3 \quad \Leftrightarrow \quad \left(\frac{x}{y}\right)^3 = 1 \quad \Leftrightarrow \quad x = y.$$

Substitution in either equation yields,

$$x^{2x} = x^3 \quad \Leftrightarrow \quad 2x = 3 \quad \Leftrightarrow \quad x = \frac{3}{2}.$$

To find other solutions of the given system, we divide its equations side by side, and we get

$$\left(\frac{x}{y}\right)^{x+y} = \left(\frac{y}{x}\right)^3 = \left(\frac{x}{y}\right)^{-3}.$$

Therefore, $x + y = -3$. Substituting this into the first equation, we obtain

$$x^{-3} = y^3 \quad \Leftrightarrow \quad (xy)^3 = 1 \quad \Leftrightarrow \quad xy = 1.$$

Hence x and y satisfy the equations

$$x + y = -3$$
$$xy = 1$$

By Example 8.4.1, the x, y are solutions of the quadratic equation

$$t^2 + 3t + 1 = 0.$$

Solving this equation we find $t_{1,2} = \frac{-3 \pm \sqrt{5}}{2}$.

Thus, the solutions of the given system are

$$(1,1), \quad \left(\frac{3}{2}, \frac{3}{2}\right), \quad \left(\frac{-3 + \sqrt{5}}{2}, \frac{-3 - \sqrt{5}}{2}\right), \quad \left(\frac{-3 - \sqrt{5}}{2}, \frac{-3 + \sqrt{5}}{2}\right).$$

Second Solution: Again we assume $x, y > 0$ and $x \neq 1$, $y \neq 1$. Taking common logarithms to both sides of both equations and using the properties of logarithms, we get the equivalent system

$$(x + y) \log x = 3 \log y$$
$$(x + y) \log y = 3 \log x.$$

Adding these equation side by side, we obtain

$$(x + y) \log(xy) = 3 \log(xy) \quad \Leftrightarrow \quad \log(xy)[(x + y) - 3] = 0.$$

This implies

$$\log(xy) = 0 \quad \text{or} \quad (x + y) - 3 = 0.$$

That is, $xy = 1$ or $x + y = 3$.

Now the given system is equivalent to two systems:

$$x^{x+y} = y^3 \qquad x^{x+y} = y^3$$
$$xy = 1, \qquad x + y = 3.$$

For the first system, the second equation implies $y = x^{-1}$. Substituting this expression of y into the first equation, we obtain

$$x^{x + \frac{1}{x}} = x^{-3} \quad \Leftrightarrow \quad x + \frac{1}{x} = -3 \quad \Leftrightarrow \quad x^2 + 3x + 1 = 0.$$

The roots of this quadratic equation are $x_1 = \frac{-3 + \sqrt{5}}{2}$ and $x_2 = \frac{-3 - \sqrt{5}}{2}$. Hence, $y_1 = x_1^{-1} = \frac{-3 - \sqrt{5}}{2}$ and $y_2 = x_2^{-1} = \frac{-3 + \sqrt{5}}{2}$.

For the second system, substituting $x + y = 3$ into the first equation, we get

$$x^3 = y^3 \iff x = y.$$

Hence, the second equation, yields $2x = 3$ or $x = \frac{3}{2}$. Thus, we obtain the same solutions as above.

9.10.2   Logarithmic Systems

Example 9.10.7. Solve the system

$$\log_4 x - \log_4 y = 1$$
$$\log_2 y^2 - \log_y x = -1$$

Solution.   First of all we must have $x, y > 0$ and $y \neq 1$. By the properties of logarithms, the first equation of the system implies

$$\log_4 \left(\frac{x}{y} \right) = 1, \iff \frac{x}{y} = 4 \iff x = 4y.$$

Substituting $x = 4y$ into the second equation and passing to base 2, we have

$$\log_2 y^2 - \log_y 4y = -1 \iff 2 \log_2 y - \log_y 4 - 1 = -1 \iff$$

$$2 \log_2 y - \frac{\log_2 4}{\log_2 y} = 0 \iff 2 \log_2 y - \frac{2}{\log_2 y} = 0.$$

Setting $t = \log_2 y$, we obtain the equation

$$2t - \frac{2}{t} = 0 \iff t^2 - 1 = 0 \iff (t - 1)(t + 1) = 0 \iff t = 1, \ t = -1.$$

Returning back to y; if $t = 1$, then $\log_2 y = 1 \iff y = 2$. If $t = -1$, then $\log_2 y = -1 \iff y = \frac{1}{2}$. Finally, using $x = 4y$, if $y = 2$, then $x = 8$. If $y = \frac{1}{2}$, then $x = 2$.

Thus, the solutions of the given system are $(8, 2)$ and $(2, \frac{1}{2})$.

Example 9.10.8. Solve the system

$$\sqrt[x]{y^2} + 10 = 11\sqrt[x]{y}$$
$$x + \log y = 1$$

Solution. We must have $y > 0$. Setting $\sqrt[x]{y} = t$, the first equation becomes $t^2 - 11t + 10 = 0$. Solving this quadratic equation, we find $t = 1$ and $t = 10$.

If $t = 1$, then $\sqrt[x]{y} = 1$, which implies $y = 1$. Substituting in the secong equation, we get $x = 1$.

If $t = 10$, then $\sqrt[x]{y} = 10$, which implies $y = 10^x$. Substituting in the second equation, we get $2x = 1$ or $x = \frac{1}{2}$, and so, $y = 10^{\frac{1}{2}}$.

Thus the solutions of the system are $(1, 1)$ and $(\frac{1}{2}, \sqrt{10})$.

Example 9.10.9. Solve the system

$$\log_3 x - \log_3 y = \log_3(x + y)$$
$$\log_4(4 - x) + \log_4 y = \log_2 x$$

Solution. First of all the logarithms appearing in the system are defined if $0 < x < 4$ and $y > 0$. We change the logarithms of the second equation to the base 2. We have

$$\frac{\log_2(4 - x)}{\log_2 4} + \frac{\log_2 y}{\log_2 4} \log_2 x \iff \log_2(4 - x) + \log_4 2y = 2\log_2 x \iff$$

$$\log_2[y(4 - x)] = \log_2 x^2 \iff y(4 - x) = x^2.$$

The first equation of the system becomes

$$\log_3\left(\frac{x}{y}\right) = \log_3(x + y) \iff \frac{x}{y} = x + y.$$

Hence, the system becomes

$$\begin{matrix} y(4 - x) = x^2 \\ \frac{x}{y} = x + y \end{matrix} \iff \begin{matrix} x(x + y) = 4y \\ x + y = \frac{x}{y}. \end{matrix}$$

Substituting $x + y = \frac{x}{y}$ into the first equation, we get $x^2 = 4y^2$ or $x^2 - 4y^2 = 0 \iff (x - 2y)(x + 2y) = 0$. Therefore $x = 2y$ and $x = -2y$.

Since $x, y > 0$ we reject the case $x = -2y$. Substituting $x = 2y$ into the second equation, we find $y = \frac{2}{3}$ and, hence $x = \frac{4}{3}$. Thus, the solution of the system is $\left(\frac{2}{3}, \frac{4}{3}\right)$.

Example 9.10.10. Solve the system

$$\log_y x + \log_x y = 2$$
$$x^2 + y = 12$$

Solution. First of all we must have $0 < x, y \neq 1$. Passing to logarithms of base x, the first equation becomes

$$\frac{1}{\log_x y} + \log_x y = 2 \quad \Leftrightarrow \quad \log_x^2 y - 2\log_x y + 1 = 0 \quad \Leftrightarrow \quad (\log_x y - 1)^2 = 0.$$

Therefore, $\log_x y = 1 \quad \Leftrightarrow \quad x = y$.

Substituting $y = x$ into the second equation of the system, we get

$$x^2 - x = 12 \quad \Leftrightarrow \quad x^2 - x - 12 = 0 \quad \Leftrightarrow \quad (x+4)(x-3) = 0.$$

Whence, $x = 3$ and $x = -4$, which we reject as $x > 0$. Thus, the solution of the system is $(3, 3)$.

Example 9.10.11. Solve the system

$$\left(\frac{x}{5}\right)^{\log 5} = \left(\frac{y}{7}\right)^{\log 7}$$
$$5^{\log y} = 7^{\log x}$$

Solution. Again we seek x, y positive. Taking common logarithms to both sides of the equations, we get

$$\log 5(\log x - \log 5) = \log 7(\log y - \log 7)$$
$$\log y \log 5 = \log x \log 7.$$

Setting $u = \log x$ and $v = \log y$, we get the system

$$\log 5 \cdot u - \log 7 \cdot v = \log^2 5 - \log^2 7$$
$$\log 7 \cdot u - \log 5 \cdot v = 0.$$

Solving this system of linear equations in u, v, we find $u = \log 5$ and $v = \log 7$. Returning to x and y, we have $\log x = \log 5 \implies x = 5$ and $\log y = \log 7 \implies y = 7$. Thus, the solution of the given system is $(5, 7)$.

Remark. Note that $5^{\log 7} = 7^{\log 5}$. This is true in general

$$a^{\log b} = b^{\log a} \quad (\text{why?})$$

Example 9.10.12. Solve the system

$$yx^{\log_y x} = x^{\frac{5}{2}}$$
$$\log_4 y \log_y (y - 3x) = 1$$

Solution. For the logarithms to be defined we seek solutions satisfying $\frac{1}{3}y > x > 0$ and $y \neq 1$. Taking logarithms of both sides of the first equation to the base y, we get

$$1 + \log_y x \log_y x = \frac{5}{2} \log_y x \quad \Leftrightarrow \quad 2\log_y^2 x - 5\log_y x + 2 = 0.$$

Setting $\log_y x = t$, we get the quadratic equation

$$2t^2 - 5t + 2 = 0.$$

Solving this equation, we find $t = 2$ and $t = \frac{1}{2}$. Returning to x, y, we have two cases:

1. If $t = 2$, then $\log_y x = 2 \Leftrightarrow x = y^2$. Since $\log_y 4 = \frac{1}{\log_4 y}$, the second equation of the system gives

$$\log_y (y - 3x) = \log_y 4 \quad \Leftrightarrow \quad y(y - 3x) = 4.$$

Substituting $x = y^2$ into the above equation, we obtain the quadratic equation

$$3y^2 - y + 4 = 0.$$

This equation has no real roots (its descriminant is $D = -47$).

2. If $t = \frac{1}{2}$, then $\log_y x = \frac{1}{2} \Leftrightarrow x = \sqrt{y} \Leftrightarrow x^2 = y$. Substituting $x^2 = y$ into the equation $y(y - 3x) = 4$, we get the quadratic equation

$$x^2 - 3x - 4 = 0.$$

This equation has roots $x = 4$ and $x = -1$ which we reject it as $x > 0$. Hence $x = 4$, and so $y = 16$. Since this value of y satisfies the condition $\frac{1}{3}y > x > 0$, the solution of the system is $(4, 16)$.

9.10.3   Exercises

Solve each exponential system.

1.

    (a) $\begin{aligned} 3^x + 2^x &= 11 \\ 3^x - 2^x &= 7 \end{aligned}$
    (b) $\begin{aligned} 3^y - 2^x &= 1 \\ 3^y + 16 \cdot 2^{-x} &= 11. \end{aligned}$

2.

    (a) $\begin{aligned} 2^x \cdot 5^y &= 2500 \\ 2^y \cdot 5^x &= 40 \end{aligned}$
    (b) $\begin{aligned} 8^{2x+1} &= 32 \cdot 4^{4y-1} \\ 5 \cdot 5^{x-y} &= 5^{2y+1}. \end{aligned}$

3.

    (a) $\begin{aligned} 3^x - 5^y &= 4 \\ 9 \cdot 3^{-x} + 5^y &= 6 \end{aligned}$
    (b) $\begin{aligned} 2^x &= 8^{y+1} \\ 9^y &= 3^{x-9}. \end{aligned}$

4.

    (a) $\begin{aligned} 2^x \cdot 4^{x+2y} &= 1024 \\ 5^{x-3y-1} &= \tfrac{1}{25} \end{aligned}$
    (b) $\begin{aligned} x^y &= y^x \\ x^x &= y^{9y}. \end{aligned}$

Solve each logarithmic system.

1.

    (a) $\begin{aligned} \log(xy) &= 4\log 2 \\ \log x \cdot \log y &= 3(\log 2)^2 \end{aligned}$
    (b) $\begin{aligned} xy &= 8 \\ \log y &= 2\log x. \end{aligned}$

2.

    (a) $\begin{aligned} 5^x - 2^y &= 1 \\ x\log 5 + y\log 2 &= \log 20 \end{aligned}$
    (b) $\begin{aligned} \log(x^2 + y^2) - 1 &= \log 13 \\ \log(x+y) - \log(x-y) &= 3\log 2. \end{aligned}$

3.

    (a) $\begin{aligned} x^y &= 3^{12} \\ y - \log_3 x &= 1 \end{aligned}$
    (b) $\begin{aligned} y &= 2x \\ y\log_x + \log 2 &= 2\log y. \end{aligned}$

4.

    (a) $\begin{aligned} 2^{\log_2 y} - \log_3 x &= 1 \\ y\log_3 x &= 2 \end{aligned}$
    (b) $\begin{aligned} \log_4(xy - 2) &= \log_2 y \\ \log_9 x^2 + \log_3(x-y) &= 1. \end{aligned}$

Appendix A

Answers to Exercises

CHAPTER 1

Exercises 1.2.2: Page 10

1. $96 = 2^5 \cdot 3$, $\quad 2520 = 2^3 \cdot 3^2 \cdot 5 \cdot 7$ **2.** 18 **3.** 540 **4.** (a) -1 (b) 0 (c) -5 (d) 1 **5.** (a) -240 (b) 2 **6.** (a) 0 (b) 10.

Exercises 1.3.2: Page 15

1. (a) $-\frac{3}{2}$ (b) $\frac{2}{3}$ **2.** (a) $\frac{13}{5}$ (b) $-\frac{1}{12}$ **3.** (a) -3 (b) $\frac{13}{24}$ **4.** (a) 1.7 (b) $3\frac{5}{8}$ **5.** (a) $\frac{9}{20}$ (b) -5 **6.** (a) $-\frac{17}{72}$ (b) $\frac{7}{6}$ **7.** (a) 1 (b) -9 **8.** (a) $\frac{1}{7}$ (b) $-5\frac{1}{2}$.

Exercises 1.3.4: Page 22

1. (a) -32 (b) $-\frac{1}{54}$ (c) $\frac{9}{25}$ (d) $\frac{1}{64}$ **2.** (a) 0.000027 (b) 1.44 (c) -125 (d) $-\frac{4}{3}$ **3.** (a) 7 (b) 4.

Review Exercises 1.7: Page 33

1. -4 **2.** 5 **3.** 65 **4.** 4 **5.** $-\frac{8}{9}$ **6.** $11\frac{2}{3}$ **7.** 3 **8.** 0.12 **9.** $\frac{1}{8}$ **10.** -18.

CHAPTER 2

Exercises 2.1.2: Page 38

1. 0   **2.** 0   **3.** -20 **4.** 10   **5.** (a) $-\frac{7}{3}$ (b) 0   **6.** $-\frac{1}{5}$ **7.** -6.

Exercises 2.2.1: Page 42

1. 1   **2.** $-\frac{1}{2}$ **3.** 3   **4.** -3 **5.** -1 **6.** -8 **7.** $\frac{4}{5}$ **8.** 3
9. $\frac{5}{14}$ **10.** 2   **11.** 18   **12.** $\frac{35}{13}$.

Exercises 2.4.3: Page 51

1. 6   **2.** 180   **3.** 16%   **4.** (a) 51 dollars and 92 cents (b) 1 hour   **5.** \$50   **6.** \$650; \$195   **7.** \$15,000   **8.** 20%
9. 38.8%   **10.** 32%   **11.** 440   **12.** 10 ounces   **13.** $2\frac{1}{4}$ liters   **14.** 4 liters   **15.** \$2,500 at 9% and \$7,500 at 14%   **16.** \$425,000

Exercises 2.5.2: Page 57

1. (a) -2 (b) 3 (c) $-\frac{3}{4}$ **2.** (a) -3 (b) $-\frac{5}{3}$ (c) $\frac{5}{3}$ **3.** $7:3$
4. (a) $x=8$ (b) $x=16$
5. (a) $x=-1$ (b) $x=20$
6. 12, 16   **7.** 160 and 400   **8.** 700 and 500   **9.** 30   **10.** 70 km
11. 30, 40, 50 inches   **12.** 18, 27, 36
13. \$200 to \$1,000; \$300 to \$9000   both   at $3\frac{1}{3}$%   **14.** 8 gallons from A, 10 gallons from B.

Exercises 2.6.2: Page 63

1. $y=-20$ **2.** $x=\pm 2\sqrt{3}/3$ **3.** -8 **4.** 225   **5.** \$1,000
6. 240 miles   **7.** $4\frac{1}{3}$ hours   **8.** 3 seconds   **9.** 4 workers   **10.** 14 days.

Appendix : Answers to Exercises 385

Exercises 2.7.1: Page 70

1. $41, 43, 45$ **2.** 4 **3.** 12 **4.** 84 **5.** 79 and 480 **6.** 4 nickels, 8 dimes, 6 quartes **7.** 4 hours **8.** 34 and, 9 **9.** In the same direction $43\frac{7}{11}$ minutes after 8; in opposite direction at $10\frac{10}{1}$ minutes after 8 **10.** 7 pears, 14 apples, 3 peaches **11.** 20 for games; 50 for practice **12.** The trains meet at $11:12\frac{12}{19}$ AM; $35\frac{10}{19}$ miles from city B.

Review Exercises 2.8: Page 71

3. (a) -2 (b) $x = 5$ **4.** (a) $x = -24$ (b) no solution **5.** $x = 6$ **6.** $x = -\frac{80}{41}$ **7.** 200 aspirin tablets; unit cost 0.0295 **8.** width 18 meters, length 25 meters **9.** 2 hours 24 minutes **10.** 95 **11.** $5\frac{5}{11}$ minutes **12.** A 25 days, B $16\frac{2}{3}$ days **13.** 13 hours. **14.** 39 miles per hour, going; 65 miles per hour, returnig **15.** 30 gallons **16.** 1 gallon the first time; 2 gallons the second time **17.** $\$1,250$ at 4%; $\$3,750$ at 6% **18.** $\$28$ **19.** $9:45$ AM **20.** 6 hours **21.** $15\frac{5}{7}$ cubic feet **22.** $\frac{15}{8}$.

CHAPTER 3

Exercises 3.2.2: Page 79

1. 0 **2.** -32 **3.** $\frac{1}{32}$ **4.** $\frac{9}{25}$ **5.** $\frac{1}{64}$ **6.** $\frac{1}{9}$ **7.** -125 **8.** 1 **9.** $\frac{1}{8}$ **10.** -608 **11.** $-\frac{3}{16}$ **12.** 1 **13.** $\frac{3}{25}$ **14.** $-\frac{124}{21}$ **15.** $\frac{1}{a^2}$ **16.** $\frac{1}{x^{11}}$ **17.** $\frac{4y^{12}}{x^{20}}$ **18.** b^2. **19.** $\frac{y^3}{25x^8}$ **20.** $\frac{x^{13}z^{17}}{y^{20}}$ **21.** (a) 8.95×10^4 (b) -2.548×10^7 (c) 5.2×10^{-4} (d) -8.9×10^{-8} **22.** (a) $5,630,000$ (b) 0.0000781 (c) -0.0032 (d) 0.53 **23.** 1.17×10^7 **24.** 1.2×10^{-22} **25.** 2.6125×10^{-17}.

Exercises 3.3.2: Page 88

1. $4\sqrt{3}$ **2.** 0 **3.** 0 **4.** $7\sqrt{3} - 2\sqrt{5}$ **5.** 30 **6.** $6a^2$ **7.** $a\sqrt{b}$ **8.** $\frac{1}{4}$ **9.** $\sqrt[3]{2}+\sqrt{3}$ **10.** $4\sqrt[3]{3}$ **11.** $-\sqrt{2}$ **12.** 2 **13.** $2\sqrt{2}-3\sqrt[3]{2}+2$ **14.** $13\sqrt[3]{9}$ **15.** $\frac{a^2}{b^2}$ **16.** $\sqrt[4]{3}$ **17.** $\sqrt[n]{a}$ **18.** $-2a$. **19.** a^2 **20.** $\frac{a-b}{\sqrt{ab}}$ **21.** $3a^3$ **22.** $\sqrt[16]{2^{15}}$.

Review Exercises 3.4: Page 89

1. -18 **2.** $\frac{23}{64}$ **3.** 2.81 **4.** $-\frac{1}{27}$ **5.** $\frac{1}{8}$ **6.** 5.3×10^{18} **7.** 180 **8.** 30
9. $a\sqrt{a}$ **10.** $21ab\sqrt[3]{a}$ **11.** 6 **12.** 0 **13.** $6\sqrt{6}$ **14.** 5 **15.** 0 **16.** $a^{-\frac{1}{12}}$
17. a^2 **18.** $\left(\frac{a}{b}\right)^{\frac{1}{9}}$ **19.** $\left(\frac{a}{b}\right)^{\frac{1}{24}}$ **20.** $\sqrt[3]{36a^2b^2}$ **21.** $\sqrt[3]{5}$ **22.** $\sqrt[4]{2}$ **23.** $3^{\frac{2}{9}} \cdot$
$2^{\frac{1}{2}}$ **24.** $5^{\frac{15}{8}}$ **25.** $5^{\frac{17}{9}}$.

CHAPTER 4

Exercises 4.1.1: Page 96

1. $-x^2$ **2.** $2x^4y^3$ **3.** $10a^2b^3c^3$ **4.** $6x^9$ **5.** $-4x^{12}$ **6.** $-4x^3$ **7.** $-2x^2$
8. $3xyz$ **9.** 1 **10.** $-144a^{11}b^7c^5$ **11.** $2x^6z$ **12.** x^5.

Exercises 4.3.1: Page 104

1. $2x^2 + 4x + 4$ **2.** $-5x^2 - 2x + 3$ **3.** $2x^4 + 4x^3 + 6x^2 + 14x + 12$
4. $x^5 + 2x^3 + x - 4$ **5.** $2x^4 - 2x^3 + x^2 - 3x - 6$ **9.** $12x^5 - 8x^3$ **11.** $-4x^5 +$
$2x^4 - 3x^2 + 9x$ **13.** $6y^2 - xy^2$ **15.** $12x^4 + 2x^3 - 29x^2 + 27x - 6$ **17.** $2x^5 -$
$12x^4 + 7x^3 + 18x^2 - 15x$ **19.** $x^3 - 6x^2 + 11x - 6$ **23.** $2x^2 - 9x +$
3 **24.** $4a^2 - 12ab + 9b^2$ **25.** $\frac{9}{4}x^4 + 12x^3y + 16x^2y^2$ **28.** $4ab$ **29.** $4ab - 4ac$
33. $4x^2 + 20x + 9$.

Exercises 4.4.3: Page 113

A. **1.** $R = 0$ **3.** $R = 0$ **5.** $R = -1$ **7.** $R = 36$ **8.** $k = -3$.

B. **1.** $-5x^2 + x - 2$ **3.** $-3a + ab - \frac{2}{3}$ **5.** $Q = 7x + 1,\ R = 0$ **7.** $Q =$
$2x^2 - x + 1,\ R = 6$ **9.** $Q = x^2 - 3x - 4,\ R = 0$ **11.** $Q = x + 5,\ R = 0$
14. $Q = \frac{1}{2}x^2 + \frac{5}{2}x + 2,\ R = 8x - 8$ **15.** $Q = x^2 + 2x - 2,\ R = 10x - 22$.

C. **3.** $Q = x^3 - x^2 + x - 1,\ R = 0$.

Exercises 4.5.1: Page 121

5. $(a-b)(x-1)^2$ **7.** $(2x+a^2)(3x+a)$ **9.** $3(x+1)(x-2)(x^2+2x+4)$
11. $(b-2y)(3ax^2+y)$ **15.** $(2x+7y)(2x-7y)$ **17.** $(4x-3)(16x^2+12x+9)$
21. $a^6(1-a)(1+a+a^2)$ **25.** $(x-3)(x^2+3x+9)(x+1)(x^2-x+1)(x-1)(x^2+x+1)$ **27.** $ab(a+b)^2(a-b)$ **29.** $(x^2-3y^2+4a^2)(x^2-3y^2-4a^2)$
30. $(x+1)(x-1)(x-2)(x^2+4x+4)$.

Exercises 4.5.3: Page 132

1. $(x+5)(x+2)$ **2.** $(x+5)(x-2)$ **5.** $(x+3)(3x-2)$ **7.** $(x-2)(2x+3)$
10. $(2x-y)(5x-6y)$ **12.** $(x(x-3)(4x+2)$ **14.** $(x-1)(x^2+3x+3)$
16. $(x+2)(x-3)(x-4)$ **17.** $(x+1)(x+2)(x+3)$ **18.** $(x-1)(2x+1)^2$
19. $(x+2)(3x-2)(5x+3)$ **21.** $(x-3)(2x+1)(3x+1)$ **23.** $(x^2+x+1)(x^2-x+1)$ **24.** $(x^2+2x+2)(x^2-2x+2)$ **25.** $(x-1)(x+2)(x^2-x+1)$ **26.** $5(x-1)^3(2x+5)$ **27.** $(2x+1)(3x-2)(x^2+x+1)$
28. $(x+2)(2x+1)(3x+2)(x^2+1)$ **29.** $(x^2+x+1)(x^2-x+1)(x^4-x^2+1)$.

Exercises 4.6.1: Page 135

A. **1.** $2x$ **2.** $3abxy$ **3.** $5abxy^2$ **4.** $2x^3y^2z^4$.

B. **1.** $(x+y)(x-y)$ **3.** $x+1$ **4.** $x+2$ **7.** $(x-1)(x-2)^2$.

Exercises 4.7.1: Page 136

A. **1.** $45x^5$ **2.** $18x^3$ **3.** $12abx^2y^2$ **4.** $45a^3b^3x^4y^3$ **5.** $60ax^4y^3z^2$.

B. **1.** $x(x+1)(x-1)$ **3.** $(x-1)(x-2)(x-3)$ **5.** $x^3(x+1)(x^3-1)$
7. $(x+1)(x+2)(x+3)(x+4)(2x-5)$ **10.** $(x-1)(x-2)(x-3)(2x-1)$
11. $x(2x+3y)(2x-3y)(2x-y)(4x+5y)$.

Exercises 4.8.2: Page 143

A. **5.** $\frac{x-3}{x-1}$ **7.** $1-y^2$ **10.** $\frac{x^2+x+2}{x-3}$ **11.** $\frac{x-1}{2x+1}$ **12.** 3.
B. **4.** $\frac{a^2}{a^2-b^2}$ **8.** $x+y$ **14.** 0 **15.** $\frac{3}{(x-1)(x-3)}$. **16.** $\frac{1}{(1-x)^2}$
17. $-\frac{8x^2+6x-9}{8x^3-27}$.

C. 3. $\frac{a}{2(a-b)}$ **4.** $\frac{1-b}{a}$ **5.** $\frac{3(x^2+xy+y^2)}{5(x-y)}$ **7.** $(x-1)(a-b)$ **9.** $\frac{x+1}{x-1}$.
D. 5. $\frac{4ab}{(a-b)^2}$ **7.** $2x^2-1$ **8.** $\frac{2}{a+b}$ **9.** 1   **10.** $\frac{2x-3}{x(x+1)}$ **11.** $3x$
12. $2x^2+2x$ **13.** $-\frac{x^2y^2}{(x-y)^2}$.

Exercises 4.9.3: Page 154

A. 1. (a) $8\sqrt{2}a^2b^3c^4$ (b) $8\sqrt[3]{2}xy^3z^4$ **7.** (a) a^4 (b) $\sqrt[3]{x^2y^2}$.
B. 2. $(3+2a)\sqrt{ax}$ **3.** 0   **4.** $4x\sqrt{x^2-1}$ **5.** 0.

C. 6. (a) $\frac{3-\sqrt{6}}{15}$ (b) $\sqrt{2}+\sqrt{3}$.

Review Exercises 4.10: Page 156

B. 6. $\frac{1}{27}(x-3)^3$ **7.** $(x-a)(x^2+ax+a^2)(x+1)(x^2-x+1)$
9. $(x+1)^2(x-3)(3x^2+2)$.
D. 2. $(x^2-1)(x^2+1)(3x-5)(x^4+x^2+1)$ **3.** $(x+1)(x-1)(x^2+1)$.

E. 5. 1   **6.** $\frac{12}{x+2}$ **7.** $\frac{1}{x+1}$ **8.** a **9.** $\frac{1}{a}$ **10.** 1

F. 1. 0   **2.** $\frac{2}{1-x}$ **3.** $\sqrt{x}$ **4.** $9a$ **5.** $\frac{x^2}{2x-1}$ **6.** 2   **7.** 1   **8.** $\sqrt{1-x}$
9. $(x-1)\sqrt{x}$ **10.** $\sqrt{1-x^2}$ **11.** $x-1$ **12.** $\frac{1-x}{\sqrt{x}}$ **13.** $2ab$
14. $-2b$ **15.** $3\sqrt{y}$.

CHAPTER 5

Exercises 5.1.1: Page 163

1. $x=9$ **2.** no solution   **3.** $x=-\frac{1}{2}$ **4.** $x=1, x=-1$ **5.**
$x=-4$ **6.** $x=$ any real number   **7.** $x=1$ **8.** $x=6$ **9.** $x=\frac{30}{22}$ **10.** $x=\frac{5}{7}$ **11.** no solution   **12.** $x=\frac{13}{59}$ **13.** $x=-\frac{26}{7}$
14. $x=-4$ **15.** no solution   **16.** $x=\frac{1}{3}$ **17.** $x=-\frac{6}{5}$ **18.** $x=\frac{a+b}{a}$.

Exercises 5.2.1: Page 167

1. (a) $x = -2, x = 4$ (b) $x = \pm\frac{3}{2}$ **2.** (a) $x = -1, x = -4$ (b) $x = 2, x = -\frac{4}{3}$ **3.** (a) $x = 2$ (b) $x = -5, x = -\frac{3}{2}$ **4.** (a) $x = -3, x = 7$ (b) $x = 5, x = 8$ **5.** (a) $x = -\frac{3}{2}$ (b) $x = -\frac{1}{2}$ **6.** (a) $x = -2$ (b) all $x \geq 2$ (c) all $x \leq 1$.

Exercises 5.3.3: Page 174

2. (a) -5 (b) $3 - i$ (c) 0 (d) 0 **3.** (a) $4 - 5i$ (b) $3 - 4i$ (c) $5 + 0i$ (d) 13 **4.** (a) $38 - 18i$ (b) $-14 + 36i$ (c) $-5 + 12i$ (d) $-2 + 2i$ **5.** (a) $1 - i$ (b) $\frac{3}{2} + \frac{1}{2}i$ (c) $1 - 2i$ (d) $3 + 7i$ **6.** $-\frac{22}{159} + \frac{5}{318}i$ **7.** $-1 + 3i$ **9.** $1, \sqrt{2}, 2, \sqrt{13}, 1$.

Exercises 5.4.1: Page 186

1. (a) $x = 0, x = 3$ (b) $x = 0, x = -\frac{3}{2}$ (c) $x = 0, x = 5$ (d) $x = \pm 4$ (e) $x = \pm 6$ **2.** (a) $x = 1, x = 5$ (b) $x = 2, x = 7$ (c) $x = -2, x = \frac{2}{3}$ (d) $x_1 = x_2 = \frac{1}{2}$ (e) $x = -\frac{1}{2} \pm \frac{3}{2}i$ (f) $x = \frac{1}{3} \pm \frac{\sqrt{3}}{3}i$ **3.** (a) $x = -\frac{\sqrt{3}}{10} \pm \frac{\sqrt{23}}{10}$ (b) $x = 1 + \sqrt{3}, x = 2 - \sqrt{3}$ because $\sqrt{13 - 4\sqrt{3}} = 2\sqrt{3} - 1$ (why?) (c) $x = -\frac{1}{2}, x = \frac{1}{4}$ (d) $x = 1, x = \frac{3}{2}$ (e) $x = \frac{1}{5}$ **4.** (a) $8x^2 + 10x + 3 = 0$ (b) $x^2 - 10x + 13 = 0$ (c) $2x^2 - 2x + 5 = 0$ **5.** $k = \frac{9}{8}$ **6.** $k = -3$ and $k = 7$ **7.** (a) $k = -1$ (b) $k = 1$ **8.** $k = 4$.

Review Exercises 5.6: Page 190

1. (a) $x = 5$ (b) $x = -\frac{3}{5}$ (c) $x = 2$ (d) $x = \frac{2}{3}$ (e) $x = 0.5$ **2.** (a) $x = 1, x = \frac{3}{2}$ (b) $x = 1, x = -\frac{5}{3}$ (c) $x = -\frac{1}{2} \pm \frac{\sqrt{5}}{2}$ (d) $x = -\frac{2}{3} \pm \frac{\sqrt{2}}{3}i$ **3.** (a) $x = \pm 3, x = \pm 4$ (b) $x = \pm 5$ (c) $x = 0, x = 2$ (d) $x = 0, x = 1$ **4.** (a) $x^2 + 4x - 21 = 0$ (b) $2x^2 - 3x + 1 = 0$ (c) Hint: first rationalize the denominators $28x^2 - 20x + 1 = 0$ (d) $x^2 - 4x + 13 = 0$ **6.** the numbers are the roots of $x^2 - 2x - 15 = 0$; $-5, 3$ **7.** $-9, 4$ **8.** $k = 2$ and $k = -\frac{22}{3}$ **10.** (a) $x = -2, x = 4; k = -1$ **11.** 117 **12.** $k = \pm 7$ **13.** $\lambda = 1, \lambda = \frac{1}{2}$ **14.** $\frac{27a^3 + 36a}{8}$ **15.** $b = c = 0$

or, $b = 1$, $c = -2$ **16.** 4, 8   **17.** 10 and 24 feet   **18.** 12, 16
19. 4 meters   **20.** 86   **21.** \$40   **22.** 50 miles per hour   **23.** 20
miles per hour going and 10 miles per hour returning   **24.** 60 km
25. 40 by 45 miles per hour   **26.** 4 hours 20 minutes   **27.** 50 miles
per hour   **28.** 30 hours   **29.** 10 cents   **30.** 25 seconds   **31.** (a)
$\frac{1}{4}$ second, 1 second; no (b) $\frac{1}{2}$ second (c) $\frac{3}{2}$ seconds.

CHAPTER 6

Exercises 6.1.5: Page 202

A. 1. $x = 0, 3, -7$ **2.** $x = 0, \frac{1}{2}, \frac{3}{4}$ **3.** $x = \pm 1, 2$ **4.** $x = 1, 1, -2$ **5.** $x = 1, 2, -3$ **6.** $x = 1, 1, -\frac{1}{2}$ **7.** $x = \pm 2, \pm 3$
8. $x = \pm \frac{3}{2}, \pm \sqrt{2}$ **9.** $x = \pm 2, \pm 4i$ **10.** $x = \pm 2, \pm \frac{1}{\sqrt{2}}$ **11.** $x = 0, \pm \sqrt{3}$ **12.** $x = \pm 2, \pm 3\sqrt{2}$.
B. 1. $x = -1, x = \frac{3 \pm \sqrt{183}}{6}$ **2.** $x = -1, 2, \frac{1}{2}$ **3.** $x = \pm 1, 3 \pm 2\sqrt{2}$ **4.** $x = 1, 1, \frac{-3 \pm \sqrt{5}}{2}$ **5.** $x = 1, 1, \pm i$ **6.** $x = -1, -1, -1, 2, \frac{1}{2}$.
C. 1. $x = -3, -1, 2$ **2.** $x = \frac{1}{2}, 1, -3$ **3.** $x = \frac{2}{3}, -1 \pm \sqrt{2}$ **4.** $x = 1, 3, \frac{-3 \pm i\sqrt{7}}{2}$ **5.** $x = 1, -2, \pm i\sqrt{3}$ **6.** $x = 1, -1, \frac{1}{2}, \frac{3}{2}$.

Exercises 6.1.8: Page 211

1. (a) $x = 2, -1 \pm i\sqrt{3}$ (b) $x = -\frac{3}{2}, \frac{3 \pm i3\sqrt{3}}{4}$ **2.** (a) $x = 0, 0, 0, \frac{1}{4}, \frac{-1 \pm i\sqrt{3}}{8}$ (b) $x = 1, \frac{1 \pm i\sqrt{3}}{4}$ **3.** (a) $x = \pm 1, \pm i$ (b) Hint: $(x^2 + 1 - \sqrt{2}x)(x^2 + 1 + \sqrt{2}x) = 0$; $x = \frac{\sqrt{2}}{2}(-1 \pm i), \frac{\sqrt{2}}{2}(1 \pm i)$
4. (a) Hint: Factoring we get

$$x^5 - 27x^2 = x^2(x^3 - 27) = x^2(x - 3)(x^2 + 3x + 9) = 0.$$

Thus, we solve $x^2 = 0$, $x - 3 = 0$ and $x^2 + 3x + 9 = 0$, etc. (b) Hint: Factoring we get

$$x^9 - x^5 + x^4 - 1 = x^5(x^4 - 1) + x^4 - 1 = (x^4 - 1)(x^5 + 1).$$

Therefore, $x^4 - 1 = 0$ and $x^5 + 1 = 0$. Hence, $(x^2 + 1)(x^2 - 1) = 0$ and $(x + 1)(x^4 - x^3 + x^2 - x + 1) = 0$. Thus, we solve the quadratic equations $x^2 + 1 = 0$ and $x^2 - 1 = 0$ whose solutions are $x = \pm i$ and $x = \pm 1$, and the equations $x + 1 = 0$, $x^4 - x^3 + x^2 - x + 1 = 0$. The first one gives $x = -1$ and the second is a $4^{th}-$ degree symmetric equation, etc. **5.** (a) Hint: set $y = x^3$ and solve $y^2 - 5y - 24 = 0$. Then solve $x^3 - 8 = 0$ and $x^3 + 3 = 0$ (b) $x = \pm 3, \pm 3i, \frac{\sqrt{2}}{2}(-1 \pm i), \frac{\sqrt{2}}{2}(1 \pm i)$
6. (a) $x = 2, 3, \pm i\sqrt{3}, \frac{1}{2} \pm \frac{\sqrt{3}}{3}i$ (b) $x = 16$ **6.** (a) $x = \frac{1}{9}$ (b) $x = 7\frac{\sqrt[3]{63}}{9}$.

Exercises 6.2.1: Page 214

1. (a) $x = 3, -\frac{1}{2}$ (b) $x = -1, 2$ **2.** (a) $x = 0$ (b) $x = \pm 1$
3. (a) $x = -\frac{6}{5}$ (b) $x = \frac{3 \pm \sqrt{13}}{2}$ **4.** (a) $x = \frac{22}{5}$ (b) $x = -1, \frac{1}{2}$
5. (a) $x = 3, -5$ (b) $x = \frac{1 \pm \sqrt{5}}{2}$ **6.** (a) $x = 2, -8$ (b) $x = 3$
7. $x = 5, \frac{5}{2}$ **8.** no solution **9.** $x = -2, \frac{1}{4}$ **10.** $x = 2, -\frac{7}{9}$.

Exercises 6.3.1: Page 219

1. (a) $x = 1$ (b) $x = 28$ **2.** (a) $x = 1$ (b) $x = -\frac{3}{2}, -\frac{5}{2}$ **3.** (a) $x = \frac{1}{4}$ (b) $x = 9$ **4.** (a) no solution (b) $x = 8$ **5.** (a) $x = 3, -1$ (b) $x = -2, -1$ **6.** (a) $x = 9$ (b) $x = -1$ **7.** (a) $x = 5$ (b) $x = \sqrt{2}, \sqrt{3}$ **8.** (a) $x = \frac{1}{9}$ (b) $x = \frac{4}{5}$.

Review Exercises 6.5: Page 228

A. **1.** $x = 0, -2, 7$ **2.** $x = 1, 1, -1$ **3.** $x = 2, -2, -3$ **4.** $x = \frac{1}{3}$, $x = \pm\sqrt{5}$ **5.** $x = 4, \frac{2}{3}, -\frac{3}{5}$ **6.** $x = \pm 1, \pm 7$ **7.** $x = \pm a, \pm i\sqrt{a}$ **8.** $x = \pm 2, \pm 2i$ **9.** $x = \pm i, \frac{3 \pm i\sqrt{7}}{2}$ **10.** $x = \frac{3 \pm i\sqrt{7}}{4}, \frac{1 \pm i\sqrt{3}}{2}$ **11.** $x = -2, 1, \pm i\sqrt{3}$ **12.** Hint: Complete the square to write the equation in the form $(x^2 + 5x)^2 + 6(x^2 + 5x) + 5 = 0$. Set $y = x^2 + 5x$, etc, $x = \frac{-5 \pm \sqrt{5}}{2}, \frac{-5 \pm \sqrt{21}}{2}$ **13.** $x = 1, -1, \frac{1}{2}, \frac{3}{2}$
15. $x = 1, 1, \frac{3 \pm \sqrt{5}}{2}$ **16.** $x = 1, 2 \pm \sqrt{3}, 3 \pm \sqrt{2}$ **17.** $x = \pm i, \pm\sqrt{\frac{-1 + i\sqrt{3}}{2}}, \pm\sqrt{\frac{-1 - i\sqrt{3}}{2}}$ **18.** Hint: Divide by x^2 and set $w =$

$x + \frac{2}{x}$; $x = \frac{\sqrt{2}-1\pm i\sqrt{5+2\sqrt{2}}}{2}$, $\frac{\sqrt{2}+1\pm i\sqrt{5+2\sqrt{2}}}{2}$ **19.** $x = 1, 2, 2, 8$.

B. 1. (a) $x = \frac{1}{3}$ (b) $x = -5$ **2.** (a) $x = -2, 1 \pm i$ (b) $x = \pm 2, \pm\sqrt{3}$ **3.** (a) $x = \pm a, \pm\frac{1}{a}$ (b) $x = 0, \pm 5\sqrt{2}$.

C. 1. (a) $x = 0, 1$ (b) $x = \pm 2$ **2.** (a) $x = 0$ (b) $x = \frac{1}{5}$

3. (a) $x = 9$ (b) $x = -1, 2$ **4.** (a) $x = 2, -\frac{5}{2}$ (b) $x = \frac{6\pm\sqrt{6}}{3}$

5. (a) $x = 1$ (b) $x = 0, 16, 81$ **6.** (a) $x = 3$ (b) $x = \frac{5}{4}$ **5.** (a) $x = \frac{16}{25}$ (b) $x = 4$.

D. 1. $x = -2, 1 \pm i$ **2.** $x = 1 \pm i, 3$ **3.** $x = 1 \pm i, 2, -2$

4. $x = 1 \pm 2i, -1, -\frac{1}{2}$ **5.** $x = 2 \pm \sqrt{2}, 1, \frac{1}{2}$ **6.** $x^4 + 12x^3 + 44x^2 + 18x - 116 = 0$.

E. 1. $x = 2, 3, -15$ **2.** $x = 1, -2, -4$ **4.** $x = 4$ **5.** $x = \frac{16}{25}$

6. $x = 9$ **7.** $x = 2$ **8.** $x = -1, 8$ **9.** $x = \frac{2}{3}$ **10.** $x = -1\frac{9}{16}$

11. $x = 0, \frac{16}{9}$ **12.** $x = -1, 7$.

CHAPTER 7

Exercises 7.2.4: Page 244

1. (a) $x > 1$ (b) $x < -1$ **2.** (a) $x > -4$ (b) $x < 3$

3. $x > 5$ **4.** $x > -\frac{43}{16}$ **5.** (a) $x > \frac{13}{8}$ (b) $x < -1$

6. (a) $x < -\frac{3}{10}$ (b) $x \in \mathbb{R}$ **7.** (a) $[5, 8)$ (b) $[\frac{1}{2}, 2)$ **8.** (a) $(8, 22]$ (b) $[2, 25]$ **9.** (a) $(2, \infty)$ (b) $(-\infty, -3)$ **10.** (a) $(4, 6)$ (b) $[-6, 2]$ **11.** (a) $x < -3$ or $x > 9$ (b) $x < -7$ or $x > 3$

12. (a) $(-\frac{7}{3}, -\frac{2}{3}] \cup (\frac{4}{3}, 3)$ (b) $(-10, -7] \cup [5, 8)$ **13.** (a) $x > 5$ or $x < -5$ (b) $x < -10$ or $x > \frac{10}{3}$ **14.** (a) $(-\frac{1}{2}, \infty)$ (b) $(-\infty, -\frac{3}{2}]$

15. (a) $x < -\frac{1}{7}$ or $x > \frac{9}{7}$ (b) $x < 2$. **16.** (a) $-3 < x < 3$ (b) $x < -2$ or $x > 4$.

Exercises 7.4.2: Page 256

1. (a) $(-\infty, 0) \cup (1, \infty)$ (b) $[0, 4]$ **2.** (a) $(-\infty, -3) \cup (1, \infty)$ (b) $2 \leq x \leq 5$ **3.** (a) $(-1, 5)$ (b) $(-\infty, -1) \cup (\frac{3}{2}, \infty)$ **4.** (a) $(-\infty, \frac{1}{2}) \cup (\frac{4}{3}, \infty)$ (b) all $x \neq \frac{1}{4}$ **5.** (a) $(2, 4) \cup (7, \infty)$ (b) $(\frac{1}{3}, 4)$

6. (a) $(-1, 2) \cup (2, 3)$ (b) $[2, \frac{5}{2}]$ and $x = -2$ **7.** (a) $x > \frac{1}{2}$ with

$x \neq 1$ (b) $\left(-1, -\frac{1}{2}\right) \cup \left(1, \frac{7}{4}\right)$ **9.** (a) $(-\infty, -2)$ (b) $x \leq -1$ or $x \geq 2$ **10.** (a) $x > -1$ with $x \neq 1$ (b) $x < -1$ or $x > 1$ **11.** (a) $k = 4$ (b) $k < 0$ or $k > 4$ (c) $0 < k < 4$.

Exercises 7.5.1: Page 260

1. (a) $(-\infty, 3) \cup (5, \infty)$ (b) $1 \leq x < 2$ **2.** (a) $(-\infty, 0) \cup \left(\frac{3}{2}, \infty\right)$ (b) $0 < x < 2$ **3.** (a) $(3, \infty)$ (b) $x < \frac{3}{2}$

4. (a) $-2 < x < -1$ or $x > 2$ (b) $-1 < x < 1 - \sqrt{3}$ or $x > 1 + \sqrt{3}$ **5.** (a) $(-\infty, -3) \cup (-3, -2] \cup (2, \infty)$ (b) $\left(-2, -\frac{1}{2}\right) \cup (1, 4)$
6. (a) $(-5, -2) \cup (3, \infty)$ (b) $(-\infty, -1] \cup (5, 6]$ **7.** (a) $(-1, \infty)$ (b) $(-\infty, 0) \cup (2, 3)$ **8.** (a) $x \leq -2$, $x = 0$, $3 \leq x < 5$ (b) $0 < x < 4$ and $x = -3$ **9.** (a) $(-\infty, 0] \cup \left[1, \frac{4}{3}\right] \cup (2, \infty)$ (b) $(-2, -1) \cup (-1, 1) \cup (3, \infty)$ **10.** (a) $(-\sqrt{2}, 0) \cup (1, \sqrt{2}) \cup (2, \infty)$ (b) $(-\infty, -1) \cup ((0, 1) \cup (2, \infty)$ **11.** (a) $\left(-\infty, -\frac{5}{2}\right) \cup (2, \infty)$ (b) $(-\infty, -3) \cup (-2, 0)$
12. (a) $(-\infty, -1) \cup (-1), 2]$ (b) $(-2, 2) \cup (2, \infty)$.

Exercises 7.6.1: Page 268

1. (a) $[5, \infty)$ (b) $\left(\frac{1}{2}, \frac{5}{2}\right)$ **2.** (a) $(3, \infty)$ (b) $[0, 2]$ **3.** (a) $\left[-2, -\frac{8}{5}\right] \cup (0, 2]$ (b) $\left[\frac{5}{2}, 3\right)$ **4.** (a) $(-\infty, 1)$ (b) $(4, 5)$ **5.** (a) $\left[-\frac{5}{2}, 2\right)$ (b) $[-14, 2)$ **6.** (a) $(-\infty, -1]$ (b) $[2, \infty)$ **7.** (a) $(-\infty, -2] \cup [2, \infty)$ (b) $\left(\frac{5 - \sqrt{13}}{6}, \infty\right)$ **8.** (a) $(-\infty, -2] \cup \left[5, \frac{74}{13}\right]$ (b) $(-\infty, -1) \cup (8, \infty)$ **9.** (a) $[3, 4) \cup (7, \infty)$ (b) $(1, \infty)$ **10.** (a) $[-2, \infty)$ (b) $1 \leq x \leq \frac{3}{2}$ **10.** (a) $\frac{12}{25} < x \leq \frac{1}{2}$ (b) $x < 0$ or $x > 1$.

Review Exercises 7.7: Page 269

1. (a) $x < 5$ (b) $x \geq 2$ **2.** (a) $x > \frac{9}{2}$ (b) $x > \frac{1}{6}$ **3.** (a) $x < -\frac{5}{4}$ or $x > 1$ (b) $\left(\frac{1}{2}, \infty\right)$ **4.** (a) $x \in \mathbb{R}$ (b) $2\frac{4}{7} < x < 4\frac{2}{3}$
5. (a) $\frac{3}{2} < x < \frac{7}{2}$ (b) $(-\infty, -7) \cup (2, \infty)$ **6.** (a) $-\frac{1}{2} \leq x \leq 3$ (b) $0 \leq x \leq 8$ **7.** (a) $(-2, -1) \cup (3, \infty)$ (b) $(-\infty, -2) \cup (2, \infty)$
8. (a) $(-\infty, -3) \cup (1, 2)$ (b) $[-3, 0] \cup [3, \infty)$ **9.** (a) $(-\infty, -1) \cup [2, 3] \cup [6, \infty)$ (b) $(-\infty, -3) \cup (-2, 1) \cup (6, \infty)$ **10.** (a) $x \in \mathbb{R}$ (b) $\left(-\frac{9}{4}, -2\right) \cup (3, \infty)$ **11.** (b) $(-3, -1) \cup \left(-\frac{1}{3}, 0\right) \cup (1, \infty)$

12. (a) $x \in \mathbb{R}$, $x \neq 2$ (b) $(-\infty, -\frac{\sqrt{7}}{2}) \cup (-1, \frac{\sqrt{7}}{2}) \cup (\frac{4}{3}, \infty)$ **13.** (a) $(-\infty, -1) \cup [4, \infty)$ (b) $(-\infty, -2) \cup (-1, 0]$ **14.** (a) $(-\infty, -1) \cup (3, 7)$ (b) $(-\infty, -2) \cup (-\frac{1}{2}, \infty)$ **15.** (a) $(-\infty, -1] \cup (0, \infty)$ (b) Set $x = \sqrt{x^2 + 1}$. All $x \in \mathbb{R}$ **16.** (a) $\frac{1 - \sqrt{13}}{2} < x < \frac{1 + \sqrt{13}}{2}$ (b) $\frac{2}{3} < x < \frac{3}{2}$, or $\frac{3}{2} < x < 4$ **17.** (a) $(-4, -1) \cup (2, 5)$ (b) $(-\infty, 0] \cup [\frac{9}{2}, \infty)$ **18.** (a) $1 < x \leq 4$ (b) $\frac{1}{2} < x < 2$, or $x > 5$ **19.** (a) $-2 \leq x < 0$ or $0 < x < \frac{8}{5}$ (b) $1 < x \leq \frac{2\sqrt{3}}{3}$ **20.** (a) $k = -3$, $k = 5$ (b) $k < -3$ or $k > 5$ (c) $-3 < k < 5$ **21.** (a) $9 < \lambda < 27$ (b) $1 \leq \lambda < 2$ **22.** $-3 < \lambda < 5$ **23.** $2 \leq \lambda \leq \frac{17}{4}$ **24.** $\lambda < 0$ **25.** $x = -2, 1$ **26.** $\frac{\sqrt{13} - 5}{2} < x \leq 1$ **27.** $x \geq 1$ **28.** $[-1, -\frac{\sqrt{3}}{2}] \cup [\frac{\sqrt{3}}{2}, 1]$ **29.** $[-\frac{1}{2}, 0) \cup (0, \frac{1}{2}]$ **30.** $x \geq \sqrt[3]{\frac{5}{4}}$ **31.** $(-\infty, 2\sqrt{2}) \cup (2 + 2\sqrt{3}, \infty)$.

CHAPTER 8

Exercises 8.1.3: Page 286

1. (a) $x = -1$, $y = -2$ (b) $x = 2$, $y = -1$ **2.** (a) $x = 3$, $y = 1$ (b) $x = 3$, $y = 0$ **3.** (a) $x = 1$, $y = -\frac{1}{2}$ (b) no solution **4.** (a) $x = -7$, $y = 5$ (b) $x = \frac{y+6}{2}$ $y \in \mathbb{R}$ **5.** (a) $x = 3 - y$, $y \in \mathbb{R}$ (b) no solution **6.** (a) $x = 2$, $y = 6$ (b) $x = \frac{5y-10}{2}$, $y \in \mathbb{R}$ **7.** (a) $x = -9$, $y = 4$ (b) $x = -1$, $y = 3$ **8.** (a) no solution (b) $x = 2$, $y = -3$ **9.** (a) $x = -1\frac{64}{49}$, $y = -\frac{38}{49}$ (b) $x = 4$, $y = 2$ **10.** (a) $x = 1$, $y = 2$, $z = 0$ (b) $x = 2$, $y = 1$, $z = \frac{1}{2}$. **11.** (a) no solution (b) $x = -7z - 1$, $y = 2z = 2$, $z \in \mathbb{R}$ **12.** (a) $x = 1$, $y = 3$, $z = -2$ (b) $x = \frac{1}{2}y + 2$, $y \in \mathbb{R}$, $z = \frac{1}{2}$ **13.** (a) $\lambda = 0$ (b) $\lambda = -3$ **14.** (a) $\lambda = 6$, $k = 4$ or $\lambda = -6$, $k = 4$ (b) $\lambda = 2$, $k = -1$ **15.** $\lambda = -7$, $\lambda = -4$.

Exercises 8.3.2: Page 301

A. 1. (a) $x = 4$, $y = 2$ (b) $x = -6$, $y = 3$, $z = 2$ **2.** (a) $x = -1$, $y = 2$, $z = -2$ (b) no solution **3.** (a) $x = -3 - z$, $y = 2 +$

$2z$, $z \in \mathbb{R}$ (b) $x = 1$, $y = 3$, $z = 1$ **4.** (a) $x_1 = -1$, $x_2 = 3$, $x_3 = -2$, $x_4 = 2$ (b) $x_1 = 4 - 2x_2 + x_4$, $x_2 \in \mathbb{R}$, $x_3 = 1 + 2x_4$, $x_4 \in \mathbb{R}$.
B. 1. (a) $x = \frac{2}{5}z$, $y = \frac{3}{5}z$, $z \in \mathbb{R}$ (b) $x = 0$, $y = 0$, $z = 0$ **2.** (a) $x = -z$, $y = z$, $z \in \mathbb{R}$ (b) $x = -zy$, $y \in \mathbb{R}$, $z = 0$ **3.** (a) $x_1 = x_4$, $x_2 = x_4$, $x_3 = -x_4$, $x_4 \in \mathbb{R}$ (b) $x_1 = 2x_3 + 8x_4$, $x_2 = -x_3 - 2x_4$, $x_3 \in \mathbb{R}$, $x_4 \in \mathbb{R}$, $x_5 = 0$.
  C. 1. When $\lambda \neq 5$, the system has no solution. For $\lambda = 5$ the system has many solutions $x = z - 4$, $y = \frac{11}{2} - 2z$, $z\mathbb{R}$ **2.** When $\lambda \neq 1$ and $\lambda \neq -2$ the system has the unique solution $x = y = z = \frac{1}{\lambda+2}$. When $\lambda = 1$, the system has many solution $x = 1 - y - z$, $y, z \in \mathbb{R}$. When $\lambda = -2$, the system has no solution.

Exercises 8.4.4: Page 313

1. (a) $x = \frac{3}{2}$, $y = \frac{1}{2}$ (b) $x = 9$, $y = -4$ and $x = -4$, $y = 9$ **2.** (a) $x = 3$, $y = 2$ and $x = -3$, $y = 2$ and $x = 2$, $y = -3$ and $x = -2$, $y = -3$ (b) $x = 11$, $y = 6$ and $x = -11$, $y = -6$ and $x = 6$, $y = 11$ and $x = -6$, $y = -11$ **3.** (a) $x = 1$, $y = 2$ and $x = \frac{33}{7}$, $y = \frac{38}{7}$ (b) $x = 1$, $y = 0$ and $x = -1$, $y = 0$ **4.** (a) $x = 0$, $y = 1$ and $x = -1$, $y = 1$ (b) $x = 2$, $y = 1$ and $x = -2$, $y = -1$ and $x = \sqrt{3}$, $y = \sqrt{3}$ and $x - \sqrt{3}$, $y = -\sqrt{3}$ **5.** (a) $x = \pm 3$, $y = \pm 2$ and $x = \pm 2$, $y = \pm\frac{1}{2}$ (b) $x = \pm 7$, $y = \pm 5$ **6.** (a) $x = \pm 2$, $y = \pm 1$ (b) $x = 7$, $y = -3$ and $x = -7$, $y = 3$ **7.** (a) Hint: multiply the first equation by 7 and the second by -3 and add the resulting equations. $x = \pm 2$, $y = \pm 1$ and $x = \pm\frac{3}{\sqrt{2}}$, $y = \pm\sqrt{2}$ (b) $x = \frac{1}{4}$, $y = -3$ and $x = \frac{27}{4}$, $y = -\frac{1}{3}$
8. (a) $x = 3$, $y = 1$ and $x = -1$, $y = -3$ (b) $x = 3$, $y = -2$
9. (a) $x = 3$, $y = -4$ and $x = 4$, $y = 3$ (b) $x = \pm 2$, $y = \pm 4$ and $x = \sqrt{6} + 2$, $y = \sqrt{6} - 2$ and $x = -\sqrt{6} + 2$, $y = -\sqrt{6} - 2$ **10.** (a) $x = \pm 3$, $y = \pm 2$ and $x = \pm 2$, $y = \pm 3$ (b) $x = 1$, $y = 2$ and $x = 2$, $y = 1$ **11.** (a) $x = 3$, $y = 2$ and $x = 2$, $y = 3$ (b) $x = 3$, $y = 2$ and $x = -3$, $y = -2$ **12.** (a) $x = 49$, $y = 16$ (b) $x = 9$, $y = 4$ and $x = y = \frac{25}{4}$ **13.** (a) $x = 4$, $y = 1$ (b) Hint: Set $t = \frac{2x-1}{y+2}$; $x = 5$, $y = 7$ **15.** (a) $x = 2$, $y = 3$ and $x = \frac{23}{4}$, $y = -\frac{9}{2}$ (b) $x = \frac{1}{2}$, $y = \frac{3}{2}$ and $x = y = \frac{11}{2}$.

Exercises 8.5.1: Page 321

1. length $30cm$, width $15cm$ **2.** length 300 feet, width 225 feet **3.** 87 **4.** 525 orchestra seats, 125 balcony seats **5.** 3 gallons from A, 5 gallons from B **6.** coffee 90 cents per pound, butter 80 cents per pound **7.** 64% in A, 32% in B **8.** 30 miles per hour and 40 miles per hour **9.** 3 miles per hour **10.** boat 20 miles per hour and river 5 miles per hour **11.** $2,345, interest rate 4.5% **12.** 20 days and 30 days **14.** 12 inches, 16 inches **15.** 39, 36 and 15 inches.

Review Exercises 8.6: Page 323

1. (a) $x = \frac{2}{3}, y = 1$ (b) $x = \frac{5-y}{2}, y \in \mathbb{R}$ **2.** (a) $x = 2, y = 3, z = -1$ (b) no solution **3.** (a) $x = y = z = 0$ (b) $x = z, y = 2 - z, z \in \mathbb{R}$ **4.** (a) $x = 2, y = -3, z = -\frac{3}{2}, w = \frac{1}{2}$ (b) $x = 3, y = 2, z = 1$ **5.** $\lambda = -4$ **6.** $\lambda = 1, \lambda = 4$ **7.** (a) $\lambda \neq \pm 2$ (b) $\lambda = -2$ (b) $\lambda = 2$ **8.** When $\lambda \neq 1$ and $\lambda \neq 3$ the system has the unique solution; $x = -1, y = \frac{\lambda-4}{\lambda-3}, z = -\frac{1}{\lambda-3}$. When $\lambda = 1$, the system has many solutions; $x = 1 - y - z, y, z \in \mathbb{R}$. When $\lambda = 3$, the system has no solution. **9.** (a) $x = 4, y = -1, x = -1, y = 4$ (b) $x = 5, y = 3$ **10.** (a) $x = 4, y = 5, x = 5, y = 4$ (b) $x = \frac{2}{3}, y = -\frac{11}{9}$ and $x = -4, y = -\frac{13}{3}$ **11.** (a) $x = 2, y = 1, x = -1, y = -2$ (b) $x = 3, y = 1, x = -1, y = -3$ and $x = 1 + i\sqrt{10}, y = -1 + i\sqrt{10}, x = 1 - i\sqrt{10}, y = -1 - i\sqrt{10}$ **12.** $\lambda = -4$ **13.** For $\lambda = 6; x_1 = 18, x_2 = 2$. For $\lambda = -\frac{6}{19}; x_1 = \frac{18}{19}, x_2 = \frac{2}{19}$ **14.** $\lambda = -\frac{125}{8}$ and $\lambda = \frac{27}{8}$ **15.** 4 miles per hour **16.** 18 feet per second;, 12 feet per second **17.** A 52%; B 22% **18.** George $4,600: Peter $600 **19.** $1,200 at 4% **20.** 4 miles per hour; 6 miles per hour **21.** 10 days **22.** 11 hours: 14 hours **23.** A 15 miles per hour; B 10 miles per hour **24.** 5, 6, 9 **25.** 33, 14, 4 **26.** $\frac{1}{2}, \frac{1}{3}$ **27.** 18 miles **28.** $\frac{7}{5}$ **29.** 6, 3 **30.** 12 and 5 feet **28.** $15cm, 12cm, 9cm$.

CHAPTER 9

Exercises 9.5: Page 338

1. (a) $\log_3 81 = 4$ (b) $\log_2 32 = 5$ (c) $\log_2(\frac{1}{8}) = -3$ (d) $\log_{\frac{1}{2}}(16) = -4$ (e) $\log_m k = p$ (f) $\ln 5 = x$ **2.** (a) $2^3 = 8$ (b) $3^{-2} = \frac{1}{9}$ (c) $(\frac{1}{5})^{-3} = 125$ (d) $(\frac{1}{3})^4 = \frac{1}{81}$ (e) $4^{-3} = \frac{1}{64}$ (f) $e^4 = x$ **3.** (a) $x = 4$ (b) $x = 3$ (c) $x = -3$ (d) $x = -2$ (e) $x = -3$ (f) $x = 2$
4. (a) $x = \frac{1}{2}$ (b) $x = \frac{1}{9}$ (c) $x = 100$ (d) $x = \frac{1}{4}$ (e) $x = 27$
5. (a) $x = 5$ (b) $x = 3$ (c) $x = \frac{1}{4}$ (d) $x = \frac{3}{2}$ (e) $x = 3$ **6.** (a) $(5, \infty)$ (b) $(-\infty, -2) \cup (2, \infty)$ (c) $(-\infty, -2) \cup (3, \infty)$ (d) $(-\infty, \infty)$ (e) $(-\infty, 0) \cup (1, \infty)$ **7.** (a) $\log_a 3 + \log_a x + 2\log_a y$ (b) $2\log_a x + 3\log_a y + \frac{1}{2}\log_a z$ (c) $\frac{1}{2}(\log_a x - \log_a y)$ (d) $5\log_a x - \log_a y - 2\log_a z$ (e) $\frac{3}{4}\log_a x + \frac{1}{2}\log_a y - \log_a z$ **8.** (a) $\log x + \frac{1}{2}\log(1 + x^2)$ (b) $3\log_a x - \log_a(2x - 1)$ (c) $2\log_a x + \frac{1}{2}\log_a(x + 1) - 3\log(x - 5)$ (d) $\frac{1}{2}\log_a(x + 1) - 2\log_a(x - 3)$ **9.** (a) $\log\left(\frac{x-2}{x+1}\right)$ (b) $\log_a[x^3(x - 1)^2]$ (c) $\log_a \frac{x(x+3)}{\sqrt{x+1}}$ (d) $-2\log(x - 1)$ (e) $\log_2[x(5x + 2)^4]$ (f) $\log y$
14. $pH \approx 3.2$ liters.   **15.** $pH \approx 4.77$ liters.   **16.** 5.01×10^{-4}.
17. 6.3×10^{-14} liters.   **18.** acid if $pH < 7$,   basic if $pH > 7$
19. 6 on Richter scale   **20.** 4.4

Exercises 9.6.3: Page 348

1. $41,410   **2.** $22,080.40   **2.** $10,010   **4.** $1,497.04
5. $694.80   **6.** $150,797.75   **7.** $12,155.61   **8.** $13,375.68   **9.** $6,849.16   **10.** $7,518.28 annual; $7,647.90 continuous   **11.** Monthly: 11.58 years or 139 months Continuously: 11.55 years or 138.6 months   **12.** 15.27 years or about

15 years and 4 months **13.** The 5.6% compounded continuously gives $1,060.62$; the 5.9% compounded monthly gives $1,057.60 **14.** 25.6% **15.** about 69.4% **16.** 33,000 years **17.** about 6.2 years **18.** about 5.1 years **19.** about 2,900 years **20.** (a) 28.4 years (b) 94.4 years **21.** (a) $Q \approx 675$ (b) 34.7 days (c) 69.3 days **22.** 40.55 hours and 69.31 hours **23.** $k \approx 0.231$; $Q(8) = 5712$, $t = 4.756$ hours **24.** $5,832$, 3.9 hours **1.** $25,198$ **26.** (a) $k \approx -0.09589$ (b) 14.5 minutes.

Exercises 9.7.1: Page 355

1. $x = 1$ **2.** $x = \frac{3}{2}$ **3.** $x = 0$ **4.** $x = -1$ **5.** $x = -\frac{1}{3}$ **6.** $x = -1, x = 3$ **7.** $x = -1, x = -2$ **8.** $x = 0$ **9.** $x = 1.7095$ **10.** $x = 0.21534$ **11.** $x = -8.2144$ **12.** $x = 0.534$ **13.** $x = 2.027$ **14.** $x = 1$ **15.** $x = 0$ **16.** $x = 1, x = 3$ **17.** $x = -1, x = 1$ **18.** $x = 1$ **19.** $x = 3$ **20.** $x = 3$ **21.** $x = 4$ **22.** $x = 0, x = 1$ **23.** $x = 0, x = -1$ **24.** $x = 2$ **25.** $x = 1$ **26.** $x = \frac{3}{2}$ **27.** $x = 0, x = 2$ **28.** $x = \frac{5}{2}$.

Exercises 9.8.1: Page 364

1. $x = 7$ **2.** $x = 2$ **3.** $x = -1, x = 4$ **4.** $x = 1$ **5.** $x = 2$ **6.** $x = 4$ **7.** $x = 10$ **8.** $x = 3, x = 4$ **9.** $x = 6, x = 12$ **10.** $x = 2$ **11.** $x = 10$ **12.** $x = 3, x = 5$ **13.** $x = \frac{9}{2}$ **14.** $x = 1, x = 100$ **15.** $x = -\frac{7}{2}, x = 3$ **16.** $x = 0.0025, x = 25$ **17.** $x = 100, x = \sqrt{10}$ **18.** $x = \frac{1}{9}, x = \frac{1}{30}$ **19.** $x = 100$ **20.** $x = 1, x = e^2$ **21.** $x = \frac{1}{e}, x = e^4$ **22.** $x = \frac{\ln 2}{\ln 2 - 1}$ **23.** $x = \frac{1}{9}, x = 9$ **24.** $x = -4$ **25.** $x = 3$ **26.** $x = 16$ **27.** $x = 3, x = \frac{5}{4}$ **28.** $x = 1, x = 15$ **29.** $x = 8, x = \frac{-3+\sqrt{3}}{3}$ **30.** $x = \frac{1}{\sqrt{2}}, x = 1, x = 4$.

Exercises 9.9.1: Page 372

Exponential Inequalities

1. $(\infty, 5)$ **2.** $(2,3)$ **3.** $(1, \infty)$ **4.** $(0,1)$ **5.** $(2, \infty)$
6. $-\log_3 2 < x < 0,$ or $\frac{1}{2} \log_3 2 < x < 1$ **7.** $(\infty, 0) \cup (1, \infty)$
8. $x \geq \frac{\ln 2}{\ln 12}$ **9.** $(0, \frac{1}{2}) \cup (4, \infty).$

Logarithmic Inequalities

1. $x > 5$ **2.** $2 < x < 4$ **3.** $-3 \leq x \leq 2$ **4.** $0 \leq x \leq 3$
5. $-3 < x < -1$ **6.** $-\frac{2}{3} < x < \frac{2}{3}$ **7.** $1 < x < 100$ **8.**
$0 < x < \frac{1}{10},$ or $x > 10$ **9.** $-\frac{17}{9} < x < -1,$ or $x > 7$ **10.**
$\frac{1}{6} < x < \frac{1}{4},$ or $x > 1.$

Exercises 9.10.3: Page 382

Exponential Systems

1. (a) $x = 2, y = 1$ (b) $x = 1, y = 1,$ or $x = 3, y = 2$ **2.** (a)
$x = 1, y = 3$ (b) $x = 0, y = 0$ **3.** (a) $x = 3, y = 1$ (b) $x = 21, y = 6$ **4.** (a) $x = 2, y = 1$ (b) $x = 3\sqrt{3}, y = \sqrt{3}.$

Logarithmic Systems

1. (a) $x = 2, y = 8,$ or $x = 8, y = 2$ (b) $x = 2, y = 4$ **2.**
(a) $x = 1, y = 2$ (b) $x = 9, y = 7$ **3.** (a) $x = 27, y = 4,$ or
$x = \frac{1}{81}, y = -3$ (b) $x = \frac{1}{2}, y = 1$ **4.** (a) $x = 3, y = 2$ (b)
$x = 3, y = 2.$

Index